AN INTRODUCTION TO HOMOLOGICAL ALGEBRA

AN INTRODUCTION TO HOMOLOGICAL ALGEBRA

BY

D. G. NORTHCOTT

PROFESSOR OF PURE MATHEMATICS IN THE
UNIVERSITY OF SHEFFIELD

CAMBRIDGE
AT THE UNIVERSITY PRESS
1962

CAMBRIDGE UNIVERSITY PRESS
Cambridge, New York, Melbourne, Madrid, Cape Town, Singapore, São Paulo, Delhi

Cambridge University Press
The Edinburgh Building, Cambridge CB2 8RU, UK

Published in the United States of America by Cambridge University Press, New York

www.cambridge.org
Information on this title: www.cambridge.org/9780521058414

First published 1960
Reprinted 1962
This digitally printed version 2008

A catalogue record for this publication is available from the British Library

ISBN 978-0-521-05841-4 hardback
ISBN 978-0-521-09793-2 paperback

CONTENTS

4. Homology functors

5. Projective and injective modules

PREFACE

The past ten years or so have seen the emergence of a new mathematical subject which now bears the name Homological Algebra. To begin with, it was the concern of a few enthusiasts in certain specialized fields but, since the publication of Cartan and Eilenberg's now famous book,† its importance for several of the main branches of pure mathematics has been generally recognized.

The young mathematician, about to start on research, will be anxious to learn about homological ideas and methods, and one of the aims of this book is to help him to get started. In trying to cater for his needs, I have imagined such a reader as being familiar with the notions of group, ring and field but still relatively inexperienced in modern algebra. For him, the account given here is self-contained save in a small number of particulars which are mentioned below, and which need not discourage him.

An introduction to homological algebra must, of necessity, be an introduction to the book of Cartan and Eilenberg, for the student who wishes to go further will need to read their work; but much of great interest and value has been achieved even more recently, and some of this later work has been given a place in the following pages. The list of contents gives a fairly detailed picture of the main topics treated, but a few additional comments may be a help.

Chapters 1–6 develop, in a leisurely manner, the results that are needed to establish and illustrate the theory of derived functors, after which follows an account of torsion and extension functors. These are the most important ones which are obtainable by the process of derivation and, in a sense, the remainder of the book is concerned with their applications. Such an application is the theory of global dimension given at the end of Chapter 7, and here are included some important results of M. Auslander on Noetherian rings that have previously been available only in the original research paper.

Chapter 9 deals with the structure of commutative Noetherian rings

† H. Cartan and S. Eilenberg, *Homological Algebra* (Princeton University Press, 1956).

of finite global dimension and represents one of the most satisfying achievements of homological methods. This, too, appears in a textbook for the first time. Here, it must be admitted, the account is not completely self-contained, but considerable care has been taken in explaining the results of Ideal Theory which are needed to supplement the purely homological arguments. This is the most ambitious chapter, and the author hopes that it will help to stimulate interest in commutative algebra. The treatment given here was found successful in a course of lectures in which the audience had no specialized knowledge of classical Ideal Theory.

Chapter 10 is an introduction to the homology and cohomology theories of monoids and groups. This, by itself, has a considerable literature and was one of the earliest branches of our subject to be developed. The chapter can be read, if desired, before Chapter 9 and does not require any specialized knowledge of Group Theory.† In deciding how far to go with this topic, I had in mind the student who might wish to acquire some general background before proceeding to the applications in some specialized field such as Class Field Theory.

Nearly all the topics covered in the following pages were included in a course of lectures given at Sheffield University. When lecturing, it is possible to digress at some length in order to explain the general plan of development and the connexions with other branches of mathematics. Also one likes to mention important results connected with what one is discussing even if there is no time for a full treatment. Some of this supplementary material, which I hope will add to the enjoyment and interest of the main text, will be found in the Notes which follow Chapter 10.

The final chapter has been much improved as the result of suggestions of J. Tate with whom I had an opportunity of discussing it. At Sheffield, I have been aided, at all stages, by my colleague H. K. Farahat. Of particular value has been his willingness to discuss points of detail and to make helpful criticisms. This work owes a great deal to his continued interest. I am also indebted to Sir William Hodge, who, when I first had the idea of writing an introduction to homological methods, encouraged me to go ahead.

† There is actually one reference to a result proved in Chapter 9, but there is no difficulty in taking this out of context.

Writing a book takes up much time and energy, and this one could never have been completed without the generous help of J. J. Kiely who typed the first draft from notes taken at lectures. I am also greatly indebted to my secretary, Mrs M. Ludbrook, for the great care and patience with which she cut innumerable exquisite stencils. To both of these I wish to express my thanks. Their strenuous efforts made it unthinkable not to finish a work to which they had contributed so much.

D. G. NORTHCOTT

SHEFFIELD
July 1958

NOTE ON CROSS-REFERENCES

If the reader is referred to a result, say to Theorem 7, and no chapter or section is specified, then he is to understand that the reference is to Theorem 7 of the chapter in which he is reading. When a reference is made in one chapter to a result which occurs in another, the section in which it will be found is also given. Thus Lemma 4 of section (5.3) means the fourth lemma in Chapter 5, and this will be found in the third subdivision of the chapter.

1

GENERALITIES CONCERNING MODULES

Notation. Λ denotes a ring, which is not necessarily commutative, with an identity element 1.

1.1 Left modules and right modules

The notion of a *ring-module* has, in recent years, come to be regarded as one of the most important in modern algebra, and the theory of ring-modules is so extensive that it includes, for example, that of vector spaces, ideals, algebras and group representations. But, in spite of the fact that this concept covers so many widely differing structures, there exists an elaborate and rich theory common to them all. Of this theory, homological algebra forms an important part just as, in topology, homology theory is a valuable system of results which is valid for many different kinds of space. In special situations one can hope to extend, in certain particulars, such a universal body of knowledge, and in this way arises the possibility of making useful applications. On the present occasion, however, these are reserved for the later sections of the book, and it is with very broadly based ideas that we shall be concerned for some time.

For the reader's convenience, we begin with the idea of a module on which the elements of a given ring Λ act as operators, it being supposed that he is familiar with the concept of a ring and also of a group. Then we shall give an account of the more elementary notions which arise out of the definition. Probably the reader will already be familiar with much that is said in the first chapter, but even so it will repay him to glance through it because the opportunity is taken to prepare the ground for the introduction of new ideas in later chapters. Also, in section (1.10), we describe a slightly unusual kind of notation which, it is hoped, will make it easier to follow some of the proofs.

We come now to the first definition. Let M be an additive abelian group, then M is called a *left Λ-module* if, for each element x of M and each element λ of Λ, there is defined a 'product' λx which belongs to M and satisfies the following axioms:

$$\text{(i)} \quad \lambda(x_1+x_2) = \lambda x_1 + \lambda x_2,$$

$$\text{(ii)} \quad (\lambda_1+\lambda_2)x = \lambda_1 x + \lambda_2 x,$$

$$\text{(iii)} \quad \lambda_1(\lambda_2 x) = (\lambda_1\lambda_2)x,$$

$$\text{(iv)} \quad 1x = x.$$

Of course, in the above, x, x_1, x_2 are arbitrary elements of M and λ, λ_1, λ_2 may vary freely in Λ while, in (iv), 1 denotes the identity element of Λ.

Right Λ-*modules* are defined similarly except that the product is written $x\lambda$ and the corresponding axioms are:

$$\text{(i)}' \quad (x_1+x_2)\lambda = x_1\lambda + x_2\lambda,$$

$$\text{(ii)}' \quad x(\lambda_1+\lambda_2) = x\lambda_1 + x\lambda_2,$$

$$\text{(iii)}' \quad (x\lambda_1)\lambda_2 = x(\lambda_1\lambda_2),$$

$$\text{(iv)}' \quad x1 = x.$$

Suppose, for the moment, that Λ is a commutative ring and that M is a left Λ-module, then we can turn M into a right Λ-module simply by putting $x\lambda = \lambda x$. Conversely, every right Λ-module can be regarded as a left Λ-module. Thus all modules over commutative rings are virtually two-sided and the distinction between left and right disappears.

In future, unless otherwise stated, we shall understand by a Λ-module a *left* Λ-*module*. However, our definitions and results will also be applicable to right modules with the appropriate formal changes. A Λ-module which comprises only the zero element will be denoted by 0.

Let M be an additive abelian group, let x be an element of M, and let k be an integer. Then kx has a well-defined meaning. Also, with an obvious notation,

$$k(x_1+x_2) = kx_1 + kx_2,$$

$$(k_1+k_2)x = k_1 x + k_2 x,$$

$$k_1(k_2 x) = (k_1 k_2)x,$$

$$1x = x.$$

Hence, if Z denotes the ring of integers, we may say that M is a Z-module.

1.2 Submodules

Let M be a (left) Λ-module and let L be a subset of M. It may happen that whenever x, x_1 and x_2 are elements of L and λ belongs to Λ, then x_1+x_2 and λx are again elements of L. Should this be so then we say that L is a *submodule* of M and that M is an *extension module* of L.

1.3 Factor modules

Let N be a submodule of the Λ-module M, then, in particular, it is a subgroup of M and therefore the cosets of N in M form the abelian group M/N. Further, when x_1 and x_2 belong to the same coset, then x_1-x_2 is an element of N and so $\lambda x_1-\lambda x_2 = \lambda(x_1-x_2)$ is also a member of N. If x is an element of M let us write $\bar{x}$ for the coset to which it belongs, then, by virtue of the above remark, we can define a 'product' $\lambda\bar{x}$, where $\lambda \in \Lambda$, by writing $\lambda\bar{x} = \overline{\lambda x}$. If this is done then M/N becomes a Λ-module called the *factor* (or *residue*) *module* of M modulo N. The mapping $x \to \bar{x}$, which carries each element into the coset to which it belongs, is called the *natural mapping* of M on to M/N, and when this natural mapping occurs in a diagram it is sometimes convenient to draw attention to it by writing

$$M \xrightarrow{\text{nat}} M/N.$$

1.4 Λ-homomorphisms

Let $f: M \to N$ be a mapping of the Λ-module M into the Λ-module N. We say that f is Λ-*linear* or that f is a Λ-*homomorphism* if

$$f(x_1+x_2) = f(x_1)+f(x_2),$$
$$f(\lambda x_1) = \lambda f(x_1),$$

where x_1, x_2 are arbitrary elements of M and λ is any element of Λ.

Remarks. (*a*) If N is a submodule of M then the 'inclusion map' $N \to M$, in which each element of N is mapped into itself, is a Λ-homomorphism.

(*b*) If N is a submodule of M then the natural map $M \to M/N$ is a Λ-homomorphism.

(*c*) Suppose that $f: M \to N$ and $g: N \to P$ are Λ-homomorphisms, then their composition $gf: M \to P$ is also a Λ-homomorphism.

(*d*) Let f_1 and f_2 be Λ-homomorphisms from M into N and for each element x of M let us write

$$f(x) = f_1(x)+f_2(x)$$

so that $f(x)$ belongs to N. Then f is a mapping of M into N and it can easily be verified that f is a Λ-homomorphism. This particular homomorphism is written f_1+f_2 and, in using this notation, we have defined 'addition' for homomorphisms of M into N. It is now a straightforward matter to verify that this set of homomorphisms, which is denoted by $\mathrm{Hom}_\Lambda(M,N)$, forms an abelian group. This group will receive a great deal of attention later.

Now suppose, for the moment, that Λ is commutative and let f belong to $\mathrm{Hom}_\Lambda(M,N)$. For each element x of M write $g(x) = \lambda f(x)$, where λ is some fixed element of Λ, then, *because* Λ *is commutative*, g also belongs to $\mathrm{Hom}_\Lambda(M,N)$. The homomorphism g is denoted by λf and, with this definition of λf, $\mathrm{Hom}_\Lambda(M,N)$ becomes a Λ-module. To summarize, we may say that *in the general (non-commutative) case* $\mathrm{Hom}_\Lambda(M,N)$ *is an additive group, but, when* Λ *is commutative, we may, if we wish, endow it with the structure of a* Λ*-module.*

(*e*) Let f, f_1, f_2 be Λ-homomorphisms $M \to N$ and let g, g_1, g_2 be Λ-homomorphisms $N \to L$. Then

(i) $g(f_1+f_2) = gf_1+gf_2$;

(ii) $(g_1+g_2)f = g_1f+g_2f$;

(iii) if Λ is a commutative ring, then $(\lambda g)f = g(\lambda f) = \lambda(gf)$, where λ is an arbitrary element of Λ.

1.5 Some different types of Λ-homomorphisms

Let $f: M \to N$ be a Λ-homomorphism.

Definition. If $f(x) \neq f(y)$ whenever $x \neq y$, then f is called a *monomorphism.*

Definition. If f maps M *on to* N, then f is called an *epimorphism.*

Definition. If f is both a monomorphism and an epimorphism, then it is said to be an *isomorphism* and we write $f: M \approx N$. In this case the inverse mapping f^{-1} is an isomorphism $N \approx M$ and f, f^{-1} are called *inverse isomorphisms.*

It is well worth noting that the Λ-homomorphisms $f: M \to N$ and $g: N \to M$ are inverse isomorphisms if and only if both gf and fg are identity maps.†

† The *identity map* of a set maps the set on to itself and leaves each element fixed.

1.6 Induced mappings

By a *pair* (M', M) will be meant a Λ-module M together with a submodule M', and by a *homomorphism* $f: (M', M) \to (N', N)$ *of pairs* will be meant a Λ-homomorphism $f: M \to N$ for which $f(M') \subseteq N'$, where $f(M')$ denotes the image of M'. Suppose that we have this situation and that x_1, x_2 are elements of M which belong to the same coset of M'. Then

$$f(x_1) - f(x_2) = f(x_1 - x_2) \in f(M') \subseteq N',$$

and so $f(x_1)$ and $f(x_2)$ belong to the same coset of N' in N. This determines a map

$$f^* : M/M' \to N/N',$$

which is easily verified to be a Λ-homomorphism. The map f^* is called the *induced map*, and it is characterized by the property that it makes the diagram

$$\begin{array}{ccc} M & \xrightarrow{f} & N \\ \downarrow & & \downarrow \\ M/M' & \xrightarrow{f^*} & N/N' \end{array}$$

commutative, the vertical maps being natural. It should be noted that if f is an epimorphism then so is f^*.

We have just referred to the idea of a *commutative diagram*, but this is a concept which requires some explanatory comment. If

$$\begin{array}{ccc} A & \xrightarrow{\phi} & B \\ \downarrow{\scriptstyle f} & & \downarrow{\scriptstyle g} \\ A' & \xrightarrow{\sigma} & B' \end{array}$$

is a square diagram of modules and homomorphisms and we say that it is commutative, then we mean that $g\phi$ and σf coincide. Similarly, the triangular diagram

$$\begin{array}{ccc} X & \xrightarrow{u} & Y \\ & {\scriptstyle w}\searrow \quad \swarrow{\scriptstyle v} & \\ & Z & \end{array}$$

is commutative when $vu = w$. Sometimes we have to deal with more complicated figures such as

$$\begin{array}{ccccc} A & \xrightarrow{\phi} & B & \xrightarrow{\psi} & C \\ \downarrow{\scriptstyle f} & & \downarrow{\scriptstyle g} & & \downarrow{\scriptstyle h} \\ A' & \xrightarrow{\sigma} & B' & \xrightarrow{\tau} & C' \end{array} \qquad (1.6.1)$$

but they will always be composed, in a simple way, out of squares and triangles. Such a diagram is said to be commutative when the *small* component squares and triangles have this property, so that (for example) this will be the case in (1.6.1) when both $\sigma f = g\phi$ and $\tau g = h\psi$. After this remark, the intention should be clear in all cases which present themselves, though the reader should observe that we do not give a definition of a commutative diagram that will apply to completely general situations.

Let us return to the consideration of induced mappings. Let $f: M \to N$ be a homomorphism which maps a submodule M' of M into the zero element of N so that $(M', M) \to (0, N)$ is a map of pairs. In this case the induced map f^* maps M/M' into N and is characterized by the fact that the diagram

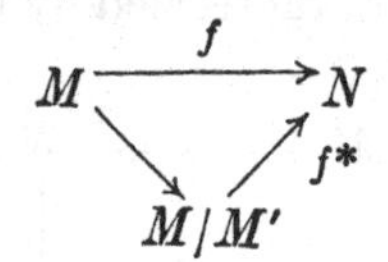

is commutative.

Suppose now that f, f_1, f_2 are homomorphisms of (M', M) into (N', N) and that g is a homomorphism of (N', N) into (P', P). Then

$$f_1+f_2 : (M', M) \to (N', N), \quad gf : (M', M) \to (P', P)$$

and

$$(f_1+f_2)^* = f_1^* + f_2^*, \quad (gf)^* = g^* f^*.$$

Furthermore, when Λ is a commutative ring and λ belongs to Λ, then

$$(\lambda f)^* = \lambda f^*.$$

1.7 Images and kernels

Let $f: M \to N$ be a Λ-homomorphism and let us write

$$\mathrm{Im}\,(f) = f(M),$$
$$\mathrm{Ker}\,(f) = f^{-1}(0),$$

so that $\mathrm{Im}\,(f)$, which is called the *image* of f, consists of all elements of the form $f(x)$, where $x \in M$, while $\mathrm{Ker}\,(f)$, the so-called *kernel* of f, is made up of all elements that are mapped into zero. In addition, we define the *coimage* of f and the *cokernel* of f by means of the formulae

$$\mathrm{Coim}\,(f) = M/\mathrm{Ker}\,(f),$$
$$\mathrm{Coker}\,(f) = N/\mathrm{Im}\,(f).$$

Let us observe that f is a monomorphism if and only if $\mathrm{Ker}\,(f) = 0$, while for f to be an epimorphism we require that $\mathrm{Coker}\,(f) = 0$.

Accordingly, f is an isomorphism when and only when Ker (f) and Coker (f) are both null modules.

Theorem 1. *Let $f: M \to N$ be an epimorphism. Then the induced map $f^*: M/\text{Ker}(f) \to N$ is an isomorphism.*

Proof. Since f is an epimorphism so is f^*; hence it remains to be shown that f^* is a monomorphism, i.e. that Ker $(f^*) = 0$. Let x be an element of M and let $\bar{x}$ be the coset of x modulo Ker (f). Then $\bar{x}$ is an entirely general element of $M/\text{Ker}(f)$. If now $\bar{x}$ belongs to Ker (f^*) then $0 = f^*(\bar{x}) = f(x)$ so that x is an element of Ker (f) and therefore $\bar{x} = 0$. Accordingly, Ker $(f^*) = 0$ and the theorem follows.

Since $M \to \text{Im}(f)$ is an epimorphism with kernel Ker (f), the theorem shows that there is an isomorphism

$$\text{Coim}(f) \approx \text{Im}(f).$$

For this reason Coim (f) does not often appear. However, in certain special situations the two concepts play genuinely different roles.

1.8 Modules generated by subsets

Let M be a Λ-module and $[u_i]_{i \in I}$ a family of elements of M, the system I of parameters being arbitrary. The subset of M, consisting of all elements which can be written in the form

$$\sum_i \lambda_i u_i,$$

where each λ_i is an element of Λ and $\lambda_i = 0$ for almost all i (that is, $\lambda_i = 0$ for all i with at most a finite number of exceptions), forms a submodule of M. This submodule is called the *submodule of M generated by* $[u_i]_{i \in I}$. If this submodule happens to coincide with M itself then $[u_i]_{i \in I}$ is called a *system of generators* of M.

Let $[u_i]_{i \in I}$ be a given system of generators of M. If now, for each element x of M, the λ_i for which

$$x = \sum_i \lambda_i u_i$$

are uniquely determined, then $[u_i]_{i \in I}$ is called a *base* of M. A module which admits a base is called *free*.†

Let F be a free Λ-module with base $[u_i]_{i \in I}$, let N be a Λ-module, and let $[v_i]_{i \in I}$ be a family of elements of N indexed with the same system I of parameters. Then there always exists a unique Λ-homomorphism $f: F \to N$ such that

$$f(u_i) = v_i,$$

† A module, which consists only of a zero element, is to be regarded as a *free* module with an *empty* base.

for all i of I. Indeed, f is defined by

$$f(\sum_i \lambda_i u_i) = \sum_i \lambda_i v_i.$$

Let $[w_i]_{i\in I}$ be a family of symbols. Consider the set of all *formal* sums $\sum_i \lambda_i w_i$, where each λ_i is an element of Λ and λ_i is zero for almost all i. For such formal sums we define addition and multiplication (by elements of Λ) in the obvious manner. If this is done the result is a Λ-module. Let us identify w_j with the formal sum $\sum_i \delta_{ij} w_i$, where $\delta_{ij} = 0$ if $i \neq j$ and $\delta_{ii} = 1$. Each element of the module has then a unique representation in the form $\sum_i \lambda_i w_i$, hence the module is free and has $[w_i]_{i\in I}$ as a base. This module is called the *free module generated by the symbols* $[w_i]_{i\in I}$.

Theorem 2. *Given any Λ-module M there exists a free module F with an epimorphism $F \to M$. If M can be generated by m elements ($0 \leqslant m < \infty$) then F can be chosen with a base of m elements.*

Proof. Let $[u_i]_{i\in I}$ be a system of generators of M,† let $[v_i]_{i\in I}$ be a similarly indexed system of distinct symbols, let F be the free module generated by $[v_i]_{i\in I}$, and let $f: F \to M$ be the Λ-homomorphism for which $f(v_i) = u_i$. Then f has the properties required by the theorem.

Theorem 3. *Let F be a free Λ-module, $p: F \to N$ a Λ-homomorphism, and $q: M \to N$ an epimorphism of Λ-modules. Then it is possible to find a Λ-homomorphism $\phi: F \to M$ such that the diagram*

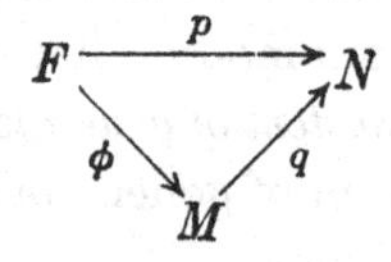

is commutative.

Proof. Let $[u_i]_{i\in I}$ be a base of F, then $p(u_i)$ is an element of N for each i. Since q is an epimorphism, we can find an element v_i of M so that $q(v_i) = p(u_i)$. But F is free, hence there exists a homomorphism $\phi: F \to M$ such that $\phi(u_i) = v_i$ and then

$$q\phi(u_i) = q(v_i) = p(u_i).$$

Thus $q\phi$ and p coincide on a base of F and hence at all elements of F. Accordingly $q\phi = p$.

† Note that we can certainly find a system of generators of M. Indeed, the set of *all* the elements of M forms such a system.

1.9 Direct products and direct sums

The notions of a direct sum of modules and a direct product of modules, which we discuss in the present section, are of fundamental importance for our theory. Let $[M_i]_{i \in I}$ be a family of Λ-modules, the set I of parameters being quite arbitrary. We consider the families $[m_i]_{i \in I}$ where, for each i, m_i is an element of M_i. For such families we define addition and multiplication (by elements of Λ) by means of the corresponding operations on individual components. This produces a Λ-module, which is written $\prod_i M_i$ and called the *direct product* of the modules M_i. From the direct product $\prod_i M_i$ we can pick out the submodule consisting of all families $[m_i]_{i \in I}$, where $m_i = 0$ for almost all i. This submodule is denoted by $\sum_i M_i$ and called the *external direct sum* of the modules of the given family. If I is finite then the direct product and external direct sum coincide.

The main features of the connexion between the M_i and their direct product can be described by means of canonical Λ-homomorphisms

$$M_j \xrightarrow{p_j} \prod_i M_i \xrightarrow{q_j} M_j \quad (j \in I).$$

Here p_j maps an element m_j into that element of $\prod_i M_i$ whose jth component is m_j and whose remaining components are zero. On the other hand, q_j maps an element of $\prod_i M_i$ into its jth component. These mappings have the property that $q_i p_i$ = identity and $q_i p_j = 0$ if $i \neq j$. For the direct sum we have similar canonical mappings

$$M_j \xrightarrow{f_j} \sum_i M_i \xrightarrow{g_j} M_j \quad (j \in I),$$

which not only satisfy $g_i f_i$ = identity and $g_j f_i = 0$ for $i \neq j$, but for which we have in addition

$$\sum_i f_i\, g_i(x) = x,$$

for all elements x of $\sum_i M_i$.

More generally suppose that $[M_i]_{i \in I}$ is a family of Λ-modules and that we have a family $[f_i]_{i \in I}$ of Λ-homomorphisms

$$f_i : M_i \to M \quad (i \in I) \tag{1.9.1}$$

of the M_i into the same module M. These determine a Λ-homomorphism $\sum_i M_i \to M$ in which an element $[m_i]_{i \in I}$ of $\sum_i M_i$ is mapped into $\sum_i f_i(m_i)$. If the homomorphism $\sum_i M_i \to M$ is an isomorphism then we say that (1.9.1) is an *injective representation of M as a direct sum* of Λ-modules.

Suppose that we have this situation. Then there exist (uniquely defined) Λ-homomorphisms

$$g_i : M \to M_i \quad (i \in I) \tag{1.9.2}$$

such that $\quad g_i f_i = \text{identity} \quad \text{and} \quad g_j f_i = 0 \quad \text{when} \quad i \neq j. \quad$ (1.9.3)

In fact if $[m_i]_{i \in I}$ belongs to $\sum_i M_i$, then the mapping g_j will carry $\sum_i f_i(m_i)$ into m_j. It will be very convenient to have a concise way of describing a situation of this kind, and therefore we shall say that

$$M_i \xrightarrow{f_i} M \xrightarrow{g_i} M_i \quad (i \in I) \tag{1.9.4}$$

is a *complete representation of M as a direct sum*. For such a complete representation we have

$$\sum_i f_i g_i(m) = m \tag{1.9.5}$$

for all elements m of M.

Next let $[N_i]_{i \in I}$ be a family of Λ-modules and suppose that we have a family $[q_i]_{i \in I}$ of homomorphisms

$$q_i : N \to N_i \quad (i \in I) \tag{1.9.6}$$

of a single module N into the various N_i. These determine a Λ-homomorphism

$$N \to \prod_i N_i,$$

in which an element n of N is mapped into $[q_i(n)]_{i \in I}$. If now the Λ-homomorphism $N \to \prod_i N_i$ is an isomorphism, then we say that (1.9.6) is a *projective representation of N as a direct product*. In such a situation there exist uniquely determined Λ-homomorphisms

$$p_i : N_i \to N \quad (i \in I) \tag{1.9.7}$$

with the properties that

$$q_i p_i = \text{identity}, \quad q_i p_j = 0 \quad \text{when} \quad i \neq j. \tag{1.9.8}$$

We say then that $\quad N_i \xrightarrow{p_i} N \xrightarrow{q_i} N_i \quad (i \in I) \quad$ (1.9.9)

is a *complete representation of N as a direct product* of Λ-modules.

Again let Y be a Λ-module and let $[Y_i]_{i \in I}$ be a family of submodules. If each element y of Y has a *unique* representation in the form

$$y = \sum_i y_i,$$

where y_i is an element of Y_i and $y_i = 0$ for almost all i, then we say that Y is the *internal direct sum* of the submodules $[Y_i]_{i \in I}$. There then exists a canonical isomorphism $\sum_i Y_i \approx Y$, between Y and the external direct sum $\sum_i Y_i$, in which $[y_i]_{i \in I}$ of $\sum_i Y_i$ is mapped into $\sum_i y_i$. We also have Λ-homomorphisms

$$Y_j \to Y \to Y_j, \tag{1.9.10}$$

with properties analogous to those described in (1.9.3) and (1.9.5). Here the first map is an inclusion map and the second carries an element of Y into the term with suffix j in its representation as a sum $\sum_i y_i$.

Theorem 4. *Let*

$$M_i \xrightarrow{f_i} M \xrightarrow{g_i} M_i \quad (i \in I) \tag{1.9.11}$$

be a system of Λ-modules and Λ-homomorphisms which satisfy

$$g_i f_i = \textit{identity} \quad \textit{and} \quad g_j f_i = 0 \quad \textit{if} \quad i \neq j.$$

Then (1.9.11) *is a complete representation of M as a direct sum if and only if we have*

$$\sum_i f_i g_i(m) = m \tag{1.9.12}$$

for all elements m of M. If I is finite, then (1.9.12) *is a necessary and sufficient condition for* (1.9.11) *to be a complete representation of M as a direct product.*

Proof. Suppose that (1.9.12) holds and consider the homomorphism $\sum_i M_i \to M$ in which $[m_i]_{i \in I}$ is mapped into $\sum_i f_i(m_i)$. We shall show that this is an isomorphism. Let us note first that, if $\sum_i f_i(m_i) = 0$, then $0 = g_j(\sum_i f_i(m_i)) = m_j$ for all j. This shows that $\sum_i M_i \to M$ is a monomorphism. Next let m be an element of M, then $\sum_i f_i g_i(m) = m$ and in particular $f_i g_i(m) = 0$ for almost all i. It therefore follows that $g_i f_i g_i(m) = g_i(m) = 0$ for almost all i, hence $[g_i(m)]_{i \in I}$ belongs to $\sum_i M_i$ and, by (1.9.12), its image in M is $\sum_i f_i g_i(m) = m$. Accordingly, we have verified that $\sum_i M_i \to M$ is both a monomorphism and an epimorphism and thereby shown that (1.9.11) is a direct sum representation. The fact that when (1.9.11) is a direct sum representation then (1.9.12) holds has already been established.

Now suppose that I is finite. Let $[m_i]_{i \in I}$ be an arbitrary element of $\prod_i M_i$, then $[m_i]_{i \in I}$ is the image of $\sum_i f_i(m_i)$ in the mapping $M \to \prod_i M_i$ determined by the g_i. Thus, in any event, $M \to \prod M_i$ is an epimorphism.

Next assume that $M \to \prod_i M_i$ is an isomorphism, that is to say, that (1.9.11) is a direct product representation. Then, since

$$g_j(m) = g_j(\sum_i f_i\, g_i(m)),$$

we see that m and $\sum_i f_i\, g_i(m)$ are elements of M with the same image in $\prod_i M_i$. Accordingly $m = \sum_i f_i\, g_i(m)$, which means that (1.9.12) holds.

Finally, assume that (1.9.12) holds. To complete the proof, we need only show that $M \to \prod_i M_i$ is a monomorphism, that is to say we must show that if $g_i(m) = 0$ for all i, then $m = 0$. But this is obvious because $\sum_i f_i\, g_i(m) = m$.

Definition. Let B be a submodule of a Λ-module M. If M is the internal direct sum of two submodules, one of which is B, then we say that B is a *direct summand* of M.

Definition. A monomorphism $L \to M$ is called *direct* if $\operatorname{Im}(L \to M)$ is a direct summand of M. An epimorphism $M \to N$ is called *direct* if its kernel is a direct summand of M. In this case we say that N is a *direct factor* of M.

1.10 Abbreviated notations

Suppose that we have a Λ-homomorphism $f : M \to N$. If there is no other homomorphism of M into N under consideration it is sometimes convenient to denote the homomorphism itself by MN. We call this the *abbreviated notation* for homomorphisms. In this notation the composition of two homomorphisms, say $L \to M$ and $M \to N$, is written as $(LM)(MN)$ or LMN. For example, the statement that the diagram

$$\begin{array}{ccc} L & \longrightarrow & M \\ \downarrow & & \downarrow \\ K & \longrightarrow & N \end{array}$$

is commutative can be expressed in the equation $LMN = LKN$. In the abbreviated notation, if m is an element of M, then its image under the homomorphism MN is written as $(m)MN$ or as mMN.

Proposition 1. *A monomorphism $L \to M$ is direct if and only if there is a Λ-homomorphism $M \to L$ such that LML = identity. An epimorphism $M \to N$ is direct if and only if there is a Λ-homomorphism $N \to M$ such that NMN = identity.*

Proof. We shall prove the first part of the proposition, the proof of the second part being similar. Assume first that $L \to M$ is direct. Put $A = \operatorname{Im}(LM)$, then there exists a submodule B of M such that $M = A + B$ (direct sum). The monomorphism $L \to M$ determines an isomorphism of L on to A or, in other words, LMA is an isomorphism, where $M \to A$ is the canonical mapping corresponding to the decomposition $M = A + B$. Let AL be the inverse isomorphism and put $ML = MAL$. Then

$$LML = (LM)(ML) = (LM)(MAL) = (LMA)(AL) = \text{identity}.$$

Conversely, suppose that we have a Λ-homomorphism ML such that $LML = $ identity. Put $A = \operatorname{Im}(LM)$ and $B = \operatorname{Ker}(ML)$. We shall show that $M = A + B$ (direct sum). Let m be an element of M then $mMLM = a$, where a is an element of A. Also

$$(m-a)\,ML = mML - mMLML = mML - mML = 0.$$

Thus $m - a = b$, where b is an element of B. This shows that every element of M can be written in the form $a + b$, where a is an element of A and b is an element of B. Now suppose that $a' + b' = 0$, where a' is an element of A and b' is an element of B. We shall show that $a' = b' = 0$, thereby proving that $M = A + B$ (direct sum). To do this we observe that $0 = (a' + b')\,ML = a'ML$. But $a' = lLM$ for a suitable element l of L, consequently $0 = a'ML = lLML = l$, and therefore $a' = lLM = 0$. Since this implies that $b' = 0$ the proof is complete.

Corollary. *Any epimorphism $M \to F$, where F is free, is direct.*

Proof. Let FF denote the identity map of F. By Theorem 3 it is possible to find a Λ-homomorphism $F \to M$ such that the diagram

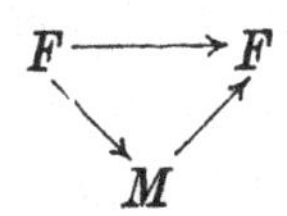

is commutative. Then $FMF = FF = $ identity and now the corollary follows from the theorem.

1.11 Sequences of Λ-homomorphisms

Let
$$L \to M \to N \tag{1.11.1}$$
be a three-term sequence of Λ-modules and Λ-homomorphisms. We shall say that (1.11.1) is a *0-sequence* if $LMN = 0$, that is, if

$$\operatorname{Im}(LM) \subseteq \operatorname{Ker}(MN).$$

Again, we shall say that (1.11.1) is *exact* if $\mathrm{Im}(LM) = \mathrm{Ker}(MN)$. As important examples let us note that $L \to M$ is a monomorphism if and only if $0 \to L \to M$ is exact, while $M \to N$ is an epimorphism if and only if $M \to N \to 0$ is exact.

More generally, a sequence

$$\ldots \to L_{n-1} \to L_n \to L_{n+1} \to L_{n+2} \to \ldots,$$

which may be finite, infinite, or semi-infinite, will be called a *0-sequence* if every triplet $L_r \to L_{r+1} \to L_{r+2}$ is a 0-sequence, and it will be called *exact* if every triplet is exact. Let us observe that if L is a submodule of M then the canonical sequence

$$0 \to L \to M \to M/L \to 0$$

is exact. Particularly important are exact sequences of the slightly more general type

$$0 \to L \to M \to N \to 0. \tag{1.11.2}$$

Definition. An exact sequence (1.11.2) is said to be *direct* or to *split* if $\mathrm{Im}(LM) = \mathrm{Ker}(MN)$ is a direct summand of M.

For example, if $A + B$ is the direct sum of A and B, then the canonical exact sequence

$$0 \to A \to (A + B) \to B \to 0$$

splits. Or again, any exact sequence of the type

$$0 \to L \to M \to F \to 0,$$

where F is free, splits. This follows from the corollary to Proposition 1.

Proposition 2. *Let* $\quad 0 \to L \to M \to N \to 0 \qquad (1.11.3)$

be a given sequence of Λ-modules and Λ-homomorphisms. Then in order that (1.11.3) *should be a split exact sequence it is necessary and sufficient that there should exist Λ-homomorphisms $M \to L$ and $N \to M$ such that*

$$\left.\begin{array}{c} LML = \text{identity}, \quad NMN = \text{identity}, \quad MLM + MNM = \text{identity}, \\ LMN = 0, \quad NML = 0. \end{array}\right\} \tag{1.11.4}$$

Remark. It should be observed that, by Theorem 4, (1.11.4) is equivalent to the statement that

$$L \to M \to L, \quad N \to M \to N$$

is a complete representation of M as a direct sum.

Proof. Assume that (1.11.3) is a split exact sequence. Then

$$\mathrm{Im}(LM) = \mathrm{Ker}(MN) = A \quad \text{(say)}$$

and $M = A + B$ (direct sum) for a suitable submodule B of M. Let MA and BM be the canonical homomorphisms associated with this direct decomposition, then LMA and BMN are isomorphisms. Denote by AL and NB the inverse isomorphisms and put $ML = MAL$, $NM = NBM$. Then

$$LML = (LM)(MAL) = (LMA)(AL) = \text{identity},$$

$$NMN = (NBM)(MN) = (NB)(BMN) = \text{identity},$$

and $NML = NBMAL = 0$, since $BMA = 0$, while $LMN = 0$ by hypothesis. Further

$$MLM = (MAL)(LMAM) = (MA)(AL.LMA)(AM) = MAM,$$

$$MNM = (MBMN)(NBM) = (MB)(BMN.NB)(BM) = MBM,$$

and therefore

$$MLM + MNM = MAM + MBM = \text{identity}$$

as required.

Conversely, let us assume that we have homomorphisms $M \to L$ and $N \to M$ for which (1.11.4) is satisfied. We first show that (1.11.3) is exact. Since $LML = \text{identity}$, LM is a monomorphism and, since $NMN = \text{identity}$, MN is an epimorphism. Also, since $LMN = 0$, $\text{Im}(LM) \subseteq \text{Ker}(MN)$. Next let m be an element of $\text{Ker}(MN)$, then

$$m = mMLM + mMNM = mMLM \subseteq \text{Im}(LM).$$

Thus $\text{Ker}(MN) \subseteq \text{Im}(LM)$, and the combined remarks show that (1.11.3) is exact. To prove that (1.11.3) splits we need now only show that $L \to M$ is direct. But this is clear from Proposition 1, because $LML = \text{identity}$.

2

TENSOR PRODUCTS AND GROUPS OF HOMOMORPHISMS

Notation. Λ denotes a ring, which is not necessarily commutative but which possesses an identity element, and Z denotes the ring of integers.

2.1 The definition of tensor products

In section (1.4) we had occasion to observe that if B and C are left Λ-modules then $\mathrm{Hom}_\Lambda(B, C)$ has the structure of an abelian group. This is an important example of a method of obtaining a new module from two given modules, and we shall have quite a lot more to say about $\mathrm{Hom}_\Lambda(B, C)$ in section (2.5). For the present, however, we shall study a construction which yields an abelian group when we are given both a right Λ-module and a left Λ-module. In this construction, which is one of the most important of modern algebra, the resulting group is known as the 'tensor product' of the two Λ-modules.

Coming now to the details, suppose that we have a right Λ-module B and a left Λ-module C. We shall use $Z(B, C)$ to designate the free Z-module generated by the set of symbols (b, c), where b belongs to B and c belongs to C. Denote by $Y(B, C)$ the smallest submodule of $Z(B, C)$ which contains all elements of each of the forms

$$\text{(i)} \quad (b_1+b_2, c)-(b_1, c)-(b_2, c),$$

$$\text{(ii)} \quad (b, c_1+c_2)-(b, c_1)-(b, c_2),$$

$$\text{(iii)} \quad (b\lambda, c)-(b, \lambda c),$$

where b, b_1, b_2 belong to B, c, c_1, c_2 belong to C and λ belongs to Λ. Put

$$Z(B, C)/Y(B, C) = B \otimes_\Lambda C,$$

and denote by $b \otimes c$ the image of (b, c) in the natural mapping

$$Z(B, C) \to Z(B, C)/Y(B, C) = B \otimes_\Lambda C.$$

By construction $B \otimes_\Lambda C$ is a Z-module. We call it the *tensor product of B and C over Λ*. Since every element of $Y(B, C)$ is mapped into zero we have

$$\text{(i)} \quad (b_1+b_2) \otimes c = b_1 \otimes c + b_2 \otimes c,$$

$$\text{(ii)} \quad b \otimes (c_1+c_2) = b \otimes c_1 + b \otimes c_2,$$

$$\text{(iii)} \quad (b\lambda) \otimes c = b \otimes (\lambda c).$$

Remarks. (*a*) If c is kept fixed, the mapping $b \to b \otimes c$ is a Z-homomorphism of B into $B \otimes_\Lambda C$, and if b is kept fixed $c \to b \otimes c$ is a Z-homomorphism of C into $B \otimes_\Lambda C$. Accordingly, if m is an integer then

$$(mb) \otimes c = m(b \otimes c) = b \otimes (mc).$$

In particular
$$0 \otimes c = 0 \quad = b \otimes 0$$
and
$$(-b) \otimes c = -(b \otimes c) = b \otimes (-c).$$

(*b*) Since every element of $Z(B, C)$ is a finite sum $\sum_i m_i(b_i, c_i)$, we see that every element of $B \otimes_\Lambda C$ is a finite sum $\Sigma m_i(b_i \otimes c_i)$. But

$$m_i(b_i \otimes c_i) = (m_i b_i) \otimes c_i,$$

hence *every element of $B \otimes_\Lambda C$ is a finite sum of elements of the form $b \otimes c$.*

2.2 Tensor products over commutative rings

Suppose, for the moment, that Λ is a commutative ring and let λ belong to Λ. There is a Z-homomorphism $Z(B, C) \to B \otimes_\Lambda C$ in which (b, c) maps into $(b\lambda) \otimes c$. In this homomorphism elements of the form

$$(b_1 + b_2, c) - (b_1, c) - (b_2, c)$$
and
$$(b, c_1 + c_2) - (b, c_1) - (b, c_2)$$

are mapped into zero. Also the image of any element of the form

$$(b\lambda', c) - (b, \lambda' c)$$

is zero, because Λ is commutative. Thus our homomorphism vanishes on $Y(B, C)$ and so there is induced a Z-homomorphism

$$B \otimes_\Lambda C \xrightarrow{(\lambda)} B \otimes_\Lambda C.$$

If x belongs to $B \otimes_\Lambda C$, we shall use λx to denote its image in this homomorphism. Then

$$\lambda(x_1 + x_2 + \ldots + x_n) = \lambda x_1 + \lambda x_2 + \ldots + \lambda x_n, \tag{2.2.1}$$

where $x_1, x_2, \ldots, x_n$ belong to $B \otimes_\Lambda C$, and

$$\lambda(b \otimes c) = (b\lambda) \otimes c. \tag{2.2.2}$$

We now assert that if λ_1, λ_2 are any elements of Λ and x belongs to $B \otimes_\Lambda C$, then

$$(\lambda_1 + \lambda_2) x = \lambda_1 x + \lambda_2 x, \tag{2.2.3}$$

$$\lambda_1(\lambda_2 x) = (\lambda_1 \lambda_2) x, \tag{2.2.4}$$

$$1x = x. \tag{2.2.5}$$

In other words, we assert that our definition of λx makes $B \otimes_\Lambda C$ into a Λ-module. But in order to prove (2.2.3), (2.2.4) and (2.2.5), we need only consider the case in which x has the form $b \otimes c$. (This follows from (2.2.1) and the fact that every element of $B \otimes_\Lambda C$ is expressible as a finite sum of terms of the form $b \otimes c$.) However, by virtue of (2.2.2), all of (2.2.3), (2.2.4) and (2.2.5) are trivial in this case.

Summary. In the general (non-commutative) case $B \otimes_\Lambda C$ is a Z-module, but when Λ is commutative it can be given the structure of a Λ-module by writing

$$\lambda(b \otimes c) = (b\lambda) \otimes c = b \otimes (\lambda c).$$

2.3 Continuation of the general discussion

We now abandon the assumption that Λ is commutative and suppose, as before, that B is a right Λ-module and C a left Λ-module. Λ itself can, of course, be regarded both as a right Λ-module and as a left Λ-module.

Theorem 1. *There is a canonical Z-isomorphism*

$$f : \Lambda \otimes_\Lambda C \approx C$$

in which $f(\lambda \otimes c) = \lambda c$. If Λ is commutative then f is a Λ-isomorphism.

Remark. There is, of course, a similar canonical isomorphism $B \otimes_\Lambda \Lambda \approx B$ for right Λ-modules.

Proof. The Z-homomorphism of $Z(\Lambda, C)$ into C in which (λ, c) is mapped into λc is easily seen to vanish on $Y(\Lambda, C)$ and so it induces a Z-homomorphism

$$f : \Lambda \otimes_\Lambda C \to C$$

in which $f(\lambda \otimes c) = \lambda c$. Since $f(1 \otimes c) = c$ for all c in C, we see that f is an epimorphism. Suppose now that x belongs to $\Lambda \otimes_\Lambda C$. Then, with an obvious notation,

$$x = \sum_i (\lambda_i \otimes c_i) = \sum_i (1\lambda_i \otimes c_i) = \sum_i (1 \otimes \lambda_i c_i) = 1 \otimes (\sum_i \lambda_i c_i),$$

or $x = 1 \otimes c$ for a suitable element c of C. If therefore

$$f(x) = f(1 \otimes c) = 0,$$

then $c = 0$ and hence $x = 0$. Thus f is not only an epimorphism but also a monomorphism, that is to say, it is an isomorphism.

Assume next that Λ is commutative, that λ belongs to Λ and that x belongs to $\Lambda \otimes_\Lambda C$. Writing x in the form $x = 1 \otimes c$ we have

$$f(\lambda x) = f(\lambda \otimes c) = \lambda c = \lambda f(x),$$

and therefore f is a Λ-isomorphism.

2.4 Tensor products of homomorphisms

As before let B and C denote right and left Λ-modules respectively and let $f: B \to B'$ and $g: C \to C'$ be Λ-homomorphisms. Then the Z-homomorphism

$$Z(B, C) \to B' \otimes_\Lambda C',$$

in which (b, c) is mapped into $f(b) \otimes g(c)$, is seen to vanish on $Y(B, C)$. It therefore induces a Z-homomorphism

$$(f \otimes g) : B \otimes_\Lambda C \to B' \otimes_\Lambda C'$$

for which $(f \otimes g)(b \otimes c) = f(b) \otimes g(c)$. The mapping $f \otimes g$ is called the *tensor product* of f and g. When Λ is commutative $f \otimes g$ is not only a Z-homomorphism but also a Λ-homomorphism.

It should be noted that if f, f_1, f_2 are Λ-homomorphisms $B \to B'$ and g, g_1, g_2 are Λ-homomorphisms $C \to C'$ then

(i) $f \otimes (g_1 + g_2) = f \otimes g_1 + f \otimes g_2$ and $f \otimes 0 = 0$,

(ii) $(f_1 + f_2) \otimes g = f_1 \otimes g + f_2 \otimes g$ and $0 \otimes g = 0$,

as may be seen by applying both sides of each equation to a general element of the form $b \otimes c$. Further, if Λ is a commutative ring and λ belongs to Λ, then

(iii) $(\lambda f) \otimes g = \lambda(f \otimes g) = f \otimes (\lambda g)$.

Theorem 2. *If $i : B \to B$ and $j : C \to C$ are identity maps then*

$$i \otimes j : B \otimes_\Lambda C \to B \otimes_\Lambda C$$

is an identity map. If $f: B \to B'$, $f' : B' \to B''$, $g : C \to C'$, $g' : C' \to C''$ are Λ-homomorphisms, then

$$(f' \otimes g')(f \otimes g) = (f'f) \otimes (g'g).$$

The proof of this theorem is completely trivial.

Corollary. *If $f : B \approx B'$, $g : C \approx C'$ are Λ-isomorphisms, then*

$$f \otimes g : B \otimes_\Lambda C \to B' \otimes_\Lambda C'$$

is an isomorphism.

Proof. Let $\phi : B' \approx B$ and $\psi : C' \approx C$ be the inverse isomorphisms. Then, by the theorem, $(f \otimes g)(\phi \otimes \psi)$ and $(\phi \otimes \psi)(f \otimes g)$ are both identity maps, and this implies, as we saw in section (1.5), that $f \otimes g$ and $\phi \otimes \psi$ are inverse isomorphisms.

Theorem 3. *Let the* Λ*-homomorphisms*

$$B_i \xrightarrow{f_i} B \xrightarrow{\phi_i} B_i \quad (i \in I) \tag{2.4.1}$$

and
$$C_j \xrightarrow{g_j} C \xrightarrow{\psi_j} C_j \quad (j \in J) \tag{2.4.2}$$

be complete representations of B and C as direct sums of Λ*-modules. Then the Z-homomorphisms*

$$B_i \otimes_\Lambda C_j \xrightarrow{f_i \otimes g_j} B \otimes_\Lambda C \xrightarrow{\phi_i \otimes \psi_j} B_i \otimes_\Lambda C_j \quad (i \in I, j \in J) \tag{2.4.3}$$

are a complete representation of $B \otimes_\Lambda C$ as a direct sum of Z-modules.

Remarks. For the notion of a complete representation see section (1.9). It should be noted that, in Theorem 3, I and J need not be finite. Also, when Λ is commutative, (2.4.3) is a complete representation of $B \otimes_\Lambda C$ as a direct sum of Λ-*modules*.

Proof. Let i, i' belong to I and j, j' to J, then

$$(\phi_{i'} \otimes \psi_{j'})(f_i \otimes g_j) = (\phi_{i'} f_i) \otimes (\psi_{j'} g_j),$$

which is an identity map if $(i', j') = (i, j)$ and is null otherwise. Accordingly, by Theorem 4 of section (1.9), we need only show that

$$\sum_{i,j} (f_i \otimes g_j)(\phi_i \otimes \psi_j) x = x$$

for each element x of $B \otimes_\Lambda C$, and this will follow if we establish it first in the case when x has the form $b \otimes c$. Assume therefore that $x = b \otimes c$. By (2.4.1) and (2.4.2),

$$b = \sum_i f_i(b_i), \quad c = \sum_j g_j(c_j),$$

where $[b_i]_{i \in I}$ belongs to $\Sigma_i B_i$ and $[c_j]_{j \in J}$ to $\Sigma_j C_j$. Consequently

$$x = b \otimes c = \sum_{i,j} f_i(b_i) \otimes g_j(c_j),$$

and therefore

$$(\phi_{i'} \otimes \psi_{j'}) x = \sum_{i,j} (\phi_{i'} f_i(b_i)) \otimes (\psi_{j'} g_j(c_j)) = b_{i'} \otimes c_{j'}$$

or, changing the notation,

$$(\phi_i \otimes \psi_j) x = b_i \otimes c_j.$$

Hence $$\sum_{i,j} (f_i \otimes g_j)(\phi_i \otimes \psi_j)x = \sum_{i,j} (f_i \otimes g_j)(b_i \otimes c_j)$$
$$= \sum_{i,j} f_i(b_i) \otimes g_j(c_j) = x.$$

This completes the proof of the theorem.

Corollary. *Let C be a free left Λ-module with base $[\gamma_i]_{i\in I}$ and let B be an arbitrary right Λ-module. Then each element of $B \otimes_\Lambda C$ has a unique representation in the form*

$$\sum_i (b_i \otimes \gamma_i),$$

where b_i belongs to B and $b_i = 0$ for almost all i.

Remark. There is, of course, a similar result when B is free and C is arbitrary.

Proof. Let b belong to B and c to C, then $c = \sum_i \lambda_i \gamma_i$, where λ_i belongs to Λ and $\lambda_i = 0$ for almost all i. Thence

$$b \otimes c = \sum_i (b \otimes \lambda_i \gamma_i) = \sum_i (b\lambda_i \otimes \gamma_i) = \sum_i (b_i \otimes \gamma_i),$$

where b_i belongs to B and $b_i = 0$ for almost all i. Thus $b \otimes c$ has a representation of the required form and therefore the same is true for every element of $B \otimes_\Lambda C$.

Suppose now that $\sum_i (b_i \otimes \gamma_i) = 0$, where b_i belongs to B and $b_i = 0$ for almost all i. We wish to show that $b_i = 0$ for *all* i. Since $[\gamma_i]_{i\in I}$ is a base of C, the inclusion maps $\Lambda\gamma_i \to C$ form an injective representation of C as a direct sum. Consequently the maps $B \otimes_\Lambda \Lambda\gamma_i \to B \otimes_\Lambda C$, to which they give rise, constitute a representation of $B \otimes_\Lambda C$ as a direct sum of Z-modules and therefore determine an isomorphism

$$\sum_i (B \otimes_\Lambda \Lambda\gamma_i) \approx B \otimes_\Lambda C.$$

In this isomorphism the element $[b_i \otimes \gamma_i]_{i\in I}$ of $\sum_i (B \otimes_\Lambda \Lambda\gamma_i)$ maps into $\sum_i (b_i \otimes \gamma_i)$ which is zero by hypothesis. Hence, for each i, $b_i \otimes \gamma_i$, as an element of $B \otimes_\Lambda \Lambda\gamma_i$, is zero. Now there is a Λ-isomorphism $\Lambda \approx \Lambda\gamma_i$ in which an element λ of Λ is mapped into $\lambda\gamma_i$ and this yields an isomorphism $B \otimes_\Lambda \Lambda \approx B \otimes_\Lambda \Lambda\gamma_i$. But, by Theorem 1, we also have a canonical isomorphism $B \approx B \otimes_\Lambda \Lambda$ and so, on combining, we obtain an isomorphism $B \approx B \otimes_\Lambda \Lambda\gamma_i$. In this isomorphism the element b_i of B maps into $b_i \otimes \gamma_i$, which is zero. Consequently $b_i = 0$ and this completes the proof.

Before proceeding to the next main result concerning tensor products it will be convenient to establish

Lemma 1. *If* $f : B \to B'$ *and* $g : C \to C'$ *are epimorphisms of* Λ-*modules, then*

$$f \otimes g : B \otimes_\Lambda C \to B' \otimes_\Lambda C'$$

is also an epimorphism.

Proof. Let b' belong to B' and c' to C'. Then $b' = f(b)$ and $c' = g(c)$ for suitable elements b in B and c in C. Thus

$$(f \otimes g)(b \otimes c) = f(b) \otimes g(c) = b' \otimes c'$$

and so $b' \otimes c'$ belongs to $\operatorname{Im}(f \otimes g)$. Consequently it follows that

$$\operatorname{Im}(f \otimes g) = B' \otimes_\Lambda C'.$$

Theorem 4. *If*

$$0 \to B' \xrightarrow{f} B \xrightarrow{g} B'' \to 0$$

and

$$0 \to C' \xrightarrow{\phi} C \xrightarrow{\psi} C'' \to 0$$

are exact sequences of Λ-*modules, then the derived sequences*

$$B' \otimes_\Lambda C \xrightarrow{f^*} B \otimes_\Lambda C \xrightarrow{g^*} B'' \otimes_\Lambda C \to 0 \tag{2.4.4}$$

and

$$B \otimes_\Lambda C' \xrightarrow{\phi^*} B \otimes_\Lambda C \xrightarrow{\psi^*} B \otimes_\Lambda C'' \to 0 \tag{2.4.5}$$

are also exact.

Remarks. Here f^* is the tensor product of f and the identity map of C while g^*, ϕ^*, ψ^* are defined similarly. Note that no assertions are made concerning the kernels of f^* and ϕ^*.

Proof. We shall only prove (2.4.4) since the proof of (2.4.5) is similar. By Lemma 1, g^* is an epimorphism. Further, since $gf = 0$, we have $g^*f^* = 0$ which shows that $\operatorname{Im}(f^*) \subseteq \operatorname{Ker}(g^*)$. Suppose now that b'' belongs to B'' and c to C. Choose an element b in B so that $g(b) = b''$. We assert that the image $q(b \otimes c)$ of $b \otimes c$ in the natural map

$$q : B \otimes_\Lambda C \to (B \otimes_\Lambda C)/\operatorname{Im}(f^*)$$

depends only on b'', c and not on the choice of b. For suppose that we have $g(b_1) = b''$ then $g(b - b_1) = 0$ and therefore $b - b_1 = f(b')$ for some b' in B'. Thus

$$b \otimes c - b_1 \otimes c = f(b') \otimes c = f^*(b' \otimes c) \in \operatorname{Im}(f^*),$$

and so $q(b \otimes c) = q(b_1 \otimes c)$ as stated.

Consider the Z-homomorphism

$$Z(B'', C) \to (B \otimes_\Lambda C)/\mathrm{Im}(f^*)$$

in which (b'', c) is mapped into $q(b \otimes c)$. This homomorphism is seen at once to vanish on $Y(B'', C)$, and so it induces a Z-homomorphism

$$u : B'' \otimes_\Lambda C \to (B \otimes_\Lambda C)/\mathrm{Im}(f^*),$$

in which $u(b'' \otimes c) = q(b \otimes c)$, where b is any element of B such that $g(b) = b''$. Again, since $\mathrm{Im}(f^*) \subseteq \mathrm{Ker}(g^*)$, we have $g^*(\mathrm{Im}(f^*)) = 0$, consequently g^* induces a Z-homomorphism

$$v : (B \otimes_\Lambda C)/\mathrm{Im}(f^*) \to B'' \otimes_\Lambda C$$

whose kernel is $\mathrm{Ker}(g^*)/\mathrm{Im}(f^*)$. If therefore we show that uv = identity, this will show that v is a monomorphism and hence that $\mathrm{Ker}(g^*)/\mathrm{Im}(f^*)$ is zero. In other words, it will follow that

$$\mathrm{Ker}(g^*) = \mathrm{Im}(f^*),$$

and this will complete the proof.

Let b belong to B and c to C and write $q(b \otimes c) = x$. Then, since every element of $(B \otimes_\Lambda C)/\mathrm{Im}(f^*)$ is a finite sum of elements having the same form as x, it will suffice to show that $uv(x) = x$. But

$$v(x) = g^*(b \otimes c) = g(b) \otimes c$$

and so $$uv(x) = u(g(b) \otimes c) = q(b \otimes c) = x.$$

This completes the proof of the theorem.

Theorem 5. *Let* $$0 \to B' \xrightarrow{f} B \xrightarrow{g} B'' \to 0$$

be an exact sequence of right Λ-modules and let F be a free left Λ-module. Then the sequence

$$0 \to B' \otimes_\Lambda F \xrightarrow{f^*} B \otimes_\Lambda F \xrightarrow{g^*} B'' \otimes_\Lambda F \to 0,$$

to which this situation gives rise, is exact. Furthermore, there is a corresponding result in which the roles of left and right modules are interchanged.

Proof. By Theorem 4, we need only show that f^* is a monomorphism. Suppose then that x belongs to $B' \otimes_\Lambda F$ and that $f^*(x) = 0$. We have to show that $x = 0$.

Let $[\gamma_i]_{i \in I}$ be a base of F then, by the corollary to Theorem 3, $x = \sum_i (b_i' \otimes \gamma_i)$, where b_i' belongs to B' and $b_i' = 0$ for almost all i. Thus

$$0 = f^*(x) = \sum_i (f(b_i') \otimes \gamma_i),$$

and therefore, by the same corollary, $f(b_i') = 0$ for all i. But f is a monomorphism; accordingly $b_i' = 0$ for all i and consequently $x = 0$.

2.5 The principal properties of $\mathrm{Hom}_\Lambda(B, C)$

We now abandon the assumption that B and C are right and left Λ-modules respectively and suppose instead that both are left Λ-modules. Then, as we saw in section (1.4), $\mathrm{Hom}_\Lambda(B, C)$ is an abelian group which, when Λ is commutative, can be regarded as a Λ-module. It will appear, from the results to be established shortly, that the tensor product $B \otimes_\Lambda C$ of the preceding sections, and $\mathrm{Hom}_\Lambda(B, C)$ of the present one are to some extent complementary concepts.

Theorem 6. *The mapping* $f : \mathrm{Hom}_\Lambda(\Lambda, B) \to B$ *defined by* $f(\phi) = \phi(1)$ *is a* Z*-isomorphism. When* Λ *is commutative* f *is a* Λ*-isomorphism.*

Proof. It is clear, with an obvious notation, that

$$f(\phi_1 + \phi_2) = f(\phi_1) + f(\phi_2),$$

hence f is a Z-homomorphism. Further, if $f(\phi) = 0$, then $\phi(1) = 0$ and so

$$\phi(\lambda) = \phi(\lambda 1) = \lambda\phi(1) = 0$$

for all λ in Λ. Accordingly $\phi = 0$, and this shows that f is a monomorphism. Next suppose that b belongs to B, then the mapping $\lambda \to \lambda b$ is a Λ-homomorphism $\phi : \Lambda \to B$. For this homomorphism $f(\phi) = \phi(1) = b$, and now the proof that f is a Z-isomorphism is complete.

Finally, if Λ is commutative and λ belongs to Λ, then

$$f(\lambda\phi) = (\lambda\phi)(1) = \lambda(\phi(1)) = \lambda f(\phi),$$

and so f is not only a Z-isomorphism but also a Λ-isomorphism. This establishes the theorem.

Consider the way in which $\mathrm{Hom}_\Lambda(B, C)$ is transformed when the modules B and C, from which it is constructed, are subjected to Λ-homomorphisms. To this end assume that $f : B' \to B$ and $g : C \to C'$ are Λ-homomorphisms and let ϕ be a variable element of $\mathrm{Hom}_\Lambda(B, C)$ so that ϕ is a Λ-linear mapping of B into C. Then $g\phi f$ is a Λ-homomorphism of B' into C' and therefore it belongs to $\mathrm{Hom}_\Lambda(B', C')$. The mapping $\phi \to g\phi f$ is now seen to be a Z-homomorphism, usually denoted by $\mathrm{Hom}(f, g)$, of $\mathrm{Hom}_\Lambda(B, C)$ into $\mathrm{Hom}_\Lambda(B', C')$. Thus we have

$$\mathrm{Hom}(f, g) : \mathrm{Hom}_\Lambda(B, C) \to \mathrm{Hom}_\Lambda(B', C'),$$

and, as is clear from the definition, *when* Λ *is a commutative ring,* $\mathrm{Hom}(f, g)$ *is a* Λ*-homomorphism of* $\mathrm{Hom}_\Lambda(B, C)$ *into* $\mathrm{Hom}_\Lambda(B', C')$ *when these are regarded as* Λ*-modules.*

There are certain elementary properties of $\mathrm{Hom}(f, g)$ which need to be noted. Let f, f_1, f_2 be Λ-homomorphisms $B' \to B$ and let g, g_1, g_2 be Λ-homomorphisms $C \to C'$. Then, by the definition,

(i) $\mathrm{Hom}(f_1+f_2, g) = \mathrm{Hom}(f_1, g) + \mathrm{Hom}(f_2, g)$ and $\mathrm{Hom}(0, g) = 0$;

(ii) $\mathrm{Hom}(f, g_1+g_2) = \mathrm{Hom}(f, g_1) + \mathrm{Hom}(f, g_2)$ and $\mathrm{Hom}(f, 0) = 0$.

Also, when Λ is commutative and λ belongs to Λ,

(iii) $\mathrm{Hom}(\lambda f, g) = \lambda\,\mathrm{Hom}(f, g) = \mathrm{Hom}(f, \lambda g)$.

Theorem 7. *If $i : B \to B$ and $j : C \to C$ are identity maps then $\mathrm{Hom}(i, j)$ is the identity map of $\mathrm{Hom}_\Lambda(B, C)$. If $f : B' \to B, f' : B'' \to B'$, $g : C \to C'$, $g' : C' \to C''$ are all Λ-homomorphisms, then*

$$\mathrm{Hom}(ff', g'g) = \mathrm{Hom}(f', g')\,\mathrm{Hom}(f, g).$$

Proof. The first assertion is trivial. Suppose now that ϕ belongs to $\mathrm{Hom}_\Lambda(B, C)$ then

$$[\mathrm{Hom}(ff', g'g)]\phi = (g'g)\phi(ff') = g'(g\phi f)f'$$
$$= [\mathrm{Hom}(f', g')\,\mathrm{Hom}(f, g)]\phi.$$

Corollary. *Λ-isomorphisms $f : B' \approx B$ and $g : C \approx C'$ give rise to an isomorphism*

$$\mathrm{Hom}(f, g) : \mathrm{Hom}_\Lambda(B, C) \approx \mathrm{Hom}_\Lambda(B', C').$$

To see this we need only observe that if $\phi : B \approx B'$ and $\psi : C' \approx C$ are the inverse isomorphisms then, by the theorem,

$$\mathrm{Hom}(f, g)\,\mathrm{Hom}(\phi, \psi) \quad \text{and} \quad \mathrm{Hom}(\phi, \psi)\,\mathrm{Hom}(f, g)$$

are identity maps.

Theorem 8. *Let*

$$B_i \xrightarrow{f_i} B \xrightarrow{g_i} B_i \quad (i \in I) \tag{2.5.1}$$

be a complete representation of B as a direct sum and let

$$C_j \xrightarrow{p_j} C \xrightarrow{q_j} C_j \quad (j \in J) \tag{2.5.2}$$

be a complete representation of C as a direct product of Λ-modules. Then

$$\mathrm{Hom}_\Lambda(B_i, C_j) \xrightarrow{\mathrm{Hom}(g_i, p_j)} \mathrm{Hom}_\Lambda(B, C) \xrightarrow{\mathrm{Hom}(f_i, q_j)} \mathrm{Hom}_\Lambda(B_i, C_j) \quad (i \in I, j \in J) \tag{2.5.3}$$

is a complete representation of $\text{Hom}_\Lambda(B, C)$ *as a direct product of Z-modules. If* Λ *is commutative the same mappings represent* $\text{Hom}_\Lambda(B, C)$ *as a direct product of* Λ*-modules.*

Proof. Since $\text{Hom}(f_{i'}, q_{j'})\,\text{Hom}(g_i, p_j) = \text{Hom}(g_i f_{i'}, q_j p_j)$ is an identity map if $(i', j') = (i, j)$ and is null otherwise, it suffices to show that the homomorphism

$$\omega : \text{Hom}_\Lambda(B, C) \rightarrow \prod_{i,j} \text{Hom}_\Lambda(B_i, C_j),$$

defined by means of the $\text{Hom}(f_i, q_j)$, is an isomorphism.

Let ϕ belong to $\text{Hom}_\Lambda(B, C)$ and suppose that $\phi \neq 0$. Then, for a suitable element b of B, $\phi(b) \neq 0$. But, since (2.5.1) is a direct sum representation,

$$b = \sum_i f_i(b_i),$$

where $[b_i]_{i \in I}$ belongs to $\sum_i B_i$. Hence, for a suitable i, $\phi f_i(b_i) \neq 0$. Now $\phi f_i(b_i)$ belongs to C and so its image under the isomorphism $C \approx \prod_j C_j$ defined by (2.5.2) is not zero. In other words, for a suitable j, $q_j \phi f_i(b_i) \neq 0$. Thus $\text{Hom}(f_i, q_j)\phi \neq 0$ and therefore $\omega(\phi) \neq 0$. Accordingly, this proves that ω is a monomorphism.

Let $[\psi_{ij}]$ belong to $\prod_{i,j} \text{Hom}_\Lambda(B_i, C_j)$. Since (2.5.1) is a direct sum representation, for each b in B, $g_i(b) = 0$ for almost all i. We may therefore write

$$\psi_j(b) = \sum_i \psi_{ij}\, g_i(b)$$

and this makes ψ_j a Λ-homomorphism of B into C_j. Let

$$\psi : B \rightarrow \prod_j C_j$$

be the Λ-homomorphism whose components are the ψ_j and let

$$\phi : B \rightarrow C$$

be the homomorphism obtained by combining $\psi : B \rightarrow \prod_j C_j$ with the isomorphism $\prod_j C_j \approx C$ given by (2.5.2). We shall complete the proof by showing that

$$\omega(\phi) = [\psi_{ij}],$$

and to do this we need only show that $q_j \phi f_i = \psi_{ij}$ for all i and j.

Let b belong to B. Then, by the definitions of ψ and ϕ,

$$[\psi_j(b)]_{j \in J} = [q_j \phi(b)]_{j \in J},$$

and so

$$q_j \phi(b) = \sum_{v \in I} \psi_{vj}\, g_v(b).$$

In this relation put $b = f_i(b_i)$, where b_i belongs to B_i, then

$$q_j\phi f_i(b_i) = \psi_{ij}(b_i),$$

and as this holds for all b_i in B_i, it follows that

$$q_j\phi f_i = \psi_{ij}.$$

This completes the proof.

To simplify the proof of the next theorem we first prove

Lemma 2. *If $f : B' \to B$ is an epimorphism and $g : C \to C'$ is a monomorphism then*

$$\mathrm{Hom}(f, g) : \mathrm{Hom}_\Lambda(B, C) \to \mathrm{Hom}_\Lambda(B', C')$$

is a monomorphism.

Proof. Suppose that ϕ belongs to $\mathrm{Hom}_\Lambda(B, C)$ and that

$$\mathrm{Hom}(f, g)\,\phi = 0.$$

We wish to show that $\phi = 0$. Let b' belong to B' then, because we are given that $g\phi f = 0$, we have $g\phi f(b') = 0$. If now b belongs to B we can, since f is an epimorphism, choose b' so that $f(b') = b$. This shows that $g\phi(b) = 0$ and hence that $\phi(b) = 0$ because g is a monomorphism. But b was an arbitrary element of B, consequently the proof that $\phi = 0$ is complete.

Theorem 9. *Let*

$$0 \to B' \xrightarrow{f} B \xrightarrow{g} B'' \to 0$$

and

$$0 \to C' \xrightarrow{\phi} C \xrightarrow{\psi} C'' \to 0$$

be exact sequences of Λ-modules. Then the sequences

$$0 \to \mathrm{Hom}_\Lambda(B'', C) \xrightarrow{g^*} \mathrm{Hom}_\Lambda(B, C) \xrightarrow{f^*} \mathrm{Hom}_\Lambda(B', C) \tag{2.5.4}$$

and

$$0 \to \mathrm{Hom}_\Lambda(B, C') \xrightarrow{\phi^*} \mathrm{Hom}_\Lambda(B, C) \xrightarrow{\psi^*} \mathrm{Hom}_\Lambda(B, C''), \tag{2.5.5}$$

to which they give rise, are also exact.

Proof. Let i and j be the identity mappings of B and C respectively. Then

$$f^* = \mathrm{Hom}(f, j), \quad g^* = \mathrm{Hom}(g, j),$$

$$\phi^* = \mathrm{Hom}(i, \phi), \quad \psi^* = \mathrm{Hom}(i, \psi),$$

consequently, by Lemma 2, g^* and ϕ^* are monomorphisms. Again

$$f^*g^* = \mathrm{Hom}(gf, j) = 0 \quad \text{and} \quad \psi^*\phi^* = \mathrm{Hom}(i, \psi\phi) = 0,$$

hence

$$\mathrm{Im}(g^*) \subseteq \mathrm{Ker}(f^*) \quad \text{and} \quad \mathrm{Im}(\phi^*) \subseteq \mathrm{Ker}(\psi^*).$$

Let p belong to $\text{Ker}(f^*)$ then $pf = f^*(p) = 0$ and so

$$p(\text{Ker}(g)) = p(\text{Im}(f)) = 0.$$

Thus $p : B \to C$ induces a homomorphism $B/\text{Ker}(g) \to C$ which makes the diagram

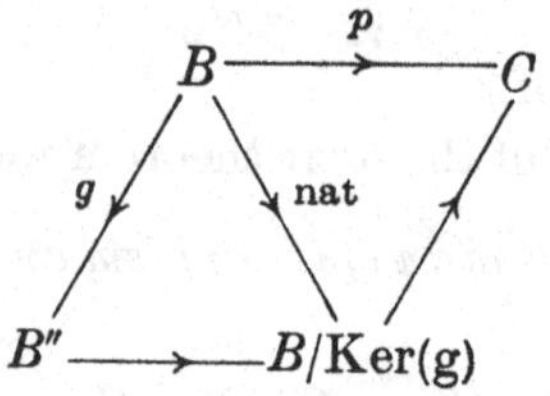

$$\begin{array}{ccc} B & \xrightarrow{p} & C \\ {\scriptstyle g}\swarrow & \searrow{\scriptstyle \text{nat}} & \nearrow \\ B'' & \longrightarrow & B/\text{Ker(g)} \end{array}$$

commutative, where $B'' \to B/\text{Ker}(g)$ is the canonical isomorphism induced by the epimorphism g. Let $q : B'' \to C$ be the combined mapping

$$B'' \to B/\text{Ker}(g) \to C.$$

Then

$$g^*(q) = qg = p,$$

and therefore p belongs to $\text{Im}(g^*)$. Accordingly

$$\text{Ker}(f^*) \subseteq \text{Im}(g^*)$$

and this establishes the exactness of (2.5.4).

Now suppose that p_1 belongs to $\text{Ker}(\psi^*)$, then

$$\psi p_1 = \psi^*(p_1) = 0$$

and so $\text{Im}(p_1) \subseteq \text{Ker}(\psi) = \text{Im}(\phi)$. Thus by restricting the range, p_1 determines a homomorphism $B \to \text{Im}(\phi)$ and, since ϕ is a monomorphism, we have an isomorphism $\text{Im}(\phi) \approx C'$. Let q_1 be the combined mapping $B \to \text{Im}(\phi) \to C'$. Then $\phi q_1 = p_1$, hence

$$p_1 = \phi q_1 = \phi^*(q_1) \in \text{Im}(\phi^*)$$

and so $\text{Ker}(\psi^*) \subseteq \text{Im}(\phi^*)$. With this the proof of the theorem is complete.

Theorem 10. *Let F be a free Λ-module and let*

$$0 \to C' \xrightarrow{\phi} C \xrightarrow{\psi} C'' \to 0$$

be an exact sequence. Then the sequence

$$0 \to \text{Hom}_\Lambda(F, C') \xrightarrow{\phi^*} \text{Hom}_\Lambda(F, C) \xrightarrow{\psi^*} \text{Hom}_\Lambda(F, C'') \to 0$$

is exact.

Proof. By Theorem 9, we need only show that ψ^* is an epimorphism. Let g belong to $\text{Hom}_\Lambda(F, C'')$, then, by Theorem 3 of section

(1.8), we can find a Λ-homomorphism $h : F \to C$ which makes the diagram

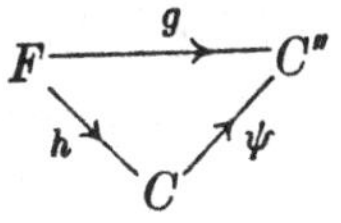

$$\begin{array}{ccc} F & \xrightarrow{g} & C'' \\ & {}_{h}\searrow \quad \nearrow_{\psi} & \\ & C & \end{array}$$

commutative. Thus $\psi^*(h) = \psi h = g$ which shows that ψ^* has the required property.

It will be observed that there is a natural association, in pairs, of results concerning tensor products and groups of homomorphisms. It is this association that we have in mind when we say loosely that $A \otimes_\Lambda C$ and $\mathrm{Hom}_\Lambda(B, C)$ are complementary concepts. However, in some respects, the association of results is at present defective. For example, the information given by Theorem 10 is less complete than that given by Theorem 5 because we cannot say for what modules C the exactness of

$$0 \to B' \to B \to B'' \to 0$$

implies the exactness of

$$0 \to \mathrm{Hom}_\Lambda(B'', C) \to \mathrm{Hom}_\Lambda(B, C) \to \mathrm{Hom}_\Lambda(B', C) \to 0.$$

This deficiency will be removed later when we introduce *projective* and *injective* modules. Also, in Chapter 8, we shall have quite a lot more to say about $A \otimes_\Lambda C$ and $\mathrm{Hom}_\Lambda(B, C)$. However, these additional results will be needed only for applications, and so we postpone their discussion in order to press on with the main developments of our subject.

3

CATEGORIES AND FUNCTORS

Notation. Λ, Λ_1, Λ_2, etc., denote rings, which need not be commutative, each having an identity element. Z denotes the ring of integers.

3.1 Abstract mappings

From now on we shall be encountering a particular type of phenomenon with such frequency that it will be worth our while to introduce a special terminology in order to describe it. This is achieved by introducing the concepts of *category* and *functor*.

In dealing with modules we are, above all, concerned with certain objects, namely, the modules themselves, and certain natural mappings of these objects, to wit, their homomorphisms. Roughly speaking, and we shall make this idea precise very shortly, a collection of 'objects' together with certain 'natural mappings' of these objects into one another constitute what is called a *category*. Modules (over a fixed ring) and their homomorphisms form one category, and, as a contrasting example, we may mention the category formed by topological spaces and continuous mappings. However, the concept is a very comprehensive one indeed and examples could be multiplied indefinitely.

Suppose now that with each object of a category is associated an object of some other category and that, in some way, a mapping of objects in the first category gives rise to a mapping of the corresponding objects in the second category. This situation, again speaking very informally, is described by saying that we have a functor from the first to the second category or that we have a *functor of a single variable*. It is not difficult to generalize this notion to obtain what are called *functors of several variables* and, in fact, $\mathrm{Hom}_\Lambda(B, C)$ and $A \otimes_\Lambda C$ can be regarded as examples of functors of two variables.

It is time, however, for these somewhat vague preliminaries to give way to a more precise discussion. When the concept of a category is analysed, it turns out that it is the 'mappings' which are important while the 'objects' play a secondary almost inessential role. But, since in most cases of interest the emphasis is reversed, we shall take an early opportunity to fit the 'objects' into the abstract scheme which is being developed.

Turning now to the details, let $\mathscr{F}$ be a set of entities, of which $f, f_1, f_2, \ldots$ will be used to indicate typical specimens, and suppose that for certain ordered pairs (f_1, f_2) there is defined a 'composition' or 'product' $f_1 f_2$, which again belongs to $\mathscr{F}$.

Definition. An element i of $\mathscr{F}$, which has the property that both $if = f$ whenever the composition if is defined and also $fi = f$ whenever fi is defined, is called an *identity* or an *identity map*.

Definition. The system $\mathscr{F}$, with its law of composition, will be called a *system of abstract maps* and the elements f of $\mathscr{F}$ will be called (abstract) *maps* if the following conditions are satisfied:

(i) if either triple product $f_1(f_2 f_3)$ or $(f_1 f_2) f_3$ is defined, then so is the other and the two of them are equal;

(ii) if $f_1 f_2$ and $f_2 f_3$ are both defined then $f_1(f_2 f_3)$ and $(f_1 f_2) f_3$ are also defined;

(iii) if f belongs to $\mathscr{F}$ then there exist identities i_1, i_2 such that fi_1 and $i_2 f$ are both defined.

Let $\mathscr{F}$ be a system of abstract maps. We make a number of observations.

(*a*) If $f_1(f_2 f_3)$ and $(f_1 f_2) f_3$ are both defined then they are equal. We denote their common value by $f_1 f_2 f_3$.

(*b*) If f belongs to $\mathscr{F}$ then the identities i_1, i_2 such that fi_1 and $i_2 f$ are defined are unique. For suppose that fi'_1 and $i'_2 f$ are also defined, where i'_1 and i'_2 are identities. Then $(fi'_1)\, i_1$ is defined and therefore, by (i), $i'_1 i_1$ is defined. But now we see that $i'_1 = i'_1 i_1 = i_1$. Similarly, we may prove that $i_2 = i'_2$.

(*c*) If i is an identity then ii is defined and $ii = i$. Indeed, we have only to show that ii is defined. But ii' is defined for some identity i', and from this it follows that $i' = ii' = i$. Accordingly ii *is* defined.

3.2 Categories

By a *category* $\mathscr{C}$ we mean a system $\mathscr{F}$ of abstract maps and a collection of objects $C, C_1, C_2, \ldots$ which are in 1-1 correspondence $C \leftrightarrow i_C$ with the identities of $\mathscr{F}$. To simplify the language, both the objects C and the maps f will be said to *belong to* $\mathscr{C}$.

Let $\mathscr{C}$ be a category and f one of its maps.

Definition. The unique objects C_1 and C_2 such that fi_{C_1} and $i_{C_2}f$ are defined, are called the *domain* and *range* of f and we indicate their connexion with f by writing $f: C_1 \to C_2$.

It should be observed that if i is the identity map corresponding to an object C then $i: C \to C$.

Proposition 1. *Let $f: C_1 \to C_2$ and $f': C_1' \to C_2'$ belong to $\mathscr{C}$ then $f'f$ is defined if and only if $C_2 = C_1'$ and when that is so $f'f: C_1 \to C_2'$.*

Proof. Let i_1, i_2, i_1', i_2' be the identities associated with C_1, C_2, C_1', C_2' respectively, and suppose first that $f'f$ is defined. Then $f'(i_2 f)$ is defined, and this shows that $f'i_2$ is defined. Therefore $i_2 = i_1'$ and so $C_2 = C_1'$. Further, $(i_2'f')f$ and $f'(fi_1)$ are defined; consequently $i_2'(f'f)$ and $(f'f)\,i_1$ are defined which proves that $f'f: C_1 \to C_2'$.

Next assume that $C_2 = C_1'$ and let $i_2 = i_1' = i$, say. Then $f'i$ and if are defined, consequently $f'(if) = f'f$ is defined.

3.3 Additive and Λ-categories

Suppose that $\mathscr{C}$ is a category and that C, C' are objects in $\mathscr{C}$. Denote by Map (C, C') the set of all maps $C \to C'$ which belong to $\mathscr{C}$. In special circumstances it may happen that, for all pairs C and C', Map (C, C') has a natural structure as an additive abelian group. Indeed, this will be the case for those categories which will concern us most. As a convenience therefore we shall indicate that we have a situation of this kind by saying that $\mathscr{C}$ is an *additive category*.

Again if, for each pair C, C', Map (C, C') has the structure of a left (right) Λ-module, then we shall say *$\mathscr{C}$ is a left (right) Λ-category*. For example, the category of modules over a *commutative* ring Λ is a Λ-category.

3.4 Equivalences

Let $\mathscr{C}$ be a category and $f: C_1 \to C_2$ a map in $\mathscr{C}$.

Definition. We say that f is an *equivalence* if there is a map $g: C_2 \to C_1$ in $\mathscr{C}$ such that $gf: C_1 \to C_1$ and $fg: C_2 \to C_2$ are identities.

Let us note that if f is an equivalence then the map g such that $gf = i_1$ and $fg = i_2$, where i_1 and i_2 are the identity maps of C_1 and C_2 respectively, is unique. For suppose we also have $g'f = i_1$ and $fg' = i_2$ then

$$g = g(i_2) = g(fg') = (gf)\,g' = i_1\,g' = g'.$$

It is therefore permissible to use f^{-1} to denote this map. Clearly f^{-1} is itself an equivalence and $(f^{-1})^{-1} = f$. We shall accordingly say that

f and f^{-1} are *inverse equivalences*. For example, any identity map i is an equivalence and it is its own inverse.

Proposition 2. *Let $f: C_1 \to C_2$ and $g: C_2 \to C_3$ be equivalences. Then gf is an equivalence and*

$$(gf)^{-1} = f^{-1}g^{-1}.$$

The proof is trivial.

3.5 The categories $\mathscr{G}_\Lambda^L$ and $\mathscr{G}_\Lambda^R$

Let Λ be a ring, then the left Λ-modules and their Λ-homomorphisms form an additive category $\mathscr{G}_\Lambda^L$ in the sense of the abstract theory. In this category domain and range have their usual meanings and an equivalence is none other than an isomorphism. Besides the category $\mathscr{G}_\Lambda^L$ we also have the additive category $\mathscr{G}_\Lambda^R$ of right Λ-modules and their Λ-homomorphisms. Now it frequently happens that for certain results it is immaterial whether our modules are all left modules or all right modules and we shall deal with this situation by speaking of the category $\mathscr{G}_\Lambda$. In other words, when we use the notation $\mathscr{G}_\Lambda$ our results will hold both when we consistently interpret $\mathscr{G}_\Lambda$ as meaning $\mathscr{G}_\Lambda^L$ and when we consistently take it to signify $\mathscr{G}_\Lambda^R$. Of course, if Λ is commutative we may identify the two categories.

3.6 Functors of a single variable

We are now ready to give a precise definition of a functor. Let $\mathscr{C}$ and $\mathscr{D}$ be two categories and suppose that with each object C of $\mathscr{C}$ there is associated an object $T(C)$ of $\mathscr{D}$, and also with each map f of $\mathscr{C}$ there is associated a map $T(f)$ of $\mathscr{D}$.

Definition. T is called a *covariant functor* from $\mathscr{C}$ to $\mathscr{D}$ when the following conditions are satisfied:

(i) if f has domain C_1 and range C_2, then $T(f)$ has domain $T(C_1)$ and range $T(C_2)$;

(ii) if f is an identity, then so is $T(f)$;

(iii) if gf is defined, then $T(gf) = T(g)\,T(f)$.

It will be observed, by virtue of (i), that if gf is defined then necessarily $T(g)\,T(f)$ is also defined.

Contravariant functors are defined similarly except that the above conditions are replaced by:

(i)′ if f has domain C_1 and range C_2, then $T(f)$ has domain $T(C_2)$ and range $T(C_1)$;

(ii)′ if f is an identity, then so is $T(f)$;

(iii)′ if gf is defined, then $T(gf) = T(f)\,T(g)$.

3.7 Functors of several variables

The notion of a functor of a single variable may be generalized without difficulty. Let $\mathscr{C}_1, \mathscr{C}_2, \ldots, \mathscr{C}_n, \mathscr{D}$ be categories and let $A_1, A_2, \ldots, A_n$ and $A'_1, A'_2, \ldots, A'_n$, etc., designate systems of n objects of which A_r, A'_r belong to $\mathscr{C}_r$. Similarly, we shall use $f_1, f_2, \ldots, f_n$ and $f'_1, f'_2, \ldots, f'_n$, etc., to denote systems of n maps of which f_r, f'_r belong to $\mathscr{C}_r$. Furthermore, let the integers $1, 2. \ldots, n$ be partitioned into two disjoint subsets I and J.

Assume now that with each set $A_1, A_2, \ldots, A_n$ of objects there is associated an object $T(A_1, A_2, \ldots, A_n)$ of $\mathscr{D}$ and that with each set $f_1, f_2, \ldots, f_n$ of maps, where $f_i : A_i \to A'_i$ $(i \in I)$ and $f_j : A'_j \to A_j$ $(j \in J)$, there is associated a map

$$T(f_1, f_2, \ldots, f_n) : T(A_1, A_2, \ldots, A_n) \to T(A'_1, A'_2, \ldots, A'_n)$$

of $\mathscr{D}$. In these circumstances we say that T *is a functor, covariant in those variables whose suffixes are in I and contravariant in those variables whose suffixes are in J*, provided that the following conditions are satisfied:

(i) if $f_1, f_2, \ldots, f_n$ are identity maps, then $T(f_1, f_2, \ldots, f_n)$ is an identity map;

(ii) whenever $f_1, f_2, \ldots, f_n$ and $f'_1, f'_2, \ldots, f'_n$ are maps such that

$$f_i : A_i \to A'_i \quad (i \in I), \qquad f_j : A'_j \to A_j \quad (j \in J)$$

and
$$f'_i : A'_i \to A''_i \quad (i \in I), \qquad f'_j : A''_j \to A'_j \quad (j \in J),$$

then
$$T(\ldots, f'_i f_i, \ldots, f_j f'_j, \ldots) = T(f'_1, \ldots, f'_n)\, T(f_1, \ldots, f_n).$$

For convenience, those suffixes which belong to I will be called *covariant suffixes*, while those in J will be termed *contravariant suffixes*.

Let us observe that if $T(A_1, A_2, \ldots, A_n)$ is a functor and we keep some of the variables fixed, then we obtain a functor of the remaining variables.

Proposition 3. *If $T(A_1, A_2, \ldots, A_n)$ is a functor and if each of $f_1, f_2, \ldots, f_n$ is an equivalence, then $T(f_1, f_2, \ldots, f_n)$ is an equivalence and the inverse equivalence is $T(f_1^{-1}, f_2^{-1}, \ldots, f_n^{-1})$.*

Proof. We have only to observe that, as an immediate consequence of the definition,

$$T(f_1, f_2, \ldots, f_n)\, T(f_1^{-1}, f_2^{-1}, \ldots, f_n^{-1})$$

and
$$T(f_1^{-1}, f_2^{-1}, \ldots, f_n^{-1})\, T(f_1, f_2, \ldots, f_n)$$

are identity maps.

Let $T(A_1, A_2, \ldots, A_n)$ be a functor and suppose that $\mathscr{C}_1, \mathscr{C}_2, \ldots, \mathscr{C}_n, \mathscr{D}$ are all additive categories in the sense of section (3.3). Further, let $f_1, f_2, \ldots, f_n$ and $g_1, g_2, \ldots, g_n$ be two sets of maps where, for each r, f_r and g_r have a common domain and a common range. If, in these circumstances, we always have

$$T(f_1, \ldots, f_r + g_r, \ldots, f_n) = T(f_1, \ldots, f_r, \ldots, f_n) + T(f_1, \ldots, g_r, \ldots, f_n),$$

where $1 \leqslant r \leqslant n$, then we say that T is an *additive functor*.

Again, if each of $\mathscr{C}_1, \mathscr{C}_2, \ldots, \mathscr{C}_n, \mathscr{D}$ is, in the terminology of section (3·3), a left (say) Λ-category and if, *in addition to being additive*, T also satisfies

$$T(f_1, \ldots, \lambda f_r, \ldots, f_n) = \lambda T(f_1, \ldots, f_r, \ldots, f_n),$$

where λ is an arbitrary element of Λ and $1 \leqslant r \leqslant n$, then T will be said to be *Λ-linear*.

3.8 Natural transformations of functors

Let $\quad T(A) = T(A_1, A_2, \ldots, A_n) \quad$ and $\quad U(A) = U(A_1, A_2, \ldots, A_n)$

be functors both (with an obvious notation) from $\mathscr{C}_1 \times \mathscr{C}_2 \times \ldots \times \mathscr{C}_n$ to $\mathscr{D}$, and suppose that both T and U are covariant in the variables A_i $(i \in I)$ and contravariant in the variables A_j $(j \in J)$. Assume now that with each set $(A) = (A_1, A_2, \ldots, A_n)$ there is associated a map $\mu_{(A)} : T(A) \to U(A)$, which belongs to $\mathscr{D}$ and has the following property:

whenever $f_1, f_2, \ldots, f_n$ are maps such that

$$f_i : A_i \to A_i' \quad (i \in I) \qquad \textit{and} \qquad f_j : A_j' \to A_j \quad (j \in J),$$

then the diagram

$$\begin{array}{ccc} T(A) & \xrightarrow{T(f)} & T(A') \\ \downarrow{\scriptstyle \mu_{(A)}} & & \downarrow{\scriptstyle \mu_{(A')}} \\ U(A) & \xrightarrow{U(f)} & U(A') \end{array}$$

is commutative. In these circumstances we say that μ is a *natural transformation* of T into U and we write

$$\mu : T \to U.$$

It is clear that, under appropriate conditions, two such natural transformations can be composed so that if we have another natural transformation $\mu' : U \to V$, then, by combining, we obtain a natural transformation $\mu'\mu : T \to V$.

Once again let $\mu : T \to U$ be a natural transformation. If, for each (A), $\mu_{(A)}$ is an equivalence in $\mathscr{D}$, then μ is called a *natural equivalence* and we write

$$\mu : T \approx U,$$

and say that T and U are *naturally equivalent* or *naturally isomorphic.* When this is the case, the $\mu_{(A)}^{-1}$ determine an *inverse equivalence* $\mu^{-1} : U \approx T$.

3.9 Functors of modules

The functors which will concern us most are of two kinds, namely, those in which the variables are modules and those in which the variables are whole diagrams. For the time being we shall consider only the former, leaving the discussion of categories of diagrams and the functors to which they give rise, until the next chapter. The results about to be obtained concerning functors of modules, will help us to familiarize ourselves with this concept and also they will prove very useful in the later chapters. The reader will find it interesting to observe how, from now on, the point of view which results from the introduction of functors will come to dominate the whole of the theory.

Consider a functor $T(A_1, A_2, \ldots, A_n)$ which is covariant in the variables A_i $(i \in I)$ and contravariant in the variables A_j $(j \in J)$. It will be supposed throughout that A_r varies in the category $\mathscr{G}_{\Lambda_r}$ and that T takes its values in $\mathscr{G}_{\Lambda}$. As explained in section (3.5), some of $\mathscr{G}_{\Lambda_1}, \mathscr{G}_{\Lambda_2}, \ldots, \mathscr{G}_{\Lambda_n}, \mathscr{G}_{\Lambda}$ will be categories of left modules and others will be categories of right modules. The situation in this respect is assumed to be known and fixed once for all, but otherwise we do not need to elucidate the position further. For brevity, we describe the present situation by saying that T is a functor of modules. Since $\mathscr{G}_{\Lambda_1}, \ldots, \mathscr{G}_{\Lambda_n}$, $\mathscr{G}_{\Lambda}$ are all additive categories, the notion of an additive functor is available on this occasion.

Proposition 4. *Let $T(A_1, A_2, \ldots, A_n)$ be an additive functor of modules. Then $T(f_1, f_2, \ldots, f_n) = 0$ whenever any of $f_1, f_2, \ldots, f_n$ is null*† *and $T(A_1, A_2, \ldots, A_n) = 0$ whenever any of the A_r is a zero module.*

Proof. Let $f_1, \ldots, f_{r-1}, f_{r+1}, \ldots, f_n$ be fixed and let f_r be a variable Λ_r-homomorphism with a fixed domain and range. Then, by the definition of an additive functor, the correspondence

$$f_r \to T(f_1, \ldots, f_r, \ldots, f_n)$$

is a Z-homomorphism and so $T(f_1, \ldots, f_r, \ldots, f_n) = 0$ if $f_r = 0$.

† A homomorphism is *null* if its domain and kernel coincide.

Now suppose that $A_r = 0$ and let i_s and o_s denote respectively the identity map of A_s and the null map of A_s into itself. Then $i_r = o_r$, hence

$$T(i_1, \ldots, i_r, \ldots, i_n) = T(i_1, \ldots, o_r, \ldots, i_n) = 0,$$

by the first part. But, by the definition of a functor, $T(i_1, \ldots, i_r, \ldots, i_n)$ is the identity map of $T(A_1, A_2, \ldots, A_n)$, hence $T(A_1, A_2, \ldots, A_n) = 0$.

Theorem 1. *Let $T(A, C)$ be an additive functor (of modules) covariant in A and contravariant in C. Further let*

$$A_p \xrightarrow{f_p} A \xrightarrow{g_p} A_p \quad (1 \leqslant p \leqslant h),$$

$$C_q \xrightarrow{\phi_q} C \xrightarrow{\psi_q} C_q \quad (1 \leqslant q \leqslant k)$$

be complete representations of A and C as direct sums of Λ_1-modules and Λ_2-modules respectively. Then

$$T(A_p, C_q) \xrightarrow{T(f_p, \psi_q)} T(A, C) \xrightarrow{T(g_p, \phi_q)} T(A_p, C_q),$$

where $1 \leqslant p \leqslant h$ and $1 \leqslant q \leqslant k$, is a complete representation of $T(A, C)$ as a direct sum of Λ-modules.

Remark. The theorem holds for a functor of any number of variables and arbitrary variances. The special case considered is sufficiently general to show that the method of proof applies to all other cases.

Proof. $T(g_{p'}, \phi_{q'}) T(f_p, \psi_q) = T(g_{p'} f_p, \psi_q \phi_{q'})$ and this is an identity map if $(p', q') = (p, q)$ and is null otherwise. Again, using the additive property of T, we have

$$\sum_{p,q} T(f_p, \psi_q) T(g_p, \phi_q) = \sum_{p,q} T(f_p g_p, \phi_q \psi_q) = T(\sum_p f_p g_p, \sum_q \phi_q \psi_q) = \text{identity},$$

hence the theorem now follows by Theorem 4 of section (1.9).

Theorem 2. *Let $T(A)$ be an additive functor of one variable and let*

$$0 \to A' \to A \to A'' \to 0$$

be a split exact sequence. If T is covariant then

$$0 \to T(A') \to T(A) \to T(A'') \to 0$$

is a split exact sequence, while if T is contravariant this is the case for

$$0 \to T(A'') \to T(A) \to T(A') \to 0.$$

Proof. By Proposition 2 of section (1.11), the assertion that

$$0 \to A' \to A \to A'' \to 0$$

is a split exact sequence is equivalent to saying that there exist Λ_1-homomorphisms $A \to A'$ and $A'' \to A$ such that

$$A' \to A \to A' \quad \text{and} \quad A'' \to A \to A''$$

is a complete representation of A as a direct sum. That being so, Theorem 2 now follows from Theorem 1.

3.10 Exact functors

As in section (3.9) we consider a functor $T(A_1, A_2, \ldots, A_n)$ of modules and use the notation of that section.

Definition. Suppose that whenever $A_r' \to A_r \to A_r''$ $(1 \leqslant r \leqslant n)$ are exact sequences, then, for each covariant suffix i and each contravariant suffix j, the sequences

$$T(A_1, \ldots, A_i', \ldots, A_n) \to T(A_1, \ldots, A_i, \ldots, A_n) \to T(A_1, \ldots, A_i'', \ldots, A_n) \tag{3.10.1}$$

and

$$T(A_1, \ldots, A_j'', \ldots, A_n) \to T(A_1, \ldots, A_j, \ldots, A_n) \to T(A_1, \ldots, A_j', \ldots, A_n) \tag{3.10.2}$$

are also exact. In such circumstances the functor T is said to be *exact*.

The procedures by which (3.10.1) and (3.10.2) are constructed are largely self-explanatory, but to avoid misunderstandings we observe that, for example, the map

$$T(A_1, \ldots, A_i', \ldots, A_n) \to T(A_1, \ldots, A_i, \ldots, A_n)$$

in (3.10.1) corresponds to the identity maps $A_r \to A_r$ $(r \neq i)$ together with the given map $A_i' \to A_i$.

Assume now that $T(A_1, A_2, \ldots, A_n)$ is an exact functor and that A_r (say) is a zero module. *We assert that* $T(A_1, \ldots, A_r, \ldots, A_n) = 0$. For let 0_r denote a zero module in $\mathscr{G}_{\Lambda_r}$, then, from the exact sequence $0_r \to 0_r \to 0_r$, we obtain an exact sequence

$$T(A_1, \ldots, 0_r, \ldots, A_n) \to T(A_1, \ldots, 0_r, \ldots, A_n) \to T(A_1, \ldots, 0_r, \ldots, A_n)$$

in which both maps are identity maps. The assertion follows at once from this.

Theorem 3. *Let* $T(A_1, A_2, \ldots, A_n)$ *be a functor and suppose that whenever*

$$0 \to A_r' \to A_r \to A_r'' \to 0 \quad (1 \leqslant r \leqslant n)$$

are exact sequences, then, for each covariant suffix i and each contravariant suffix j, the sequences

$$0 \to T(A_1, \ldots, A'_i, \ldots, A_n) \to T(A_1, \ldots, A_i, \ldots, A_n) \to T(A_1, \ldots, A''_i, \ldots, A_n) \to 0$$

and

$$0 \to T(A_1, \ldots, A''_j, \ldots, A_n) \to T(A_1, \ldots, A_j, \ldots, A_n) \to T(A_1, \ldots, A'_j, \ldots, A_n) \to 0$$

are also exact. Then T is an exact functor.

Proof. Since we may consider the behaviour in each variable separately, we may suppose that we are dealing with a functor $T(A)$ of a single variable. We have, of course, to treat both the case in which T is covariant and also the case in which T is contravariant, but, for brevity, we shall give details only in the latter situation.

Let $A' \to A \to A''$ be an exact sequence, then we have to show that

$$T(A'') \to T(A) \to T(A')$$

is exact. Put

$$A_1 = \operatorname{Ker}(A' \to A), \quad A_2 = \operatorname{Im}(A' \to A) = \operatorname{Ker}(A \to A'')$$

and

$$A_3 = \operatorname{Im}(A \to A''),$$

then

$$0 \to A_1 \to A' \to A_2 \to 0,$$

$$0 \to A_2 \to A \to A_3 \to 0$$

and

$$0 \to A_3 \to A'' \to A''/A_3 \to 0$$

are all exact. Consequently, by hypothesis, these give rise to exact sequences

$$0 \to T(A_2) \to T(A') \to T(A_1), \tag{3.10.3}$$

$$T(A_3) \to T(A) \to T(A_2), \tag{3.10.4}$$

$$T(A'') \to T(A_3) \to 0. \tag{3.10.5}$$

But $A' \to A$ is the combined map $A' \to A_2 \to A$, consequently, by the definition of a functor, $T(A) \to T(A')$ is the combined map

$$T(A) \to T(A_2) \to T(A').$$

Moreover, by (3.10.3), $T(A_2) \to T(A')$ is a monomorphism and therefore

$$\operatorname{Ker}(T(A) \to T(A')) = \operatorname{Ker}(T(A) \to T(A_2)) = \operatorname{Im}(T(A_3) \to T(A))$$

by (3.10.4). Again $A \to A''$ is the combined map $A \to A_3 \to A''$ and so $T(A'') \to T(A)$ is the combined map $T(A'') \to T(A_3) \to T(A)$. Now (3.10.5) shows that $T(A'') \to T(A_3)$ is an epimorphism, accordingly

$\operatorname{Im}(T(A'') \to T(A)) = \operatorname{Im}(T(A_3) \to T(A))$. Combining these results we find that

$$\operatorname{Ker}(T(A) \to T(A')) = \operatorname{Im}(T(A'') \to T(A)),$$

so that

$$T(A'') \to T(A) \to T(A')$$

is exact as required.

Theorem 3 prepares the way for the generalizations of the next section.

3.11 Left exact and right exact functors

Very few of the functors that we shall deal with are exact, but the most important of all (for our purposes) are partially exact. It is this notion of partial exactness that we shall now investigate. As in sections (3.9) and (3.10), $T(A_1, A_2, \ldots, A_n)$ always denotes a functor of modules.

Definition. Suppose that whenever

$$0 \to A'_r \to A_r \to A''_r \to 0 \quad (1 \leqslant r \leqslant n) \tag{3.11.1}$$

are exact sequences, then, for each covariant suffix i and contravariant suffix j, the sequences

$$T(A_1, \ldots, A'_i, \ldots, A_n) \to T(A_1, \ldots, A_i, \ldots, A_n) \to T(A_1, \ldots, A''_i, \ldots, A_n) \to 0 \tag{3.11.2}$$

and

$$T(A_1, \ldots, A''_j, \ldots, A_n) \to T(A_1, \ldots, A_j, \ldots, A_n) \to T(A_1, \ldots, A'_j, \ldots, A_n) \to 0 \tag{3.11.3}$$

are also exact. In these circumstances T is called a *right exact* functor.

The definition of a *left exact* functor is similar except that (3.11.2) and (3.11.3) have to be replaced by exact sequences

$$0 \to T(A_1, \ldots, A'_i, \ldots, A_n) \to T(A_1, \ldots, A_i, \ldots, A_n) \to T(A_1, \ldots, A''_i, \ldots, A_n) \tag{3.11.4}$$

and

$$0 \to T(A_1, \ldots, A''_j, \ldots, A_n) \to T(A_1, \ldots, A_j, \ldots, A_n) \to T(A_1, \ldots, A'_j, \ldots, A_n). \tag{3.11.5}$$

From Theorem 3 and the remark immediately preceding it, it now follows at once that we have

Proposition 5. *A functor T is exact if and only if it is both left exact and right exact.*

Before we proceed to record certain facts about right exact functors we mention, in passing, that there is another kind of partially exact functor which is called a *half exact* functor. However, this concept is not important for us and we shall make no further reference to it.

3.12 Properties of right exact functors

The next two results deal with useful facts concerning right exact functors. The second of these (Theorem 4), which is about the effect of simultaneous variation of all the variables, may be regarded as a generalization of Proposition 6. However, it is convenient to have both the proposition and the theorem stated separately.

Proposition 6. *Let $T(A_1, A_2, \ldots, A_n)$ be a right exact functor of modules and suppose that*

$$A'_i \to A_i \to A''_i \to 0 \quad \textit{and} \quad 0 \to A'_j \to A_j \to A''_j$$

are exact sequences, i and j being typical covariant and contravariant suffixes respectively. Then all the sequences

$$T(A_1, \ldots, A'_i, \ldots, A_n) \to T(A_1, \ldots, A_i, \ldots, A_n) \to T(A_1, \ldots, A''_i, \ldots, A_n) \to 0$$

and

$$T(A_1, \ldots, A''_j, \ldots, A_n) \to T(A_1, \ldots, A_j, \ldots, A_n) \to T(A_1, \ldots, A'_j, \ldots, A_n) \to 0$$

are exact.

Proof. As in the proof of Theorem 3, we may suppose, without real loss of generality, that we are dealing with a contravariant functor $T(A)$ of a single variable. Let $0 \to A' \to A \to A''$ be an exact sequence, then we have to show that

$$T(A'') \to T(A) \to T(A') \to 0$$

is exact. Put $A^* = \operatorname{Im}(A \to A'')$, then

$$0 \to A' \to A \to A^* \to 0 \quad \text{and} \quad 0 \to A^* \to A'' \to A''/A^* \to 0$$

are exact consequently, since T is a right exact functor,

$$T(A^*) \to T(A) \to T(A') \to 0 \quad \text{and} \quad T(A'') \to T(A^*) \to 0$$

are both exact. Now $A \to A''$ is the combined map $A \to A^* \to A''$, consequently $T(A'') \to T(A)$ is the combined map $T(A'') \to T(A^*) \to T(A)$, and, as we have seen, $T(A'') \to T(A^*)$ is an epimorphism. Accordingly,

$$\operatorname{Im}(T(A'') \to T(A)) = \operatorname{Im}(T(A^*) \to T(A)) = \operatorname{Ker}(T(A) \to T(A')).$$

Our combined remarks show that

$$T(A'') \to T(A) \to T(A') \to 0$$

is exact and so the proof is complete.

Theorem 4. *Let $T(A_1, A_2, \ldots, A_n)$ be a right exact functor and suppose that*

$$A'_i \to A_i \to A''_i \to 0, \quad 0 \to A''_j \to A_j \to A'_j \tag{3.12.1}$$

are exact sequences, where i and j range over all covariant and contravariant suffixes respectively. Then

$$\sum_{r=1}^{n} T(A_1, \ldots, A_r', \ldots, A_n) \xrightarrow{\phi} T(A_1, \ldots, A_n) \xrightarrow{\psi} T(A_1'', \ldots, A_n'') \to 0 \quad (3.12.2)$$

is an exact sequence.

Remarks. It should be noted that in (3.12.1) we have departed from our usual custom in regard to notation. Further, in (3.12.2) the sum is to be understood to be a *direct sum* and the mapping ϕ is the one determined by the separate homomorphisms

$$T(A_1, \ldots, A_r', \ldots, A_n) \to T(A_1, \ldots, A_r, \ldots, A_n).$$

Proof. We have $\psi = T(f_1, f_2, \ldots, f_n)$, where each f_i is an epimorphism and each f_j is a monomorphism. Now

$$T(f_1, f_2, \ldots, f_n) = T(f_1, i_2, \ldots, i_n)\, T(i_1, f_2, \ldots, f_n),$$

where $i_1, i_2, \ldots, i_n$ are certain identity mappings. Proceeding in this way, we can express $T(f_1, f_2, \ldots, f_n)$ as a composition of n maps of the form

$$T(i_1^{(r)}, \ldots, f_r, \ldots, i_n^{(r)}),$$

where once again $i_1^{(r)}$, $i_2^{(r)}$, etc., are identity maps on certain modules. But, because of the right exactness of T and the special properties of the f_r, each of the

$$T(i_1^{(r)}, \ldots, f_r, \ldots, i_n^{(r)})$$

is an epimorphism. Thus $T(f_1, f_2, \ldots, f_n)$ is the composition of n epimorphisms and therefore it is itself an epimorphism.

It remains for us to prove that $\operatorname{Im}(\phi) = \operatorname{Ker}(\psi)$, and this will be done by induction on n. If $n = 1$ the result follows by Proposition 6. Accordingly, we suppose that $n > 1$ and that the required result has been established for right exact functors in $n-1$ variables. Consider the diagram

$$\begin{array}{ccccccc}
T(A_1', A_2, \ldots, A_n) & \xrightarrow{u} & T(A_1, A_2, \ldots, A_n) & \xrightarrow{v} & T(A_1'', A_2, \ldots, A_n) & \longrightarrow & 0 \\
\downarrow{\scriptstyle p'} & & \downarrow{\scriptstyle p} & & \downarrow{\scriptstyle p''} & & \\
T(A_1', A_2'', \ldots, A_n'') & \xrightarrow{u''} & T(A_1, A_2'', \ldots, A_n'') & \xrightarrow{v''} & T(A_1'', A_2'', \ldots, A_n'') & \longrightarrow & 0 \\
\downarrow & & \downarrow & & \downarrow & & \\
0 & & 0 & & 0 & &
\end{array}$$

In the first place, this diagram is commutative because of the way in which functors transform compositions of mappings; in the second place, by Proposition 6, the rows are exact.† Moreover, the columns are exact, by virtue of the observations which showed that ψ is an epimorphism. Finally, the mapping ψ itself is given by $\psi = p''v = v''p$. It follows from this that

$$\operatorname{Ker}(\psi) = \operatorname{Im}(u) + \operatorname{Ker}(p),$$

where by $\operatorname{Im}(u)+\operatorname{Ker}(p)$ we mean the smallest submodule of $T(A_1, A_2, \ldots, A_n)$ which contains both $\operatorname{Im}(u)$ and $\operatorname{Ker}(p)$. Indeed, if $x \in \operatorname{Ker}(\psi)$, then

$$p(x) \in \operatorname{Ker}(v'') = \operatorname{Im}(u'') = \operatorname{Im}(u''p') = \operatorname{Im}(pu) = p(\operatorname{Im}(u)),$$

and therefore $x \in \operatorname{Ker}(p) + \operatorname{Im}(u)$. Thus $\operatorname{Ker}(\psi) \subseteq \operatorname{Im}(u) + \operatorname{Ker}(p)$ and the opposite inclusion is obvious.

Again, by the inductive hypothesis applied to T with the first variable fixed,

$$\sum_{r=2}^{n} T(A_1, A_2, \ldots, A'_r, \ldots, A_n) \xrightarrow{h} T(A_1, A_2, \ldots, A_n) \xrightarrow{p} T(A_1, A''_2, \ldots, A''_n)$$

is an exact sequence, hence $\operatorname{Ker}(p) = \operatorname{Im}(h)$ and therefore

$$\operatorname{Ker}(\psi) = \operatorname{Im}(h) + \operatorname{Im}(u).$$

But the image of the map

$$\phi : \sum_{r=1}^{n} T(A_1, \ldots, A'_r, \ldots, A_n) \to T(A_1, \ldots, A_n)$$

is clearly the sum of the images of

$$\sum_{r=2}^{n} T(A_1, A_2, \ldots, A'_r, \ldots, A_n) \xrightarrow{h} T(A_1, A_2, \ldots, A_n)$$

and $$T(A'_1, A_2, \ldots, A_n) \xrightarrow{u} T(A_1, A_2, \ldots, A_n),$$

consequently $$\operatorname{Im}(\phi) = \operatorname{Im}(h) + \operatorname{Im}(u) = \operatorname{Ker}(\psi).$$

This completes the proof.

We record in the corollary a fact which was established when we showed that ψ is an epimorphism.

Corollary. *Let $T(A_1, A_2, \ldots, A_n)$ be a right exact functor and let $f_1, f_2, \ldots, f_n$ be maps such that, for each covariant suffix i, f_i is an epimorphism and for each contravariant suffix j, f_j is a monomorphism. Then $T(f_1, f_2, \ldots, f_n)$ is an epimorphism.*

† The fact that the rows have been broken, to facilitate printing, should be ignored.

There are, of course, results for left exact functors which correspond to Proposition 6 and Theorem 4. We shall now state these but we shall not give proofs since these do not involve any essentially new principles.

Proposition 7. *Let $T(A_1, A_2, \ldots, A_n)$ be a left exact functor and let*

$$0 \to A'_i \to A_i \to A''_i \quad (i \in I) \qquad \text{and} \qquad A'_j \to A_j \to A''_j \to 0 \quad (j \in J)$$

be exact sequences, where as usual I and J denote the set of covariant and contravariant suffixes respectively. Then the sequences

$$0 \to T(A_1, \ldots, A'_i, \ldots, A_n) \to T(A_1, \ldots, A_i, \ldots, A_n) \to T(A_1, \ldots, A''_i, \ldots, A_n) \quad (i \in I)$$

and

$$0 \to T(A_1, \ldots, A''_j, \ldots, A_n) \to T(A_1, \ldots, A_j, \ldots, A_n) \to T(A_1, \ldots, A'_j, \ldots, A_n) \quad (j \in J)$$

are also exact.

Theorem 5. *Let $T(A_1, A_2, \ldots, A_n)$ be a left exact functor and suppose that*

$$0 \to A'_i \to A_i \to A''_i \quad \text{and} \quad A''_j \to A_j \to A'_j \to 0$$

are exact sequences, where i and j range over all covariant and contravariant suffixes respectively. Then, with a similar notation to that of Theorem 4,

$$0 \to T(A'_1, \ldots, A'_n) \xrightarrow{\psi} T(A_1, \ldots, A_n) \xrightarrow{\phi} \sum_{r=1}^{n} T(A_1, \ldots, A''_r, \ldots, A_n)$$

is an exact sequence.

This time the fact that ψ is a monomorphism gives the

Corollary. *Let $T(A_1, A_2, \ldots, A_n)$ be a left exact functor and let $f_1, f_2, \ldots, f_n$ be maps such that, for each covariant suffix i, f_i is a monomorphism and, for each contravariant suffix j, f_j is an epimorphism. Then $T(f_1, f_2, \ldots, f_n)$ is a monomorphism.*

3.13 $A \otimes_\Lambda C$ and $\mathrm{Hom}_\Lambda(B, C)$ as functors

We shall now restate some of the main results of Chapter 2 in the language of functor theory. Theorems 6 and 7 are derived respectively from Theorem 4 of section (2.4) and Theorem 9 of section (2.5).

Theorem 6. *$A \otimes_\Lambda C$ is an additive right exact functor which is covariant in both variables.*

It will be recalled that in the general (non-commutative) case $A \otimes_\Lambda C$ varies in $\mathscr{G}_Z$, but, when Λ is commutative, we may, if we wish, regard the functor as having values in $\mathscr{G}_\Lambda$.

Theorem 7. $\mathrm{Hom}_\Lambda(B,C)$ *is an additive left exact functor contravariant in B and covariant in C.*

As in the last theorem, this functor takes values in $\mathscr{G}_Z$ or $\mathscr{G}_\Lambda$ according as to whether we are considering the non-commutative or the commutative case.

Again, by Theorem 5 of section (2.4), *if either A or C is a free module then $A \otimes_\Lambda C$ becomes an exact functor of the other variable.* Also by Theorem 10 of section (2.5), *when B is free* $\mathrm{Hom}_\Lambda(B,C)$ *is an exact functor of C.* These latter results will be refined later.

4

HOMOLOGY FUNCTORS

Notation. Λ denotes a ring, not necessarily commutative, with an identity element.

4.1 Diagrams over a ring

It is convenient to introduce certain terminology relating to diagrams of modules and homomorphisms and, in order to achieve greater precision in this, we shall begin by analysing the notion of a diagram to a greater extent than we have done previously.

Let $[D_i]_{i\in I}$ be a family of left Λ-modules and let Ω be a set of ordered pairs of elements of I. Suppose now that with each element (i,j) of Ω there is associated a Λ-homomorphism

$$\phi_{ij} : D_i \to D_j. \tag{4.1.1}$$

Then the complete system (D_i, ϕ_{ij}) of modules and homomorphisms will be called a *diagram over* Λ. We shall use a capital letter in bold type to denote a diagram and if $\mathbf{D}$ denotes the diagram just described then we shall indicate the details of its composition by writing

$$\mathbf{D} = [D_i, \phi_{ij}, \quad i \in I, \ (i,j) \in \Omega].$$

Two diagrams
$$\mathbf{D} = [D_i, \phi_{ij}, \quad i \in I, \ (i,j) \in \Omega] \tag{4.1.2}$$
and
$$\mathbf{D}' = [D'_i, \phi'_{ij}, \quad i \in I, \ (i,j) \in \Omega], \tag{4.1.3}$$

which share the same labels I and Ω for their constituent modules and homomorphisms, will be said to be *similar*. Frequently we shall have occasion to consider a set of mutually similar diagrams. If one of these happens to be such that all its component modules are null, then it will be convenient to denote this diagram by $\mathbf{0}$.

So far we have spoken only of diagrams of left Λ-modules, and indeed for the rest of the chapter the phrase *diagram over* Λ will be restricted to mean diagrams of this kind. Of course, one can also consider diagrams of right Λ-modules, and to these our results will apply after any necessary minor modifications have been made.

4.2 Translations of diagrams

Let $\mathbf{D}$ and $\mathbf{D}'$ be two similar diagrams, the details of their composition being as in (4.1.2) and (4.1.3). Then by a *map* or *translation*†

$$\mathbf{f} : \mathbf{D} \to \mathbf{D}' \tag{4.2.1}$$

will be meant a family $[f_i]_{i \in I}$ of Λ-homomorphisms

$$f_i : D_i \to D'_i$$

such that for each pair (j, k) in Ω the diagram

$$\begin{array}{ccc} D_j & \xrightarrow{\phi_{jk}} & D_k \\ \downarrow{\scriptstyle f_j} & & \downarrow{\scriptstyle f_k} \\ D'_j & \xrightarrow{\phi'_{jk}} & D'_k \end{array}$$

is commutative. It is clear that translations

$$\mathbf{f} : \mathbf{D} \to \mathbf{D}' \quad \text{and} \quad \mathbf{f}' : \mathbf{D}' \to \mathbf{D}''$$

can be combined to give a translation

$$\mathbf{f}'\mathbf{f} : \mathbf{D} \to \mathbf{D}'',$$

and that two translations

$$\mathbf{f}_1 : \mathbf{D} \to \mathbf{D}' \quad \text{and} \quad \mathbf{f}_2 : \mathbf{D} \to \mathbf{D}'$$

can be 'added', by adding their components, so as to produce a translation

$$\mathbf{f}_1 + \mathbf{f}_2 : \mathbf{D} \to \mathbf{D}'.$$

Indeed, if we take as a set of *objects* a collection of mutually similar diagrams and as *maps* the set of all translations of one into another (including the translations of diagrams into themselves), then the resulting system is an additive category. Such a category of diagrams and translations will be called a *translation category*.

If the basic ring Λ is commutative then we can go further. For, given any translation

$$\mathbf{f} : \mathbf{D} \to \mathbf{D}'$$

and any λ in Λ, the family $[\lambda f_i]_{i \in I}$ is a translation

$$\lambda\mathbf{f} : \mathbf{D} \to \mathbf{D}'.$$

In brief we may say, in the general case, that translation categories are additive, but, when Λ is commutative, every translation category may be regarded as a Λ-category in the sense of section (3.3).

† Small letters in bold type will be used to denote translations of diagrams.

We conclude by describing some further terminology. Let

$$\mathbf{D} \to \mathbf{D}' \to \mathbf{D}''$$

be a sequence of diagrams and translations. Then this sequence will be called a 0-*sequence* (*exact sequence*) if, for each i,

$$D_i \to D_i' \to D_i''$$

is a 0-sequence (exact sequence). To take another example, we shall be particularly concerned with sequences of the form

$$\mathbf{0} \to \mathbf{D}' \to \mathbf{D} \to \mathbf{D}'' \to \mathbf{0}.$$

Such a sequence will be called a *split exact sequence* if

$$0 \to D_i' \to D_i \to D_i'' \to 0$$

is a split exact sequence for every i.

4.3 Images and kernels as functors

We shall now make a detailed study of the translation properties of a simple one-arrow diagram

$$M_1 \xrightarrow{\phi} M_2.$$

Assume then that we have a translation

$$\begin{array}{ccc} M_1 & \xrightarrow{\phi} & M_2 \\ \downarrow{\scriptstyle f_1} & & \downarrow{\scriptstyle f_2} \\ M_1' & \xrightarrow{\phi'} & M_2' \end{array}$$

of this diagram and let x belong to $\mathrm{Ker}(\phi)$ and m_1 to M_1. Then

$$\phi'(f_1(x)) = f_2(\phi(x)) = 0$$

and
$$f_2(\phi(m_1)) = \phi'(f_1(m_1)) \in \mathrm{Im}(\phi'),$$

which shows that

$$f_1(\mathrm{Ker}(\phi)) \subseteq \mathrm{Ker}(\phi') \quad \text{and} \quad f_2(\mathrm{Im}(\phi)) \subseteq \mathrm{Im}(\phi').$$

Accordingly, using the notation of section (1.6),

$$f_1 : (\mathrm{Ker}(\phi), M_1) \to (\mathrm{Ker}(\phi'), M_1')$$

and
$$f_2 : (\mathrm{Im}(\phi), M_2) \to (\mathrm{Im}(\phi'), M_2'),$$

thereby giving rise to homomorphisms

$$\mathrm{Ker}(\phi) \to \mathrm{Ker}(\phi'), \quad \mathrm{Im}(\phi) \to \mathrm{Im}(\phi'),$$

$$\mathrm{Coim}(\phi) \to \mathrm{Coim}(\phi'), \quad \mathrm{Coker}(\phi) \to \mathrm{Coker}(\phi'),$$

and to commutative diagrams

$$\begin{array}{ccccccc} \mathrm{Ker}(\phi) & \longrightarrow & M_1 & \longrightarrow & \mathrm{Coim}(\phi) & \longrightarrow & M_2 \\ \downarrow & & \downarrow & & \downarrow & & \downarrow \\ \mathrm{Ker}(\phi') & \longrightarrow & M_1' & \longrightarrow & \mathrm{Coim}(\phi') & \longrightarrow & M_2' \end{array}$$

and

$$\begin{array}{ccccccc} M_1 & \longrightarrow & \mathrm{Im}(\phi) & \longrightarrow & M_2 & \longrightarrow & \mathrm{Coker}(\phi) \\ \downarrow & & \downarrow & & \downarrow & & \downarrow \\ M_1' & \longrightarrow & \mathrm{Im}(\phi') & \longrightarrow & M_2' & \longrightarrow & \mathrm{Coker}(\phi'). \end{array}$$

Moreover, the canonical isomorphism $\mathrm{Coim}(\phi) \approx \mathrm{Im}(\phi)$ of section (1.7), together with the commutative properties of the diagram

$$\begin{array}{ccccc} \mathrm{Coim}(\phi) & \longleftarrow & M_1 & \longrightarrow & \mathrm{Im}(\phi) \\ \downarrow & & \downarrow & & \downarrow \\ \mathrm{Coim}(\phi') & \longleftarrow & M_1' & \longrightarrow & \mathrm{Im}(\phi') \end{array}$$

shows that

$$\begin{array}{ccc} \mathrm{Coim}(\phi) & \longrightarrow & \mathrm{Im}(\phi) \\ \downarrow & & \downarrow \\ \mathrm{Coim}(\phi') & \longrightarrow & \mathrm{Im}(\phi') \end{array}$$

is also commutative. These remarks, in a slightly amplified form, may be restated as

Theorem 1. Ker (ϕ), Im (ϕ), Coker (ϕ) *and* Coim (ϕ) *are all covariant additive functors from the translation category of all diagrams* (*over* Λ), *which are similar to*

$$M_1 \xrightarrow{\phi} M_2, \tag{4.3.1}$$

to the category of Λ*-modules. Further, by means of these functors, every translation of* (4.3.1) *induces a translation of the canonically augmented diagram*

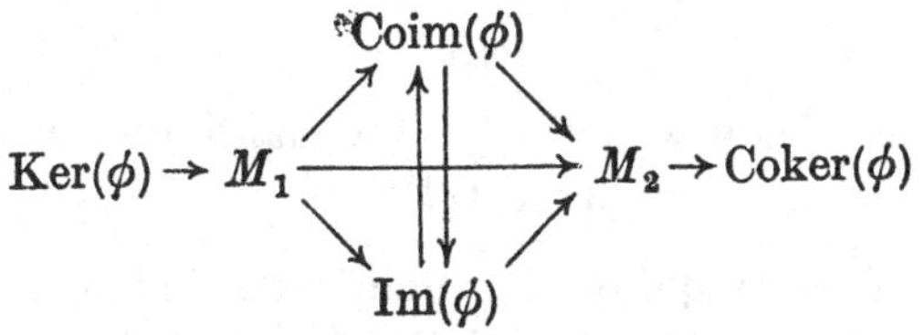

$$\begin{array}{ccccc} & & \mathrm{Coim}(\phi) & & \\ & \nearrow & \uparrow\downarrow & \searrow & \\ \mathrm{Ker}(\phi) \rightarrow M_1 & & \longrightarrow & & M_2 \rightarrow \mathrm{Coker}(\phi) \\ & \searrow & & \nearrow & \\ & & \mathrm{Im}(\phi) & & \end{array} \tag{4.3.2}$$

Remarks. The theorem shows that Coim (ϕ) and Im (ϕ) are naturally equivalent functors in the sense of section (3.8). It may be noted that the diagram (4.3.2) may itself be regarded as a covariant additive functor of the diagram (4.3.1). Further, when Λ is commutative, all these functors are not only additive but also Λ-linear.

We take the opportunity to record, in Propositions 1, 2 and their corollaries, some other translation properties of simple diagrams. Whenever convenient we shall employ the abbreviated notation of section (1.10) without further explanatory comment.

Proposition 1. *Let*

$$\begin{array}{ccccccc} A' & \longrightarrow & A & \longrightarrow & A'' & \longrightarrow & 0 \\ & & \downarrow & & & & \\ & & D & & & & \end{array} \tag{4.3.3}$$

be a diagram over Λ *in which the row is exact and* $A'AD = 0$. *Then there exists a unique* Λ*-homomorphism* $A'' \to D$ *for which*

$$\begin{array}{ccccccc} A' & \to & A & \to & A'' & \to & 0 \\ & & \downarrow & \swarrow & & & \\ & & D & & & & \end{array} \tag{4.3.4}$$

is commutative. Further, every translation of (4.3.3) *is necessarily also a translation of* (4.3.4).

Proof. Let $$X = \operatorname{Im}(A' \to A) = \operatorname{Ker}(A \to A'').$$
Then, since $A'AD = 0$, it follows that $A \to D$ carries X into zero. Thus $A \to D$ induces a map $A/X \to D$ for which

$$\begin{array}{ccc} & A & \\ \swarrow & & \searrow \\ D & \longleftarrow & A/X \end{array}$$

is commutative. But $A \to A''$ is an epimorphism with kernel X, and so we have an isomorphism $A'' \to A/X$ for which

$$\begin{array}{ccc} A & \longrightarrow & A'' \\ \searrow & & \swarrow \\ & A/X & \end{array}$$

is commutative. If now we put $A'' \to D$ equal to the combined map

$$A'' \to A/X \to D$$

then $AA''D = AD$ as required. The uniqueness follows from the commutative character of (4.3.4) and the fact that $A \to A''$ is an epimorphism. The statement about translations is obvious.

Corollary. *Let*

$$\begin{array}{ccccccc} A' & \longrightarrow & A & \longrightarrow & A'' & \longrightarrow & 0 \\ \downarrow & & \downarrow & & & & \\ B' & \longrightarrow & B & \longrightarrow & B'' & & \end{array} \tag{4.3.5}$$

be a commutative diagram in which the upper row is exact and the lower row is a 0-sequence. Then there exists a unique Λ*-homomorphism* $A'' \to B''$ *such that*

$$\begin{array}{ccccccc} A' & \longrightarrow & A & \longrightarrow & A'' & \longrightarrow & 0 \\ \downarrow & & \downarrow & & \downarrow & & \\ B' & \longrightarrow & B & \longrightarrow & B'' & & \end{array} \tag{4.3.6}$$

is a commutative diagram. Further, every translation of (4.3.5) *is a translation of* (4.3.6).

Proof. $$A'A(ABB'') = A'ABB'' = A'B'BB'' = 0,$$

consequently ABB'' satisfies the condition imposed on $A \to D$ in the proposition and now the corollary follows at once.

Proposition 2. *Let*

$$\begin{array}{ccccccc} & & & & D & & \\ & & & & \downarrow & & \\ 0 & \longrightarrow & A' & \longrightarrow & A & \longrightarrow & A'' \end{array} \tag{4.3.7}$$

be a diagram over Λ *in which the row is exact and* $DAA'' = 0$. *Then there exists a unique* Λ*-homomorphism* $D \to A'$ *such that*

$$\begin{array}{ccccccc} & & & & D & & \\ & & & \swarrow & \downarrow & & \\ 0 & \to & A' & \to & A & \to & A'' \end{array} \tag{4.3.8}$$

is commutative. Further, any translation of (4.3.7) *is also a translation of* (4.3.8).

Proof. Put $$X = \text{Im}\,(A' \to A) = \text{Ker}\,(A \to A'').$$

Then, since $DAA'' = 0$, it follows that $D \to A$ maps D into X and thereby gives rise to a commutative diagram

$$\begin{array}{ccccc} & & & & D \\ & & & \swarrow & \downarrow \\ A' & \to & X & \to & A \end{array}$$

But $A' \to X$ is an isomorphism. Hence, if XA' denotes the inverse isomorphism and we put $DA' = DXA'$, then DA' has the required properties. The other assertions are obvious.

Corollary. *Let*

$$\begin{array}{ccccccc} & & B' & \longrightarrow & B & \longrightarrow & B'' \\ & & & & \downarrow & & \downarrow \\ 0 & \longrightarrow & A' & \longrightarrow & A & \longrightarrow & A'' \end{array} \tag{4.3.9}$$

be a commutative diagram in which the upper row is a 0*-sequence and the lower row is exact. Then there exists a unique* Λ*-homomorphism* $B' \to A'$ *such that*

$$\begin{array}{ccccccc} & & B' & \longrightarrow & B & \longrightarrow & B'' \\ & & \downarrow & & \downarrow & & \downarrow \\ 0 & \longrightarrow & A' & \longrightarrow & A & \longrightarrow & A'' \end{array} \tag{4.3.10}$$

is commutative. Further, every translation of (4.3.9) *is also a translation of* (4.3.10).

This follows immediately from the proposition.

4.4 Homology functors

Let

$$(\mathbf{A}) \quad A_2 \to A_1 \to A_0 \tag{4.4.1}$$

be a three-term 0-sequence, then $\operatorname{Im}(A_2A_1) \subseteq \operatorname{Ker}(A_1A_0)$, and so, by means of the inclusion map, we obtain a one-arrow diagram

$$\operatorname{Im}(A_2A_1) \to \operatorname{Ker}(A_1A_0). \tag{4.4.2}$$

Further, by Theorem 1, each translation of (4.4.1) into a similar 0-sequence determines a translation of (4.4.2) and thereby (4.4.2) becomes a covariant additive functor of (4.4.1) defined on the translation category of three-term 0-sequences. Let us write

$$H(\mathbf{A}) = H(A_2 - A_1 - A_0) = \operatorname{Ker}(A_1A_0)/\operatorname{Im}(A_2A_1)$$

so that $H(\mathbf{A})$ is the cokernel of (4.4.2). Then each translation of $\mathbf{A}$ into a similar three-term 0-sequence, now determines first a translation of (4.4.2) and, in consequence of this, a Λ-homomorphism of $H(\mathbf{A})$. Indeed, $H(\mathbf{A})$ has become a functor of $\mathbf{A}$.

Definition. The additive covariant functor $H(\mathbf{A})$, which is defined on the translation category of three-term 0-sequences and whose values lie in the category of Λ-modules, is called the *homology functor*.

Let us observe, in passing, that when Λ is commutative $H(\mathbf{A})$ will not only be additive but also Λ-linear.

Assume now that $\mathbf{A}$ and $\mathbf{B}$ are three-term 0-sequences and that we have a translation

$$\begin{array}{ccccccc} (\mathbf{A}) & & A_2 & \longrightarrow & A_1 & \longrightarrow & A_0 \\ \downarrow & & \downarrow & & \downarrow & & \downarrow \\ (\mathbf{B}) & & B_2 & \longrightarrow & B_1 & \longrightarrow & B_0. \end{array}$$

Then the associated Λ-homomorphism $H(\mathbf{A}) \to H(\mathbf{B})$ is simply that induced by the map

$$(\operatorname{Im}(A_2A_1), \operatorname{Ker}(A_1A_0)) \to (\operatorname{Im}(B_2B_1), \operatorname{Ker}(B_1B_0))$$

of pairs, where $\text{Ker}\,(A_1 A_0) \to \text{Ker}\,(B_1 B_0)$ is obtained from $A_1 \to B_1$ by restriction of the range and domain. Accordingly, it follows that *if* $A_1 \to B_1$ *is such that* $\text{Ker}\,(A_1 A_0)$ *is mapped into* $\text{Im}\,(B_2 B_1)$ *then* $H(\mathbf{A}) \to H(\mathbf{B})$ *is null.* In particular, $H(\mathbf{A}) \to H(\mathbf{B})$ is null whenever $A_1 \to B_1$ is null.

In the theorem which follows **A**, **B** and **C** denote three-term 0-sequences

$$A_2 \to A_1 \to A_0, \quad B_2 \to B_1 \to B_0 \quad \text{and} \quad C_2 \to C_1 \to C_0$$

respectively.

Theorem 2. *If translations*

$$\mathbf{A} \xrightarrow{\mathfrak{f}} \mathbf{B} \xrightarrow{\mathfrak{g}} \mathbf{C} \tag{4.4.3}$$

are such that (i) $B_2 \to C_2$ *is an epimorphism,*

(ii) $A_1 \to B_1 \to C_1$ *is exact,*

(iii) $A_0 \to B_0$ *is a monomorphism,*

then

$$H(\mathbf{A}) \xrightarrow{H(\mathfrak{f})} H(\mathbf{B}) \xrightarrow{H(\mathfrak{g})} H(\mathbf{C}) \tag{4.4.4}$$

is exact. In particular, (4.4.4) *is exact if*

$$0 \to \mathbf{A} \xrightarrow{\mathfrak{f}} \mathbf{B} \xrightarrow{\mathfrak{g}} \mathbf{C} \to 0$$

is exact.

Proof. We are concerned with a commutative diagram

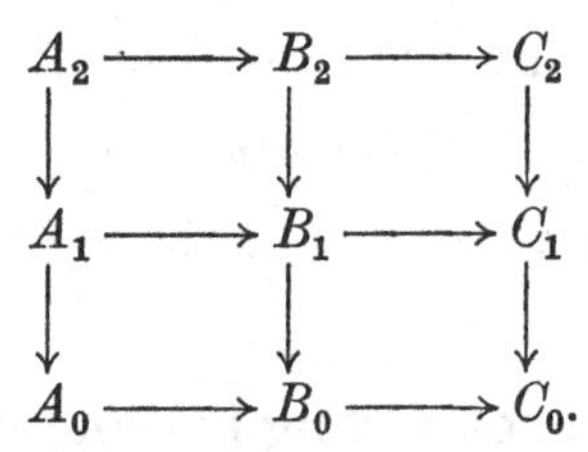

By the functor properties of $H(\mathbf{A})$,

$$H(\mathfrak{g})\,H(\mathfrak{f}) = H(\mathfrak{g}\mathfrak{f}).$$

Now $\mathfrak{g}\mathfrak{f} : \mathbf{A} \to \mathbf{C}$, and in this translation $A_1 \to C_1$ is a null map, accordingly, by the remark just preceding the theorem, $H(\mathbf{A}) \to H(\mathbf{C})$ is null. Thus $H(\mathfrak{g})\,H(\mathfrak{f}) = 0$ and so

$$\text{Im}\,(H(\mathfrak{f})) \subseteq \text{Ker}\,(H(\mathfrak{g})).$$

Suppose now that β belongs to $H(\mathbf{B})$ and that $H(\mathfrak{g})\,\beta = 0$, so that β is a typical element of $\text{Ker}\,(H(\mathfrak{g}))$. Since

$$\beta \in H(\mathbf{B}) = \text{Ker}\,(B_1 B_0)/\text{Im}\,(B_2 B_1),$$

it can be represented by an element b_1 of $\operatorname{Ker}(B_1B_0)$. Further, $H(\mathbf{g})\beta$ is represented by $b_1B_1C_1$, and, since $H(\mathbf{g})\beta = 0$, this means that $b_1B_1C_1$ belongs to $\operatorname{Im}(C_2C_1)$. We may therefore write

$$b_1B_1C_1 = c_2C_2C_1,$$

where c_2 belongs to C_2. But, by hypothesis, $B_2 \to C_2$ is an epimorphism hence $c_2 = b_2B_2C_2$, for a suitable element b_2 of B_2. We now have

$$b_1B_1C_1 = c_2C_2C_1 = b_2B_2C_2C_1 = b_2B_2B_1C_1,$$

which shows that

$$b_1 - b_2B_2B_1 \in \operatorname{Ker}(B_1C_1) = \operatorname{Im}(A_1B_1),$$

since $A_1 \to B_1 \to C_1$ is exact. Let us therefore write

$$b_1 - b_2B_2B_1 = a_1A_1B_1,$$

where a_1 belongs to A_1. We now assert that a_1 belongs to $\operatorname{Ker}(A_1A_0)$. For

$$a_1A_1A_0B_0 = a_1A_1B_1B_0 = b_1B_1B_0 - b_2B_2B_1B_0 = 0 - 0 = 0,$$

and hence $a_1A_1A_0 = 0$, since $A_0 \to B_0$ is a monomorphism. Since a_1 belongs to $\operatorname{Ker}(A_1A_0)$, a_1 represents an element α of

$$H(\mathbf{A}) = \operatorname{Ker}(A_1A_0)/\operatorname{Im}(A_2A_1).$$

Further, $H(\mathbf{f})\alpha$ is represented by $a_1A_1B_1 = b_1 - b_2B_2B_1$. But b_1 and $b_1 - b_2B_2B_1$ represent the same element of

$$H(\mathbf{B}) = \operatorname{Ker}(B_1B_0)/\operatorname{Im}(B_2B_1),$$

namely, β, because they differ by an element of $\operatorname{Im}(B_2B_1)$. Accordingly

$$\beta = H(\mathbf{f})\alpha \in \operatorname{Im}(H(\mathbf{f})).$$

This shows that $\operatorname{Ker}(H(\mathbf{g}))$ is contained in $\operatorname{Im}(H(\mathbf{f}))$ and now the proof is complete.

4.5 The connecting homomorphism

In this section we consider a commutative diagram

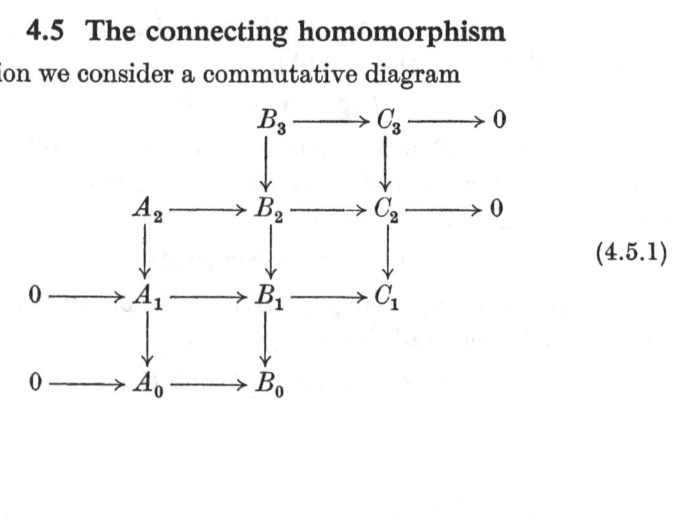

(4.5.1)

of Λ-modules and Λ-homomorphisms, in which the rows are exact and the columns are 0-sequences, and we shall show that it is possible to define, in a natural way, a Λ-homomorphism

$$\Delta : H(C_3 - C_2 - C_1) \to H(A_2 - A_1 - A_0). \tag{4.5.2}$$

Since the details of the construction of Δ will concern us for some time, we shall set these out in a form that will facilitate the making of future references.

The construction of Δ. Suppose that γ belongs to

$$H(C_3 - C_2 - C_1) = \operatorname{Ker}(C_2 C_1)/\operatorname{Im}(C_3 C_2).$$

Choose a representative c_2 from $\operatorname{Ker}(C_2 C_1)$ for γ, then, since $B_2 \to C_2$ is an epimorphism, we can find b_2 in B_2 such that $b_2 B_2 C_2 = c_2$. Put $b_1 = b_2 B_2 B_1$, then b_1 belongs to

$$\operatorname{Ker}(B_1 C_1) = \operatorname{Im}(A_1 B_1),$$

(for $b_1 B_1 C_1 = b_2 B_2 B_1 C_1 = b_2 B_2 C_2 C_1 = c_2 C_2 C_1 = 0$), and so we can find a_1 in A_1 such that $a_1 A_1 B_1 = b_1$. But a_1 belongs to $\operatorname{Ker}(A_1 A_0)$, (for $a_1 A_1 A_0 B_0 = a_1 A_1 B_1 B_0 = b_1 B_1 B_0 = b_2 B_2 B_1 B_0 = 0$ and $A_0 B_0$ is a monomorphism), and so it represents an element α of

$$\operatorname{Ker}(A_1 A_0)/\operatorname{Im}(A_2 A_1) = H(A_2 - A_1 - A_0).$$

We put $\Delta(\gamma) = \alpha$. This completes the construction of Δ.

We first show that $\Delta(\gamma)$ is uniquely defined by the construction. Indeed, suppose that in selecting c_2, b_2, b_1, a_1 we had made different choices and obtained instead c_2^*, b_2^*, b_1^*, a_1^*. Then, since c_2 and c_2^* both represent γ, $c_2 - c_2^*$ belongs to $\operatorname{Im}(C_3 C_2)$ and therefore $c_2 - c_2^* = c_3 C_3 C_2$ for some element c_3 in C_3. But $B_3 \to C_3$ is an epimorphism and so $c_3 = b_3 B_3 C_3$ for a suitable element b_3 of B_3. Now

$$(b_2 - b_2^* - b_3 B_3 B_2) B_2 C_2 = c_2 - c_2^* - b_3 B_3 B_2 C_2 = c_2 - c_2^* - c_3 C_3 C_2 = 0,$$

consequently $b_2 - b_2^* - b_3 B_3 B_2$ belongs to $\operatorname{Ker}(B_2 C_2) = \operatorname{Im}(A_2 B_2)$. We may therefore write $b_2 - b_2^* - b_3 B_3 B_2 = a_2 A_2 B_2$, where a_2 belongs to A_2. Further

$$\begin{aligned}(a_1 - a_1^* - a_2 A_2 A_1) A_1 B_1 &= b_1 - b_1^* - a_2 A_2 A_1 B_1 \\ &= (b_2 - b_2^* - a_2 A_2 B_2) B_2 B_1 = b_3 B_3 B_2 B_1 = 0.\end{aligned}$$

But $A_1 B_1$ is a monomorphism and so $a_1 - a_1^* - a_2 A_2 A_1 = 0$. Accordingly a_1 and a_1^* belong to the same coset modulo $\operatorname{Im}(A_2 A_1)$ and hence represent the same element in $\operatorname{Ker}(A_1 A_0)/\operatorname{Im}(A_2 A_1)$; that is, in $H(A_2 - A_1 - A_0)$. This establishes the uniqueness of the construction.

Suppose now that $\overline{\gamma}$ also belongs to $H(C_3-C_2-C_1)$ and that λ is in Λ. An easy verification shows that

$$\Delta(\gamma+\overline{\gamma}) = \Delta(\gamma)+\Delta(\overline{\gamma})$$

and

$$\Delta(\lambda\gamma) = \lambda\Delta(\gamma).$$

Thus Δ is a Λ-homomorphism. It is called the *connecting homomorphism.*

We next consider the translation properties of the connecting homomorphism. Suppose then that we have a translation of (4.5.1) into another diagram with analogous properties. The modules of the new diagram will be denoted by A'_i, B'_j, C'_k and the components of the translation will be written as $A_iA'_i$, $B_jB'_j$, $C_kC'_k$. Let γ belong to $H(C_3-C_2-C_1)$, let $\Delta(\gamma) = \alpha$ and let c_2, b_2, b_1, a_1 be as in the construction of $\Delta(\gamma)$ described above. Put

$$c'_2 = c_2C_2C'_2, \quad b'_2 = b_2B_2B'_2, \quad b'_1 = b_1B_1B'_1 \quad \text{and} \quad a'_1 = a_1A_1A'_1.$$

Since c_2 belongs to $\operatorname{Ker}(C_2C_1)$ and a_1 to $\operatorname{Ker}(A_1A_0)$, it follows that c'_2 belongs to $\operatorname{Ker}(C'_2C'_1)$ and a'_1 to $\operatorname{Ker}(A'_1A'_0)$, consequently they represent, respectively, elements γ' and α' of $H(C'_3-C'_2-C'_1)$ and $H(A'_2-A'_1-A'_0)$. Further, γ' and α' are the images of γ and α under the respective homomorphisms

$$H(C_3-C_2-C_1)\rightarrow H(C'_3-C'_2-C'_1)$$

and

$$H(A_2-A_1-A_0)\rightarrow H(A'_2-A'_1-A'_0).$$

Again, since $c_2 = b_2B_2C_2$, $b_1 = b_2B_2B_1$ and $b_1 = a_1A_1B_1$, we also have $c'_2 = b'_2B'_2C'_2$, $b'_1 = b'_2B'_2B'_1$ and $b'_1 = a'_1A'_1B'_1$. But this shows that $\Delta(\gamma') = \alpha'$ and therefore the diagram

$$\begin{array}{ccc} H(C_3-C_2-C_1) & \xrightarrow{\Delta} & H(A_2-A_1-A_0) \\ \downarrow & & \downarrow \\ H(C'_3-C'_2-C'_1) & \xrightarrow{\Delta} & H(A'_2-A'_1-A'_0) \end{array}$$

is commutative.

These results are brought together in

Theorem 3. *If*

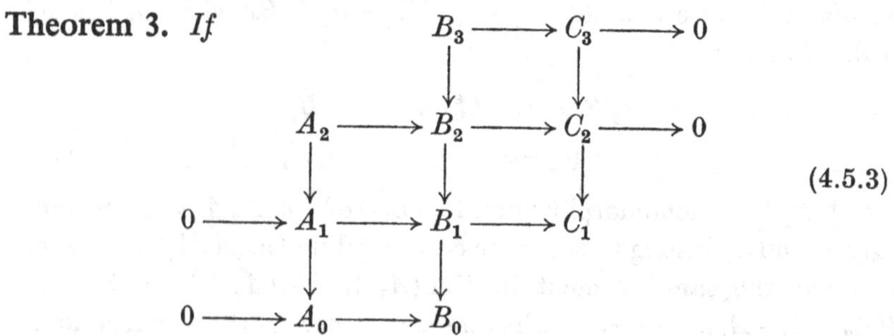

(4.5.3)

is a commutative diagram of Λ-modules and Λ-homomorphisms, in which the rows are exact and the columns are 0-sequences, then there exists a canonical Λ-homomorphism

$$H(C_3-C_2-C_1)\overset{\Delta}{\rightarrow}H(A_2-A_1-A_0) \tag{4.5.4}$$

defined by the above construction. This homomorphism is such that every translation of (4.5.3) *into a diagram with corresponding properties, determines a translation of* (4.5.4).

The diagram (4.5.3), besides enabling us to define the connecting homomorphism Δ, also yields Λ-homomorphisms

$$H(B_3-B_2-B_1)\overset{f}{\rightarrow}H(C_3-C_2-C_1)$$

and
$$H(A_2-A_1-A_0)\overset{g}{\rightarrow}H(B_2-B_1-B_0).$$

Concerning the whole system we have the following result:

Theorem 4. *If the hypotheses are the same as in Theorem* 3, *then the sequence*

$$H(B_3-B_2-B_1)\overset{f}{\rightarrow}H(C_3-C_2-C_1)\overset{\Delta}{\rightarrow}H(A_2-A_1-A_0)\overset{g}{\rightarrow}H(B_2-B_1-B_0)$$

is exact.

Proof. The proof will be in four stages.

(i) $\text{Im}(f) \subseteq \text{Ker}(\Delta)$. Let β of $H(B_3-B_2-B_1)$ be represented by the element b_2 of $\text{Ker}(B_2B_1)$. Then $f(\beta)$ is represented by $b_2B_2C_2 = c_2$ (say). But $b_2B_2B_1 = 0 = 0A_1B_1$ and this shows that $\Delta f(\beta)$ is represented by the zero element of A_1. Thus $\Delta f(\beta) = 0$ and this shows that $\text{Im}(f) \subseteq \text{Ker}(\Delta)$.

(ii) $\text{Ker}(\Delta) \subseteq \text{Im}(f)$. Suppose that γ belongs to $H(C_3-C_2-C_1)$ and that $\Delta(\gamma) = 0$. Let c_2, b_2, b_1, a_1 be as in the construction of $\Delta(\gamma)$. Then, since $\Delta(\gamma) = 0$, we see that a_1 belongs to $\text{Im}(A_2A_1)$. Put $a_1 = a_2A_2A_1$, where a_2 belongs to A_2. Now

$$(b_2-a_2A_2B_2)\,B_2B_1 = b_1-a_2A_2A_1B_1 = b_1-a_1A_1B_1 = 0,$$

consequently $b_2-a_2A_2B_2$ belongs to $\text{Ker}(B_2B_1)$ and therefore it represents an element β of $H(B_3-B_2-B_1)$. Further, $f(\beta)$ is represented by $(b_2-a_2A_2B_2)\,B_2C_2 = b_2B_2C_2 = c_2$. Thus $f(\beta) = \gamma$ and γ belongs to $\text{Im}(f)$. It follows that $\text{Ker}(\Delta) \subseteq \text{Im}(f)$.

(iii) $\text{Im}(\Delta) \subseteq \text{Ker}(g)$. Let γ belong to $H(C_3-C_2-C_1)$ and let c_2, b_2, b_1, a_1 be as in the construction of $\Delta(\gamma)$. Then $g\Delta(\gamma)$ is represented by $a_1A_1B_1 = b_1 = b_2B_2B_1$ which belongs to $\text{Im}(B_2B_1)$. Accordingly $g\Delta(\gamma) = 0$ and from this we see that $\text{Im}(\Delta) \subseteq \text{Ker}(g)$.

(iv) $\operatorname{Ker}(g) \subseteq \operatorname{Im}(\Delta)$. Let α belong to $H(A_2 - A_1 - A_0)$ and suppose that $g(\alpha) = 0$. Let a_1 in $\operatorname{Ker}(A_1 A_0)$ represent α then $g(\alpha)$ is represented by $a_1 A_1 B_1 = b_1$ (say). But $g(\alpha) = 0$; consequently b_1 belongs to $\operatorname{Im}(B_2 B_1)$ and therefore we may write $b_1 = b_2 B_2 B_1$. Put $c_2 = b_2 B_2 C_2$ then

$$c_2 C_2 C_1 = b_2 B_2 C_2 C_1 = b_2 B_2 B_1 C_1 = b_1 B_1 C_1 = a_1 A_1 B_1 C_1 = 0.$$

Accordingly c_2 represents an element γ of $H(C_3 - C_2 - C_1)$ and, by the construction of Δ, $\Delta(\gamma)$ is represented by a_1. Hence $\alpha = \Delta(\gamma) \subseteq \operatorname{Im}(\Delta)$ and the required relation $\operatorname{Ker}(g) \subseteq \operatorname{Im}(\Delta)$ is established.

Theorems 2, 3 and 4 have various special forms of which we shall now give two examples.

Proposition 3. *Let*

$$\begin{array}{ccccc} A & \longrightarrow & B & \longrightarrow & C \\ \downarrow{\scriptstyle f} & & \downarrow{\scriptstyle g} & & \downarrow{\scriptstyle h} \\ A' & \longrightarrow & B' & \longrightarrow & C' \end{array}$$

be a commutative diagram over Λ with exact rows. If $A' \to B'$ is a monomorphism then the sequence

$$\operatorname{Ker}(f) \to \operatorname{Ker}(g) \to \operatorname{Ker}(h) \tag{4.5.5}$$

is exact, while if $B \to C$ is an epimorphism then

$$\operatorname{Coker}(f) \to \operatorname{Coker}(g) \to \operatorname{Coker}(h) \tag{4.5.6}$$

is exact.

Proof. The exactness of (4.5.5) follows by applying Theorem 2 to the diagram

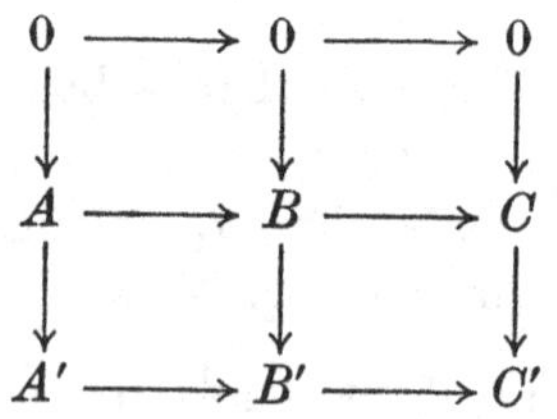

while that of (4.5.6) comes from applying the same theorem to

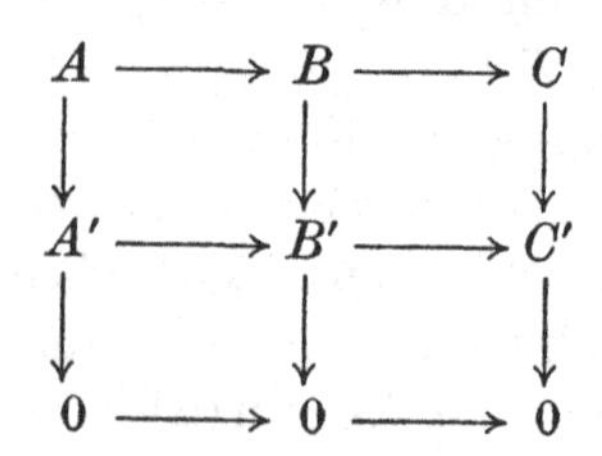

Proposition 4. *Let*

$$\begin{array}{ccccccc} & & A & \longrightarrow & B & \longrightarrow & C & \longrightarrow 0 \\ & & \downarrow f & & \downarrow g & & \downarrow h & \\ 0 & \longrightarrow & A' & \longrightarrow & B' & \longrightarrow & C' & \end{array} \qquad (4.5.7)$$

be a commutative diagram over Λ *with exact rows. Then we have a canonical exact sequence*

$$\operatorname{Ker}(f) \to \operatorname{Ker}(g) \to \operatorname{Ker}(h) \to \operatorname{Coker}(f) \to \operatorname{Coker}(g) \to \operatorname{Coker}(h). \qquad (4.5.8)$$

Further, every translation of (4.5.7) *into a diagram with corresponding properties determines a translation of* (4.5.8).

Proof. If we apply Theorem 4 to the diagram

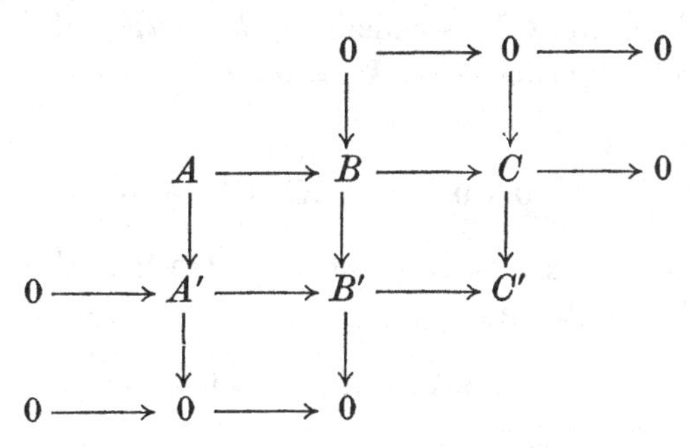

and combine the results with those of Proposition 3, then the exactness of (4.5.8) follows. The translation property is a consequence of Theorem 3 taken in conjunction with the functor properties of kernels and cokernels.

4.6 Complexes

By a *complex* **X** we mean a 0-sequence

$$\cdots \longrightarrow X^{n-1} \xrightarrow{d^{n-1}} X^n \xrightarrow{d^n} X^{n+1} \longrightarrow \cdots, \qquad (4.6.1)$$

over Λ, infinite in both directions. The Λ-homomorphism $X^n \to X^{n+1}$, which we have denoted by d^n, is called the *boundary homomorphism* or the *differentiation homomorphism*. The fact that (4.6.1) is a 0-sequence is then expressed by the relation $d^{n+1}d^n = 0$, which holds for all integers n. Put

$$H^n(\mathbf{X}) = H(X^{n-1} - X^n - X^{n+1}).$$

These Λ-modules are called the *homology modules* of the complex $\mathbf{X}$. Each translation $\mathbf{X} \to \mathbf{X}'$ of $\mathbf{X}$ into another complex $\mathbf{X}'$ then determines, for each n, a Λ-homomorphism $H^n(\mathbf{X}) \to H^n(\mathbf{X}')$ and in this way $H^n(\mathbf{X})$ becomes an additive covariant functor of $\mathbf{X}$ defined on the translation category of complexes.

It is convenient to write

$$X_n = X^{-n}, \quad d_n = d^{-n}, \quad H_n(\mathbf{X}) = H^{-n}(\mathbf{X}).$$

If this is done then (4.6.1) takes the form

$$\cdots \longrightarrow X_{n+1} \xrightarrow{d_{n+1}} X_n \xrightarrow{d_n} X_{n-1} \longrightarrow \cdots \tag{4.6.2}$$

and we have $d_n d_{n+1} = 0$ and $H_n(\mathbf{X}) = H(X_{n+1} - X_n - X_{n-1})$.

Definition. A complex $\mathbf{X}$ is called a *right complex* if $X^n = 0$ for all $n < 0$; it is called a *left complex* if $X_n = 0$ for all $n < 0$.

A typical right complex has the form

$$\cdots \to 0 \to 0 \to X^0 \to X^1 \to X^2 \to \cdots$$

and for such a complex $H^n(\mathbf{X}) = 0$ if $n < 0$. On the other hand, a left complex may be written as

$$\cdots \to X_2 \to X_1 \to X_0 \to 0 \to 0 \to \cdots,$$

and, for this case, $H_n(\mathbf{X}) = 0$ if $n < 0$.

Theorem 5. *Let*
$$0 \to \mathbf{X}' \to \mathbf{X} \to \mathbf{X}'' \to 0 \tag{4.6.3}$$
be an exact sequence of complexes and translations. Then there results a canonical exact sequence

$$\cdots \to H_{n+1}(\mathbf{X}') \to H_{n+1}(\mathbf{X}) \to H_{n+1}(\mathbf{X}'') \xrightarrow{\Delta} H_n(\mathbf{X}') \to H_n(\mathbf{X})$$
$$\to H_n(\mathbf{X}'') \xrightarrow{\Delta} H_{n-1}(\mathbf{X}') \to \cdots. \tag{4.6.4}$$

Further, every translation of (4.6.3), *that is, every commutative pattern*

$$\begin{array}{ccccccccc} 0 & \longrightarrow & \mathbf{X}' & \longrightarrow & \mathbf{X} & \longrightarrow & \mathbf{X}'' & \longrightarrow & 0 \\ & & \downarrow & & \downarrow & & \downarrow & & \\ 0 & \longrightarrow & \mathbf{Y}' & \longrightarrow & \mathbf{Y} & \longrightarrow & \mathbf{Y}'' & \longrightarrow & 0 \end{array}$$

of complexes and translations in which the rows are exact, determines a translation of (4.6.4).

Proof. We are concerned with a commutative diagram

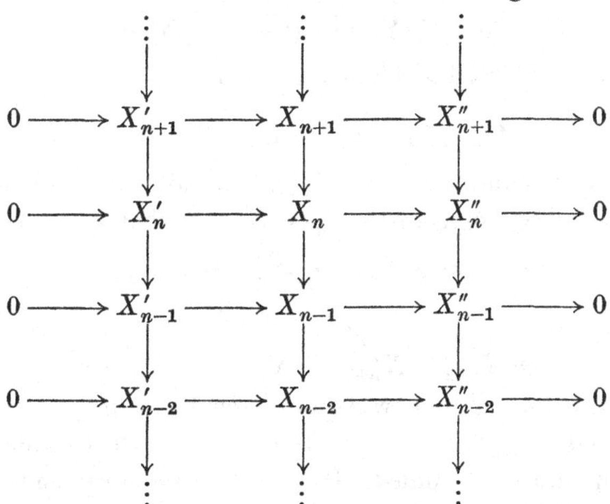

in which the rows are exact and the columns are 0-sequences. The exactness of (4.6.4) follows from Theorems 2 and 4, while the translation property is a consequence of Theorem 3 taken in conjunction with the functor properties of the $H_n(\mathbf{X})$.

The sequence (4.6.4) is called the *exact homology sequence* of (4.6.3). In order to familiarize ourselves with its structure we note that if we raise all the suffixes then it becomes

$$\cdots \to H^{n-1}(\mathbf{X}') \to H^{n-1}(\mathbf{X}) \to H^{n-1}(\mathbf{X}'') \xrightarrow{\Delta} H^n(\mathbf{X}') \to H^n(\mathbf{X}) \xrightarrow{} H^n(\mathbf{X}'') \xrightarrow{\Delta} H^{n+1}(\mathbf{X}') \to \cdots. \quad (4.6.5)$$

Should it happen that $\mathbf{X}'$, $\mathbf{X}$, $\mathbf{X}''$ are all left complexes then the non-trivial part of (4.6.4) is

$$\cdots \to H_1(\mathbf{X}') \to H_1(\mathbf{X}) \to H_1(\mathbf{X}'') \xrightarrow{\Delta} H_0(\mathbf{X}') \to H_0(\mathbf{X}) \to H_0(\mathbf{X}'') \to 0,$$

whereas if they are all right complexes the essential part of (4.6.5) is

$$0 \to H^0(\mathbf{X}') \to H^0(\mathbf{X}) \to H^0(\mathbf{X}'') \xrightarrow{\Delta} H^1(\mathbf{X}') \to H^1(\mathbf{X}) \to H^1(\mathbf{X}'') \to \cdots.$$

Theorem 6. *Let* $\quad 0 \to \mathbf{X}' \to \mathbf{X} \to \mathbf{X}'' \to 0$

be an exact sequence of complexes. Then if any two of $\mathbf{X}'$, $\mathbf{X}$, $\mathbf{X}''$ *are exact so is the third.*

Proof. Since every complex is a 0-sequence it is clear what we mean by saying that a complex is exact. Indeed, a complex is exact if and only if all its homology modules are null. Suppose now that two of $\mathbf{X}'$, $\mathbf{X}$, $\mathbf{X}''$ are exact and that $\mathbf{Y}$ is the third complex. Then the exact homology sequence has the form

$$\cdots \to 0 \to H_n(\mathbf{Y}) \to 0 \to 0 \to H_{n-1}(\mathbf{Y}) \to \cdots,$$

and so $H_n(\mathbf{Y}) = 0$ for all n. Thus $\mathbf{Y}$ is exact.

4.7 Homotopic translations

Let $\mathbf{X}$ and $\mathbf{X}'$ be complexes over Λ and suppose that, for each n, we are given a Λ-homomorphism $X_n \to X'_{n+1}$ so that we have, in fact, mappings

$$\begin{array}{ccccccccc} \cdots \to & X_{n+2} & \to & X_{n+1} & \to & X_n & \to & X_{n-1} & \to \cdots \\ & & \swarrow & & \swarrow & & \swarrow & & \\ \cdots \to & X'_{n+2} & \to & X'_{n+1} & \to & X'_n & \to & X'_{n-1} & \to \cdots \end{array}$$

We emphasize that these new Λ-homomorphisms may be quite arbitrary, so that, in particular, no requirements as to commutative properties have to be fulfilled. If now Λ-homomorphisms $X_n \to X'_n$ are defined by putting

$$X_n X'_n = X_n X'_{n+1} X'_n + X_n X_{n-1} X'_n, \tag{4.7.1}$$

then $\quad X_n X'_n X'_{n-1} = X_n X_{n-1} X'_n X'_{n-1} = X_n X_{n-1} X'_{n-1},$

which shows that the homomorphisms $X_n \to X'_n$ constitute a translation of $\mathbf{X}$ into $\mathbf{X}'$. A translation which can be obtained in this way will be called an *inessential translation*. It is a simple matter to verify that the inessential translations of $\mathbf{X}$ into $\mathbf{X}'$ form a subgroup of the group of *all* translations of $\mathbf{X}$ into $\mathbf{X}'$, and that, when Λ is commutative, this subgroup is a Λ-module.

Definition. Two translations of $\mathbf{X}$ into $\mathbf{X}'$, which differ by an in essential translation, will be said to be *homotopic*.

In terms of this definition, an inessential translation of $\mathbf{X}$ into $\mathbf{X}'$ is one which is homotopic to the translation all of whose component homomorphisms are null mappings. Instead, therefore, of saying that a translation is inessential we can equally well say that it is *null homotopic*.

Assume now that we have a null homotopic translation of $\mathbf{X}$ into $\mathbf{X}'$ defined (say) by (4.7.1) and suppose that x_n belongs to Ker $(X_n X_{n-1})$. Then $x_n X_n X'_n = x_n X_n X'_{n+1} X'_n \in \text{Im}\,(X'_{n+1} X'_n)$, and so, by the remark just preceding Theorem 2, the associated Λ-homomorphism

$$H(X_{n+1} - X_n - X_{n-1}) \to H(X'_{n+1} - X'_n - X'_{n-1})$$

is null. From this observation one obtains

Theorem 7. *If two translations of* $\mathbf{X}$ *into* $\mathbf{X}'$ *are homotopic, then the corresponding* Λ*-homomorphisms* $H_n(\mathbf{X}) \to H_n(\mathbf{X}')$ *coincide.*

5

PROJECTIVE AND INJECTIVE MODULES

Notation. Λ denotes a ring, not necessarily commutative, with an identity element, and Z denotes the ring of integers. Unless otherwise stated the term *Λ-module* should be taken to mean *left Λ-module*, but corresponding results for right Λ-modules can be obtained by making the obvious formal changes. The abbreviated notation of section (1.10) will be used wherever convenient.

5.1 Projective modules

As will be seen shortly, the notion of a projective module is a generalization of the concept of a free module, and it has the great advantage that it possesses a natural dual, the so-called *injective module* which will be studied in the next section. The formal definition of a projective module is as follows:

Definition. A Λ-module P is said to be *Λ-projective* if given any diagram

$$\begin{array}{ccccc} & & P & & \\ & & \downarrow & & \\ A & \longrightarrow & B & \longrightarrow & 0 \end{array}$$

over Λ, in which the row is exact, it is always possible to find a Λ-homomorphism $P \to A$ such that $PAB = PB$.

Proposition 1. *Let*

$$\begin{array}{ccccc} & & P & & \\ & & \downarrow & & \\ A & \longrightarrow & B & \longrightarrow & C \end{array}$$

be a diagram over Λ, in which P is Λ-projective, $A \to B \to C$ is exact and $PBC = 0$. Then there exists a Λ-homomorphism $P \to A$ such that $PAB = PB$.

Proof. Let $X = \text{Im}(AB) = \text{Ker}(BC)$ then, since $PBC = 0$, $P \to B$ maps P into X and thereby gives rise to a diagram

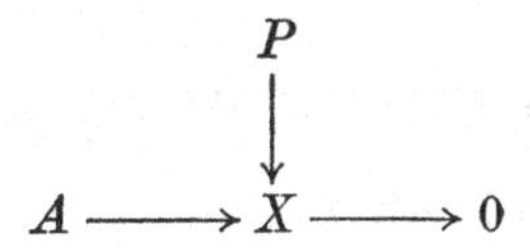

in which $A \to X \to 0$ is exact. Accordingly, since P is projective, there exists a Λ-homomorphism $P \to A$ such that $PAX = PX$, and this homomorphism clearly has the required property.

As an immediate consequence we have the

Corollary. *Let*

$$\begin{array}{ccccc} P & \longrightarrow & L & \longrightarrow & M \\ & & \downarrow & & \downarrow \\ A & \longrightarrow & B & \longrightarrow & C \end{array}$$

be a commutative diagram over Λ, in which P is Λ-projective, $PLM = 0$ and $A \to B \to C$ is exact. Then there exists a Λ-homomorphism $P \to A$ for which the diagram

$$\begin{array}{ccccc} P & \longrightarrow & L & \longrightarrow & M \\ \downarrow & & \downarrow & & \downarrow \\ A & \longrightarrow & B & \longrightarrow & C \end{array}$$

is commutative.

Theorem 1. *If P is a Λ-projective module then every epimorphism $M \to P$ is direct and every exact sequence*

$$0 \to L \to M \to P \to 0$$

splits.

Proof. The projective property of P shows that there exists a Λ-homomorphism $P \to M$ such that PMP is an identity map and from this the directness of $M \to P$ follows from Proposition 1 of section (1.10). This proves the first assertion and the second is an immediate consequence of the first.

Proposition 2. *Every direct summand of a Λ-projective module P is again Λ-projective.*

Proof. Let $P = M + N$ (direct sum) and let

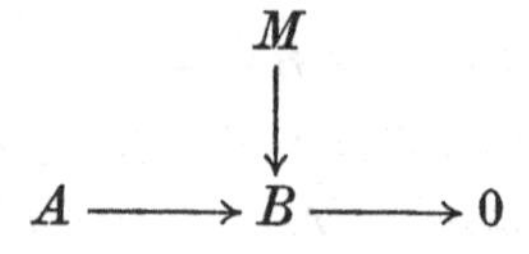

be a diagram in which the row is exact. If now $P \to M$ and $M \to P$ are canonical mappings associated with the direct decomposition, then we can find a Λ-homomorphism $P \to A$ such that $PAB = PMB$. Put $MA = MPA$ then

$$MAB = MPAB = MPMB = MB.$$

This shows that M is Λ-projective.

Proposition 3. *A direct sum $M = \sum_{i \in I} M_i$ of Λ-modules is Λ-projective if and only if each M_i is Λ-projective.*

Proof. Assume, to begin with, that each M_i is Λ-projective and let

$$\begin{array}{ccccc} & & M & & \\ & & \downarrow h & & \\ A & \xrightarrow{g} & B & \longrightarrow & 0 \end{array}$$

be a diagram in which the row is exact. For each i we have a canonical homomorphism $\phi_i : M_i \to M$ and therefore, by the projective character of M_i, we can find a Λ-homomorphism $f_i : M_i \to A$ such that $gf_i = h\phi_i$. With the aid of the f_i we can now construct a Λ-homomorphism $f : M \to A$. Let $m = [m_i]_{i \in I}$ be an element of M, then

$$gf(m) = g(\sum_i f_i(m_i)) = \sum_i gf_i(m_i) = \sum_i h\phi_i(m_i) = h(\sum_i \phi_i(m_i)) = h(m).$$

But m was an arbitrary element of M; consequently it has been shown that $gf = h$. This proves that M is Λ-projective.

To establish the converse, we assume now that M is Λ-projective. Then each M_i is isomorphic to a direct summand of M and so is Λ-projective by Proposition 2.

There is an intimate connexion between projective modules and free modules, the details of which are given by

Theorem 2. *A Λ-module is Λ-projective if and only if it is a direct summand of a Λ-free module.*

Proof. Let P be a Λ-projective module, then there exists a Λ-free module F for which there is an epimorphism $F \to P$. By Theorem 1, the kernel of this epimorphism is a direct summand of F and therefore P must be isomorphic to a direct summand of F. This implies, however, that P itself is a direct summand of a suitably constructed free module.

To see the converse we note that, by Theorem 3 of section (1.8), every Λ-free module is Λ-projective; consequently, by Proposition 2, every direct summand of a Λ-free module is projective.

It is now a simple matter to give an example of a projective module which is not free. Let Λ be the ring which consists of the integers taken module 6 and let A and B be the Λ-modules generated by the residues of 2 and 3 respectively. It is readily seen that Λ is the direct sum of A and B. But Λ is itself a Λ-free module because it admits its identity element as a base, consequently, by Theorem 2, A is Λ-projective. However, it is not Λ-free, because every non-null Λ-free module contains at least 6 elements, whereas A has only 3 elements.

Theorem 3. *A given Λ-module M is Λ-projective if and only if $\mathrm{Hom}_\Lambda(M,N)$ is an exact functor of N.*

Proof. Suppose first that $\mathrm{Hom}_\Lambda(M,N)$ is an exact functor of N and let

$$\begin{array}{ccccc} & & M & & \\ & & \downarrow & & \\ A & \longrightarrow & B & \longrightarrow & 0 \end{array}$$

be a diagram over Λ in which the row is exact, then it follows that

$$\mathrm{Hom}_\Lambda(M,A) \to \mathrm{Hom}_\Lambda(M,B) \to 0$$

is exact. Consequently there is a Λ-homomorphism $M \to A$, that is, an element of $\mathrm{Hom}_\Lambda(M,A)$, whose image in $\mathrm{Hom}_\Lambda(M,B)$ is the given mapping MB. In other words, $MAB = MB$, and this shows that M is Λ-projective.

Next assume that P is Λ-projective and let

$$0 \to N' \to N \to N'' \to 0$$

be an exact sequence. The proof will be complete if we show that

$$0 \to \mathrm{Hom}_\Lambda(P,N') \to \mathrm{Hom}_\Lambda(P,N) \to \mathrm{Hom}_\Lambda(P,N'') \to 0$$

is also exact and, referring back to Theorem 10 of section (2.5), we find that this has already been established if P is Λ-free. However, the original argument works equally well in the present case because it depends essentially on the fact that free modules are necessarily projective.

Theorem 4. *If P is a right Λ-projective module, then $P \otimes_\Lambda N$ is an exact functor on the category of left Λ-modules N. If P' is a left Λ-projective module, then $M \otimes_\Lambda P'$ is an exact functor on the category of right Λ-modules M.*

Proof. We shall only establish the first assertion, since the second is proved similarly. Let

$$0 \to N' \to N \to N'' \to 0$$

be an exact sequence. Then, by Theorem 3 of section (3.10), it will be sufficient to prove that

$$0 \to P \otimes_\Lambda N' \to P \otimes_\Lambda N \to P \otimes_\Lambda N'' \to 0$$

is exact. But $M \otimes_\Lambda N$ is a right exact functor, and so we need only show that $P \otimes_\Lambda N' \to P \otimes_\Lambda N$ is a monomorphism. By Theorem 2, P is a direct summand of a Λ-free module F, and so we may construct a split exact sequence of the form

$$0 \to P \to F \to L \to 0.$$

Consider the commutative diagram

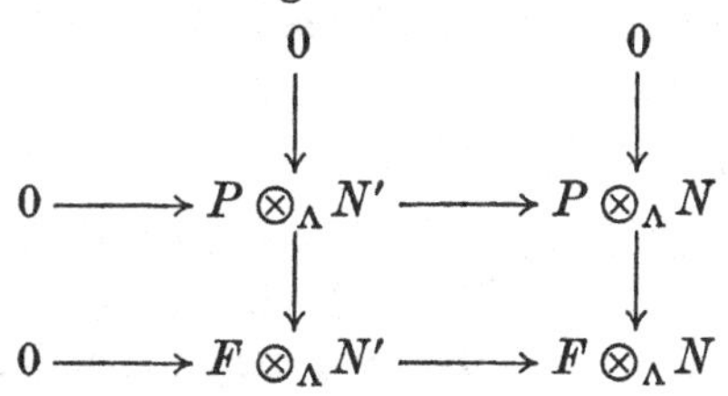

Here, by Theorem 2 of section (3·9), the columns are exact, while the bottom row is exact by Theorem 5 of section (2.4). But this implies that $P \otimes_\Lambda N' \to P \otimes_\Lambda N$ is a monomorphism and so the proof is complete.

We already know that every Λ-module is a homomorphic image of a Λ-free module, and we know that Λ-free modules are projective. This observation leads us to the following result, which is important enough to be stated as a theorem:

Theorem 5. *Given a Λ-module M there exists a projective module P for which there is an epimorphism $P \to M$.*

5.2 Injective modules

The concept of an injective module, which is of comparatively recent origin, is an essential one for our subject if we are to preserve the illuminating duality which runs through it. Λ-injective modules are defined as follows:

Definition. A Λ-module Q is said to be Λ-*injective* if given any diagram

$$\begin{array}{ccccc} 0 & \longrightarrow & A & \longrightarrow & B \\ & & \downarrow & & \\ & & Q & & \end{array}$$

with $0 \to A \to B$ exact, it is always possible to find a Λ-homomorphism $B \to Q$ such that $ABQ = AQ$.

This definition may be restated as follows:

A Λ-module Q is Λ-injective if, whenever A is a submodule of a Λ-module B, every Λ-homomorphism $A \to Q$ can be extended to a Λ-homomorphism $B \to Q$.

Proposition 4. *Let*

$$\begin{array}{ccccc} A & \longrightarrow & B & \longrightarrow & C \\ & & \downarrow & & \\ & & Q & & \end{array}$$

be a diagram over Λ, *in which* Q *is* Λ*-injective,* $A \to B \to C$ *is exact and* $ABQ = 0$. *Then there exists a* Λ*-homomorphism* $C \to Q$ *such that* $BCQ = BQ$.

Proof. Put $\operatorname{Im}(AB) = \operatorname{Ker}(BC) = X$. Then $B \to Q$ maps every element of X into zero and so induces a Λ-homomorphism $B/X \to Q$. Further, since $B \to C$ has kernel X, it yields an exact sequence $0 \to B/X \to C$, and now the injective character of Q shows that there exists a Λ-homomorphism $C \to Q$ such that

$$\begin{array}{ccccc} 0 & \longrightarrow & B/X & \longrightarrow & C \\ & & \downarrow & \swarrow & \\ & & Q & & \end{array}$$

is commutative. This new homomorphism has the required property.

As an immediate application we have the

Corollary. *Let*

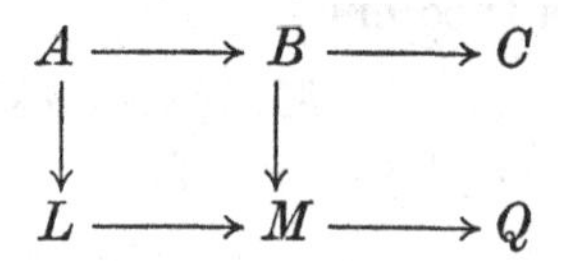

be a commutative diagram over Λ, *in which* Q *is* Λ*-injective,* $A \to B \to C$ *is exact and* $LMQ = 0$. *Then there exists a* Λ*-homomorphism* $C \to Q$ *for which the diagram*

$$\begin{array}{ccccc} A & \longrightarrow & B & \longrightarrow & C \\ \downarrow & & \downarrow & & \downarrow \\ L & \longrightarrow & M & \longrightarrow & Q \end{array}$$

is commutative.

Injective modules, like their projective counterparts, possess certain splitting properties. These are set out in

Theorem 6. *If* Q *is* Λ*-injective then every monomorphism* $Q \to L$ *is direct and every exact sequence*

$$0 \to Q \to L \to M \to 0$$

splits.

Proof. The injective character of Q shows that there exists a Λ-homomorphism $L \to Q$ such that $QLQ = \text{identity}$. Consequently, by Proposition 1 of section (1.10), $Q \to L$ is direct and now the rest of the theorem is obvious.

Proposition 5. *If M is a direct factor of a Λ-injective module Q, then M is itself Λ-injective.*

Proof. The argument, which is dual to that of Proposition 2, runs as follows. Let

$$\begin{array}{ccccc} 0 & \longrightarrow & A & \longrightarrow & B \\ & & \downarrow & & \\ & & M & & \end{array}$$

be a diagram over Λ, with $0 \to A \to B$ exact. By hypothesis, there exists an epimorphism $Q \to M$, which is direct. Consequently we can find a Λ-homomorphism $M \to Q$ such that MQM = identity. Again, since Q is Λ-injective, there exists a Λ-homomorphism $B \to Q$ such that $ABQ = AMQ$. Put $BM = BQM$; then

$$ABM = ABQM = AMQM = AM.$$

This completes the proof.

The analogue of Proposition 3 is

Proposition 6. *A direct product $M = \prod_{i \in I} M_i$ of Λ-modules is Λ-injective if and only if each M_i is Λ-injective.*

Proof. Assume first that each M_i is Λ-injective and let

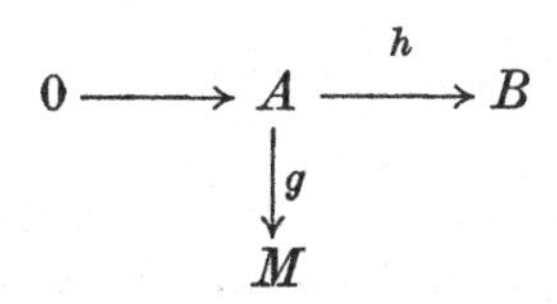

be a diagram over Λ, with $0 \to A \to B$ exact. For each i, let $\phi_i : M \to M_i$ be the usual canonical homomorphism, then, since M_i is Λ-injective, we can find a Λ-homomorphism $f_i : B \to M_i$ such that $f_i h = \phi_i g$. Further, the homomorphisms f_i can be combined to yield a homomorphism $f : B \to M$. Let a belong to A, then

$$\phi_i fh(a) = f_i h(a) = \phi_i g(a),$$

and as this holds for all i in I, it follows that $fh(a) = g(a)$. Accordingly, $fh = g$, and we have shown that M is Λ-injective.

To see the converse assume that M is Λ-injective. Since $\phi_i : M \to M_i$ is a direct epimorphism, M_i is a direct factor of M and so it is Λ-injective by virtue of Proposition 5.

A useful characterization of injective modules is given by

Theorem 7. *A Λ-module N is Λ-injective if and only if* $\mathrm{Hom}_\Lambda(M, N)$ *is an exact functor of M.*

Proof. Assume first that $\mathrm{Hom}_\Lambda(M, N)$ is an exact functor of M and let

$$\begin{array}{ccccc} 0 & \longrightarrow & A & \longrightarrow & B \\ & & \downarrow & & \\ & & N & & \end{array}$$

be a diagram over Λ with $0 \to A \to B$ exact. Then

$$\mathrm{Hom}_\Lambda(B, N) \to \mathrm{Hom}_\Lambda(A, N)$$

is an epimorphism and so the given homomorphism $A \to N$, regarded as an element of $\mathrm{Hom}_\Lambda(A, N)$, is the image, under

$$\mathrm{Hom}_\Lambda(B, N) \to \mathrm{Hom}(A, N),$$

of some element $B \to N$ of $\mathrm{Hom}_\Lambda(B, N)$. But this means simply that $ABN = AN$, and so N has been shown to be injective.

Next suppose that N is Λ-injective and let

$$0 \to A \to B \to C \to 0$$

be an exact sequence. It will now suffice to show that

$$0 \to \mathrm{Hom}_\Lambda(C, N) \to \mathrm{Hom}_\Lambda(B, N) \to \mathrm{Hom}_\Lambda(A, N) \to 0$$

is also exact. However, we know, in any case, that $\mathrm{Hom}_\Lambda(M, N)$ is a left exact functor, consequently it will be enough to prove that $\mathrm{Hom}_\Lambda(B, N) \to \mathrm{Hom}_\Lambda(A, N)$ is an epimorphism. But this is clear, for, by the injective property of N, given any homomorphism $A \to N$ there exists a homomorphism $B \to N$ such that $ABN = AN$.

It has already been observed that every Λ-module is a homomorphic image of a Λ-projective module. The corresponding result for injective modules asserts that *every Λ-module is a submodule of a Λ-injective module.* As compared with the former result this is comparatively difficult to prove. What we shall do is state the result in question as Theorem 8 and devote the next section to developing ideas which will yield a proof.

Theorem 8. *Given any Λ-module M there exists a Λ-injective module Q for which there is a monomorphism $M \to Q$.*

5.3 An existence theorem for injective modules

Let R be the additive group of rational numbers and put $\Omega = R/Z$, where, as usual, Z denotes the integers. We shall begin by showing that Ω is Z-injective and for this we need a lemma.

Lemma 1. *Let H be a submodule of a Z-module G and let $\phi : H \to \Omega$ be a homomorphism. If now γ belongs to G then ϕ can be extended to a homomorphism $\phi' : H' \to \Omega$, where H' is the submodule generated by H and γ. Further, if γ does not belong to H, then we can arrange that $\phi'(\gamma) \neq 0$.*

Proof. We assume that γ does not belong to H (for otherwise the assertions are trivial) and we distinguish two cases.

Case 1. This is the case in which, for every integer $m \neq 0$, $m\gamma$ is not an element of H. In this situation, each element h' of H' has a *unique* representation in the form $h' = h + n\gamma$, where h belongs to H and n is an integer. If therefore ω is an arbitrary element of Ω we obtain the required extension by writing $\phi'(h') = \phi(h) + n\omega$. Further, by choosing $\omega \neq 0$, we can secure that $\phi'(\gamma) = \omega \neq 0$.

Case 2. In the remaining case there is a smallest positive integer m such that $m\gamma$ belongs to H. Let $\phi(m\gamma) = \omega_1$, then we can choose ω in Ω so that $m\omega = \omega_1$. Since γ does not belong to H, the integer m must be at least 2 and therefore we can arrange that $\omega \neq 0$. Each element h' of H' has a *unique* representation in the form $h' = h + n\gamma$, where h belongs to H and $0 \leqslant n < m$. Put $\phi'(h') = \phi(h) + n\omega$, then it is clear that, as a mapping, ϕ' extends ϕ. We now assert that ϕ' is a homomorphism. To see this let $x = h + n\gamma$ and $\bar{x} = \bar{h} + \bar{n}\gamma$, where $0 \leqslant n < m$ and $0 \leqslant \bar{n} < m$, be two elements of H'. Then either $n + \bar{n} < m$ or $n + \bar{n} \geqslant m$. In the former case it is trivial that

$$\phi'(x + \bar{x}) = \phi'(x) + \phi'(\bar{x}),$$

while if $n + \bar{n} \geqslant m$ then

$$\begin{aligned} \phi'(x+\bar{x}) &= \phi'(h + \bar{h} + m\gamma + (n + \bar{n} - m)\gamma) \\ &= \phi(h + \bar{h} + m\gamma) + (n + \bar{n} - m)\omega \\ &= \phi(h) + \phi(\bar{h}) + \omega_1 + (n + \bar{n} - m)\omega = \phi(h) + n\omega + \phi(\bar{h}) + \bar{n}\omega \\ &= \phi'(x) + \phi'(\bar{x}) \end{aligned}$$

The proof of the lemma is therefore complete.

From this result we can deduce, by a straightforward application of Zorn's lemma,† that the homomorphism $H \to \Omega$ can be extended to a homomorphism $G \to \Omega$. To be more explicit, let $\mathfrak{S}$ be the set of all pairs (K, ψ), where K is a submodule of G containing H and ψ is a homomorphism $K \to \Omega$ extending ϕ. $\mathfrak{S}$ can be partially ordered by writing $(K, \psi) \leqslant (K', \psi')$ if K is contained in K' and ψ' agrees with ψ on K. The partial ordering makes $\mathfrak{S}$ what is called an *inductive system*. For suppose that $[(K_i, \psi_i)]_{i \in I}$ is a family of pairs with the property that, whenever i_1, i_2 are in I, either

$$(K_{i_1}, \psi_{i_1}) \leqslant (K_{i_2}, \psi_{i_2}) \quad \text{or} \quad (K_{i_2}, \psi_{i_2}) \leqslant (K_{i_1}, \psi_{i_1}).$$

Then the union K of the K_i is a submodule of G and there is a homomorphism $K \to \Omega$, which, for each i, agrees with ψ_i on K_i. Zorn's lemma now shows that there is a pair, (K^*, ψ^*) say, which is maximal in $\mathfrak{S}$ and, by Lemma 1, this can only happen when $K^* = G$. Accordingly $\psi^* : G \to \Omega$ is the required extension of $H \to \Omega$.

The above remarks constitute a proof of the following theorem.

Theorem 9. *The additive group Ω, consisting of the rational numbers taken modulo the integers, is a Z-injective module.*

The module Ω will be used to discuss injective modules over arbitrary rings. To this end, let M be a *left* Λ-module and let us write

$$\mathrm{Hom}_Z(M, \Omega) = M^*. \tag{5.3.1}$$

Then M^* is an additive group and it will now be shown that it can be endowed with certain additional structure. In fact, let f belong to M^*, let λ belong to Λ and let us write, for each m in M, $g(m) = f(\lambda m)$. It is clear that the mapping $m \to g(m)$ is a group homomorphism $M \to \Omega$ and therefore an element $f\lambda$ (say) of M^*. With this definition of 'multiplication' M^*, as is easily verified, becomes a *right* Λ-module.

Next we consider the effect of a Λ-*homomorphism* $\phi : M \to M_1$ of left Λ-modules. This determines a Z-homomorphism

$$\phi^* : \mathrm{Hom}_Z(M_1, \Omega) \to \mathrm{Hom}_Z(M, \Omega),$$

so that, in other symbols, $\phi^* : M_1^* \to M^*$. Let f_1 belong to M_1^* and let λ belong to Λ, then

$$\phi^*(f_1) = f_1 \phi, \quad \phi^*(f_1 \lambda) = (f_1 \lambda)\phi.$$

Accordingly, for any m in M,

$$\begin{aligned}[\phi^*(f_1\lambda)]\,m = [(f_1\lambda)\,\phi]\,m &= [f_1\lambda]\,\phi(m) = f_1[\lambda\phi(m)] \\ &= f_1[\phi(\lambda m)] = [\phi^*(f_1)]\,(\lambda m) = [\phi^*(f_1)\,\lambda]\,m.\end{aligned}$$

† If this is unfamiliar to the reader he should consult the Notes on Chapter 5.

Thus $\phi^*(f_1\lambda) = \phi^*(f_1)\lambda$, and so we see that $\phi^* : M_1^* \to M^*$ is not only a Z-homomorphism but also a Λ-homomorphism.

Lemma 2. *M^* is an additive contravariant functor of M (from the category of left Λ-modules to the category of right Λ-modules) and, as a functor, it is exact.*

Proof. The fact that M^* is an additive functor is an immediate consequence of the remarks which have already been made.† Let

$$0 \to M_1 \to M \to M_2 \to 0$$

be an exact sequence of Λ-modules and Λ-homomorphisms. Then the lemma will be established if we show that

$$0 \to \mathrm{Hom}_Z(M_2, \Omega) \to \mathrm{Hom}_Z(M, \Omega) \to \mathrm{Hom}_Z(M_1, \Omega) \to 0$$

is also exact. But this follows at once from Theorem 7 and the fact that Ω is Z-injective.

So far we have considered $\mathrm{Hom}_Z(M, \Omega)$ only when M is a *left* Λ-module. Suppose now that N is a *right* Λ-module, then, in an analogous manner, $\mathrm{Hom}_Z(N, \Omega)$ can be given the structure of a *left* Λ-module. In order to avoid complicating the notation, we shall denote this module by N^* and then we find that N^* is an additive, exact, contravariant functor of N.

Let us now return to the case of a left Λ-module M and let us consider $(M^*)^*$, which we shall denote by M^{**}. This is, of course, also a left Λ-module. Suppose now that m is a fixed element of M and f is a variable element of M^*, then the mapping $f \to f(m)$ is a Z-homomorphism $\alpha(m) : M^* \to \Omega$. Thus $\alpha(m)$ belongs to M^{**} and by construction $[\alpha(m)]f = f(m)$ for each f in M^*. Clearly, if m_1, m_2 are in M, then

$$\alpha(m_1 + m_2) = \alpha(m_1) + \alpha(m_2).$$

Also, if λ belongs to Λ and f belongs to M^*, then

$$[\alpha(\lambda m)]f = f(\lambda m) = [f\lambda]m = [\alpha(m)](f\lambda) = [\lambda\alpha(m)]f.$$

This shows that $\alpha(\lambda m) = \lambda\alpha(m)$ and, combining our observations, we see that the mapping $m \to \alpha(m)$ is a Λ-homomorphism $M \to M^{**}$.

Lemma 3. *The Λ-homomorphism $\alpha : M \to M^{**}$ is a monomorphism.*

Proof. Suppose that $\alpha(m) = 0$, then $0 = [\alpha(m)]f = f(m)$ for all f in M^*. We wish to show that $m = 0$. Assume the contrary; then, by

† More generally, for an arbitrary Z-module Θ, the above construction enables us to regard $\mathrm{Hom}_Z(M, \Theta)$ as an additive contravariant functor from $\mathscr{G}_\Lambda^L$ to $\mathscr{G}_\Lambda^R$.

Lemma 1, the trivial mapping $0 \to \Omega$ can be extended to a Z-homomorphism $Zm \to \Omega$, in which the image of m is not zero. Since Ω is Z-injective, we can extend this new mapping to a Z-homomorphism $f: M \to \Omega$, and this will have the property that $f(m) \neq 0$. However, we know already that such a situation is impossible.

We need one further result and then we shall have all that we require to prove Theorem 8.

Lemma 4. *If Λ is regarded as a right Λ-module, then Λ^* is an injective left Λ-module.*

Proof. Let A be a left Λ-module and B one of its submodules, and suppose that a Λ-homomorphism $f: B \to \Lambda^*$ is given. We want to show that f can be extended to a Λ-homomorphism $\bar{f}: A \to \Lambda^*$.

Suppose that $b \in B$, then $f(b) \in \mathrm{Hom}_Z(\Lambda, \Omega)$. Putting $\psi(b) = [f(b)]1$ we see that the mapping $b \to \psi(b)$ is a Z-homomorphism $\psi: B \to \Omega$ consequently, since Ω is Z-injective, it has an extension to a Z-homomorphism $\bar{\psi}: A \to \Omega$.

For each $a \in A$ define $\bar{f}(a) \in \mathrm{Hom}_Z(\Lambda, \Omega)$ by

$$[\bar{f}(a)]\lambda = \bar{\psi}(\lambda a), \tag{5.3.2}$$

where $\lambda \in \Lambda$; then, when $b \in B$,

$$[\bar{f}(b)]\lambda = \psi(\lambda b) = [f(\lambda b)]1 = [\lambda f(b)]1 = [f(b)]\lambda, \tag{5.3.3}$$

and now it is seen that the mapping $a \to \bar{f}(a)$ is a Z-homomorphism $\bar{f}: A \to \Lambda^*$ extending f. But if $a \in A$ and λ, λ' are in Λ, we have

$$[\bar{f}(\lambda a)]\lambda' = \bar{\psi}(\lambda'\lambda a) = [\bar{f}(a)]\lambda'\lambda = [\lambda\bar{f}(a)]\lambda'. \tag{5.3.4}$$

Accordingly $\bar{f}(\lambda a) = \lambda\bar{f}(a)$, and therefore $\bar{f}$ is a Λ-homomorphism and not merely a Z-homomorphism. This completes the proof.

Corollary. *If F is a Λ-free right module then F^* is a Λ-injective left module.*

Proof. Since F is free, it is a direct sum of modules each of which is isomorphic to Λ and therefore we have a complete representation

$$\Lambda \xrightarrow{f_i} F \xrightarrow{g_i} \Lambda \quad (i \in I) \tag{5.3.5}$$

of F as a direct sum. From this we obtain, using Theorem 8 of section (2.5), a direct product representation

$$\Lambda^* \xrightarrow{\phi_i} F^* \xrightarrow{\psi_i} \Lambda^* \quad (i \in I), \tag{5.3.6}$$

for $F^* = \mathrm{Hom}_Z(F, \Omega)$. More precisely, the result quoted shows that (5.3.6) gives a Z-isomorphism $F^* \approx \prod_i \Lambda^*$. But, since we know that the mappings in (5.3.6) are Λ-homomorphisms, it follows that this isomorphism is, in fact, a Λ-isomorphism. Hence to establish the corollary we need only show that $\prod_i \Lambda^*$ is Λ-injective. But we already know that Λ^* is Λ-injective and so the required result follows from Proposition 6.

We come now to the demonstration for which the present section has been developed.

***Proof of Theorem* 8.** Let M be an arbitrary left Λ-module; then M^* is a right Λ-module. We can therefore construct an exact sequence $F \to M^* \to 0$, where F is a right Λ-free module. Next, by Lemma 2, the sequence $0 \to M^{**} \to F^*$, to which it gives rise, is also exact and so $M^{**} \to F^*$ is a Λ-monomorphism. However, by Lemma 3, we have a canonical monomorphism $M \to M^{**}$, and, so, by combining, we obtain a Λ-monomorphism $M \to F^*$. But, by the last corollary, F^* is Λ-injective, consequently this completes the proof.

5.4 Complexes over a module

The facts, that we have just established concerning projective and injective modules, will now be used to introduce the concept of a *resolution* of a module. We recall that by a complex $\mathbf{X}$ of Λ-modules is meant a 0-sequence

$$\cdots \to X^{n-1} \to X^n \to X^{n+1} \to \cdots, \tag{5.4.1}$$

which is infinite in both directions. The complex $\mathbf{X}$ is said to be a *projective complex* if all the X^n are Λ-projective, and to be *injective* when they are all Λ-injective modules.

Let A be a Λ-module and let us denote by $\mathbf{A}$ the complex

$$\cdots \to 0 \to 0 \to A \to 0 \to 0 \to \cdots. \tag{5.4.2}$$

Here it is to be understood that $A^n = 0$ if $n \neq 0$ and $A^0 = A$, so that $\mathbf{A}$ is both a right complex and a left complex. $\mathbf{A}$ will be called the *complex associated with* A. We note that $\mathbf{A}$ is projective (injective) when A is Λ-projective (Λ-injective).

Definition. By a *right complex over* A we mean a right complex $\mathbf{Y}$ together with a translation $\boldsymbol{\epsilon} : \mathbf{A} \to \mathbf{Y}$ called the *augmentation translation*.

Suppose that we have this situation; then in view of the degenerate character of **A** only one component of $\boldsymbol{\epsilon}$ is non-trivial and that is the so-called *augmentation homomorphism* $\epsilon : A \to Y^0$. Since $\boldsymbol{\epsilon}$ is a translation, the diagram

$$\begin{array}{ccc} A & \longrightarrow & 0 \\ \downarrow{\scriptstyle \epsilon} & & \downarrow \\ Y^0 & \longrightarrow & Y^1 \end{array}$$

is commutative and this shows that

$$0 \to A \xrightarrow{\epsilon} Y^0 \to Y^1 \to Y^2 \to \cdots \qquad (5.4.3)$$

is a 0-sequence. Conversely every 0-sequence (5.4.3) determines a right complex over A.

Definition. By a *left complex over* A is meant a left complex **X** together with an *augmentation translation* $\boldsymbol{\epsilon} : \mathbf{X} \to \mathbf{A}$.

This time the non-trivial part of $\boldsymbol{\epsilon}$ is the augmentation homomorphism $\epsilon : X_0 \to A$ and

$$\cdots \to X_2 \to X_1 \to X_0 \xrightarrow{\epsilon} A \to 0 \qquad (5.4.4)$$

is a 0-sequence. On the other hand, every 0-sequence (5.4.4) determines, in an obvious manner, a left complex over A.

Let **Y** be a right complex over A and **X** a left complex over A.

Definition. If the sequences (5.4.3) and (5.4.4) are exact, then we say that **Y** is an *acyclic right complex over* A and **X** is called an *acyclic left complex over* A.

Among the acyclic complexes over A those of particular interest are the so-called projective and injective resolutions. These are defined as follows.

Definition. An acyclic projective left complex over A is called a *projective resolution* of A. An *injective resolution* of A is an acyclic injective right complex over A.

In other words, a projective resolution of A is determined by an exact sequence

$$\cdots \to P_2 \to P_1 \to P_0 \to A \to 0, \qquad (5.4.5)$$

where each P_i is Λ-projective, while an injective resolution is associated with an exact sequence

$$0 \to A \to Q^0 \to Q^1 \to \cdots, \qquad (5.4.6)$$

where Q^0, Q^1, etc., are all Λ-injective.

Suppose now that we have an exact sequence

$$\cdots \to F_2 \to F_1 \to F_0 \to A \to 0, \tag{5.4.7}$$

where each F_i is Λ-free. Then this determines a special kind of pro jective resolution of A, which is called a *free resolution.*

Now, by Theorem 2 of section (1.8), every module is a homomorphic image of a Λ-free module; consequently, given A, we can always construct an exact sequence (5.4.7). This establishes

Theorem 10. *Every Λ-module possesses free resolutions. In particular, for each Λ-module, there exist projective resolutions.*

Theorem 11. *Every Λ-module possesses injective resolutions.*

Proof. Let A be a Λ-module then, by Theorem 8, we can construct an exact sequence $0 \to A \to Q^0$, where Q^0 is Λ-injective. By the same theorem we can construct an exact sequence

$$0 \to Q^0/\mathrm{Im}\,(AQ^0) \to Q^1,$$

where Q^1 is again Λ-injective. Denote by $Q^0 \to Q^1$ the combined homomorphism

$$Q^0 \xrightarrow{\text{nat.}} Q^0/\mathrm{Im}\,(AQ^0) \to Q^1$$

then $0 \to A \to Q^0 \to Q^1$ is exact. Proceeding in this way we build up an infinite exact sequence

$$0 \to A \to Q^0 \to Q^1 \to Q^2 \to \cdots,$$

where all the Q^i are Λ-injective, and this establishes the theorem.

5.5 Properties of resolutions of modules

The present section is devoted to establishing some basic facts concerning projective and injective resolutions. It will be found that, for each theorem involving the former, there is a corresponding theorem for injective resolutions. Further, the proofs for corresponding theorems are similar in the sense that we can change the arguments for one situation into those applicable to the other, simply by using the duality which exists between projective and injective modules. Accordingly, only one result of each pair will be proved. Usually the one chosen will be that concerned with left complexes because there is a notational convenience in having indices attached at the bottom of our symbols.

Let $\mathbf{X}$ be a left complex over A, $\mathbf{X}'$ a left complex over A' and suppose that $f: A \to A'$ is a Λ-homomorphism. This homomorphism determines a translation

$$\begin{array}{ccccccccccc} \cdots \longrightarrow & 0 & \longrightarrow & 0 & \longrightarrow & A & \longrightarrow & 0 & \longrightarrow & 0 & \longrightarrow \cdots \\ & \downarrow & & \downarrow & & \downarrow f & & \downarrow & & \downarrow & \\ \cdots \longrightarrow & 0 & \longrightarrow & 0 & \longrightarrow & A' & \longrightarrow & 0 & \longrightarrow & 0 & \longrightarrow \cdots \end{array}$$

of $\mathbf{A}$ into $\mathbf{A}'$, that is a translation $\mathbf{f}$ of the associated complexes.

Definition. A translation $\mathbf{X} \to \mathbf{X}'$ is said to be *over the Λ-homomorphism $f: A \to A'$* if

$$\begin{array}{ccc} \mathbf{X} & \longrightarrow & \mathbf{A} \\ \downarrow & & \downarrow \mathbf{f} \\ \mathbf{X}' & \longrightarrow & \mathbf{A}' \end{array}$$

is a commutative diagram of complexes and translations.

Thus a translation $\mathbf{X} \to \mathbf{X}'$ over f is determined by Λ-homomorphisms $X_0 \to X_0'$, $X_1 \to X_1'$, etc., which make

$$\begin{array}{ccccccccc} \cdots \longrightarrow & X_2 & \longrightarrow & X_1 & \longrightarrow & X_0 & \longrightarrow & A & \longrightarrow 0 \\ & \downarrow & & \downarrow & & \downarrow & & \downarrow f & \\ \cdots \longrightarrow & X_2' & \longrightarrow & X_1' & \longrightarrow & X_0' & \longrightarrow & A' & \longrightarrow 0 \end{array}$$

commutative.

Similarly, if $\mathbf{Y}$ and $\mathbf{Y}'$ are right complexes over A and A' respectively, then a translation $\mathbf{Y} \to \mathbf{Y}'$ is said to be over $f: A \to A'$ if

$$\begin{array}{ccc} \mathbf{A} & \longrightarrow & \mathbf{Y} \\ \downarrow & & \downarrow \\ \mathbf{A}' & \longrightarrow & \mathbf{Y}' \end{array}$$

is commutative and such a translation is determined by Λ-homomorphisms $Y^n \to Y'^n$ $(n = 0, 1, 2, \ldots)$ which make

$$\begin{array}{ccccccccc} 0 \longrightarrow & A & \longrightarrow & Y^0 & \longrightarrow & Y^1 & \longrightarrow & Y^2 & \longrightarrow \cdots \\ & \downarrow f & & \downarrow & & \downarrow & & \downarrow & \\ 0 \longrightarrow & A' & \longrightarrow & Y'^0 & \longrightarrow & Y'^1 & \longrightarrow & Y'^2 & \longrightarrow \cdots \end{array}$$

a commutative diagram.

Theorem 12. Let $\mathbf{P}$ *be a projective left complex over* A *and* $\mathbf{X}$ *an acyclic left complex over* A'. *Then, given a Λ-homomorphism $f: A \to A'$, there exists a translation $\mathbf{P} \to \mathbf{X}$ over f and any two such translations are homotopic.*†

† It is convenient to express this by saying that the translation is uniquely determined to 'within a homotopy'.

Proof. In the diagram

$$\begin{array}{ccccc} P_0 & \longrightarrow & A & & \\ & & \Big\downarrow f & & \\ X_0 & \longrightarrow & A' & \longrightarrow & 0 \end{array}$$

P_0 is projective and the lower row is exact; hence we can find a homomorphism $P_0 \to X_0$ such that $P_0 X_0 A' = P_0 A A'$. Suppose now that we have constructed $P_0 X_0, P_1 X_1, \ldots, P_m X_m$ so as to make

$$\begin{array}{ccccccccccc} P_m & \longrightarrow & P_{m-1} & \longrightarrow & \cdots & \longrightarrow & P_0 & \longrightarrow & A & \longrightarrow & 0 \\ \Big\downarrow & & \Big\downarrow & & & & \Big\downarrow & & \Big\downarrow f & & \\ X_m & \longrightarrow & X_{m-1} & \longrightarrow & \cdots & \longrightarrow & X_0 & \longrightarrow & A' & \longrightarrow & 0 \end{array} \tag{5.5.1}$$

commutative, then, since in

$$\begin{array}{ccccc} P_{m+1} & \longrightarrow & P_m & \longrightarrow & P_{m-1} \\ & & \Big\downarrow & & \Big\downarrow \\ X_{m+1} & \longrightarrow & X_m & \longrightarrow & X_{m-1} \end{array}$$

the upper row is a 0-sequence and the lower row is exact, it follows, by the corollary to Proposition 1, that we can extend (5.5.1) by the addition of a homomorphism $P_{m+1} X_{m+1}$ without disturbing its commutative properties. Thus the first assertion follows by induction.

To prove the second assertion we suppose that $A \to A'$ is a null map and show that $\mathbf{P} \to \mathbf{X}$ is null homotopic. In other words, we must establish the existence of homomorphisms $P_0 X_1, P_1 X_2, P_2 X_3, \ldots$ such that $P_0 X_1 X_0 = P_0 X_0$ and $P_n X_{n+1} X_n + P_n P_{n-1} X_n = P_n X_n$ for $n \geqslant 1$. The existence of $P_0 X_1$ follows by consideration of the diagram

$$\begin{array}{ccccc} & & P_0 & & \\ & & \Big\downarrow & & \\ X_1 & \longrightarrow & X_0 & \longrightarrow & A' \end{array}$$

and the fact $P_0 X_0 A' = P_0 A A' = 0$. Suppose now that $P_0 X_1, P_1 X_2, \ldots, P_m X_{m+1}$ have all been constructed with the required properties. We wish to establish the existence of a homomorphism $P_{m+1} X_{m+2}$ such that $P_{m+1} X_{m+2} X_{m+1} + P_{m+1} P_m X_{m+1} = P_{m+1} X_{m+1}$; and this will follow, by Proposition 1, from the diagram

$$\begin{array}{ccccc} & & P_{m+1} & & \\ & & \Big\downarrow {\scriptstyle P_{m+1} X_{m+1} - P_{m+1} P_m X_{m+1}} & & \\ X_{m+2} & \longrightarrow & X_{m+1} & \longrightarrow & X_m \end{array}$$

provided we show that $(P_{m+1}X_{m+1}-P_{m+1}P_mX_{m+1})X_{m+1}X_m = 0$. But

$$\begin{aligned}(P_{m+1}X_{m+1}&-P_{m+1}P_mX_{m+1})X_{m+1}X_m\\ &= P_{m+1}P_mX_m-P_{m+1}P_mX_{m+1}X_m = P_{m+1}P_m(P_mX_m-P_mX_{m+1}X_m)\\ &= P_{m+1}P_m(P_mP_{m-1}X_m)\end{aligned}$$

by the inductive hypothesis. This completes the proof, since

$$P_{m+1}P_mP_{m-1} = 0.$$

Corollary. *If* **P** *and* **P**′ *are projective resolutions of* A *and* A' *respectively, then given any* Λ*-homomorphism* $f: A \to A'$ *there exists a translation* $\mathbf{P} \to \mathbf{P}'$ *over* f *and any two such translations are homotopic.*

The dual theorem and its corollary will now be stated without proof.

Theorem 13. *Let* **Q** *be an injective right complex over* A *and* **Y** *an acyclic right complex over* A'. *Then given any* Λ*-homomorphism* $f: A' \to A$ *there exists a translation* $\mathbf{Y} \to \mathbf{Q}$ *over* f *and any two such translations are homotopic.*

Corollary. *If* **Q** *and* **Q**′ *are injective resolutions of* A *and* A' *respectively, then given a* Λ*-homomorphism* $f: A' \to A$ *there exists a translation* $\mathbf{Q}' \to \mathbf{Q}$ *over* f *and any two such translations are homotopic.*

5.6 Properties of resolutions of sequences

Let
$$0 \to A' \to A \to A'' \to 0$$
be a sequence of Λ-modules and homomorphisms and let $\mathbf{X}'$, $\mathbf{X}$ and $\mathbf{X}''$ be left complexes over A', A and A'' respectively. Then a sequence

$$\mathbf{0} \to \mathbf{X}' \to \mathbf{X} \to \mathbf{X}'' \to \mathbf{0}$$

of translation is said to be *over*

$$0 \to A' \to A \to A'' \to 0$$

if $\mathbf{X}' \to \mathbf{X}$ is over $A' \to A$ and $\mathbf{X} \to \mathbf{X}''$ is over $A \to A''$. A similar terminology is used in connexion with sequences

$$\mathbf{0} \to \mathbf{Y}' \to \mathbf{Y} \to \mathbf{Y}'' \to \mathbf{0},$$

where $\mathbf{Y}'$, $\mathbf{Y}$ and $\mathbf{Y}''$ are right complexes over A', A and A'' respectively.

Definition. A *projective resolution* of an exact sequence

$$0 \to A' \to A \to A'' \to 0 \tag{5.6.1}$$

is an exact sequence
$$\mathbf{0} \to \mathbf{P}' \to \mathbf{P} \to \mathbf{P}'' \to \mathbf{0} \tag{5.6.2}$$

(in the sense of section (4.2)) over (5.6.1), where $\mathbf{P}'$, $\mathbf{P}$ and $\mathbf{P}''$ are projective resolutions of A', A and A'' respectively.

Injective resolutions of (5.6.1) are defined by replacing 'projective' by 'injective' wherever the former occurs in the above definition.

Theorem 14. *Let*
$$0 \to A' \to A \to A'' \to 0 \tag{5.6.3}$$
be an exact sequence, let $\mathbf{X}'$, $\mathbf{X}$ *and* $\mathbf{X}''$ *be left complexes over* A', A *and* A'' *respectively, and let*
$$\mathbf{0} \to \mathbf{X}' \to \mathbf{X} \to \mathbf{X}'' \to \mathbf{0} \tag{5.6.4}$$
be an exact sequence over (5.6.3). *If now* $\mathbf{X}'$ *and* $\mathbf{X}''$ *are acyclic over* A' *and* A'', *then* $\mathbf{X}$ *is acyclic over* A. *Furthermore, if* $\mathbf{X}'$ *and* $\mathbf{X}''$ *are projective resolutions of* A' *and* A'', *then* $\mathbf{X}$ *will be a projective resolution of* A.

This time, for a change, we shall prove the dual theorem, which reads thus:

Theorem 15. *Let*
$$0 \to A' \to A \to A'' \to 0 \tag{5.6.5}$$
be an exact sequence, let $\mathbf{Y}'$, $\mathbf{Y}$ *and* $\mathbf{Y}''$ *be right complexes over* A', A *and* A'' *respectively, and let*
$$\mathbf{0} \to \mathbf{Y}' \to \mathbf{Y} \to \mathbf{Y}'' \to \mathbf{0} \tag{5.6.6}$$
be an exact sequence over (5.6.5). *If now* $\mathbf{Y}'$ *and* $\mathbf{Y}''$ *are acyclic over* A' *and* A'', *then* $\mathbf{Y}$ *is acyclic over* A. *Furthermore, if* $\mathbf{Y}'$ *and* $\mathbf{Y}''$ *are injective resolutions of* A' *and* A'', *then* $\mathbf{Y}$ *must be an injective resolution of* A.

Proof. By hypothesis we have a commutative diagram

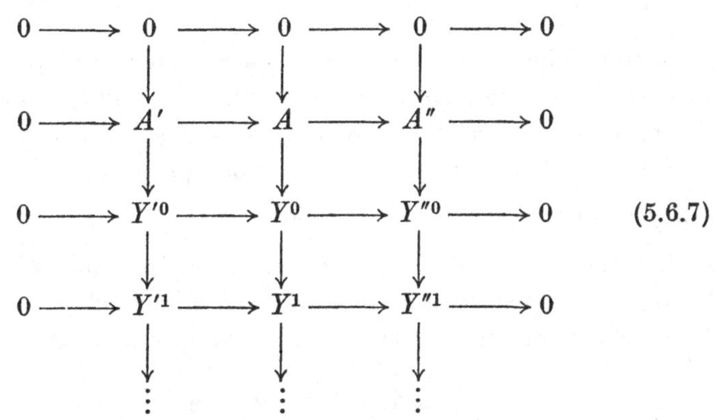

(5.6.7)

in which the rows are exact and the columns are 0-sequences. Suppose that $\mathbf{Y}'$ and $\mathbf{Y}''$ are acyclic over A' and A'', then the extreme columns are exact and therefore it follows,† by Theorem 6 of section (4.6), that the centre column is exact. This shows that $\mathbf{Y}$ is acyclic over A.

† More precisely (5.6.7) must be extended upwards by means of null modules before Theorem 6 of section (4.6) is directly applicable.

Now assume that $\mathbf{Y}'$ and $\mathbf{Y}''$ are injective resolutions of A' and A''. Then to complete the proof it is enough to show that each Y^n is Λ-injective. But

$$0 \to Y'^n \to Y^n \to Y''^n \to 0 \tag{5.6.8}$$

is exact and Y'^n is injective; consequently, by Theorem 6, (5.6.8) splits. Hence Y^n is isomorphic to $Y'^n + Y''^n$ (direct sum), and therefore, by Proposition 6, Y^n is Λ-injective.

Theorems 14 and 15 were concerned with the properties of exact sequences

$$\mathbf{0} \to \mathbf{X}' \to \mathbf{X} \to \mathbf{X}'' \to \mathbf{0} \tag{5.6.9}$$

and

$$\mathbf{0} \to \mathbf{Y}' \to \mathbf{Y} \to \mathbf{Y}'' \to \mathbf{0} \tag{5.6.10}$$

over a given exact sequence

$$0 \to A' \to A \to A'' \to 0.$$

We now turn to the question of the existence of sequences such as (5.6.9) and (5.6.10) satisfying certain prescribed conditions.

Theorem 16. *Let*

$$0 \to A' \xrightarrow{\phi} A \xrightarrow{\psi} A'' \to 0 \tag{5.6.11}$$

be an exact sequence, let $\mathbf{X}''$ *be a projective left complex over* A'' *and* $\mathbf{X}'$ *an acyclic left complex over* A'. *Then it is possible to find a left complex* $\mathbf{X}$ *over* A *which can be embedded in a split exact sequence*

$$\mathbf{0} \to \mathbf{X}' \to \mathbf{X} \to \mathbf{X}'' \to \mathbf{0}$$

over (5.6.11).

Remarks. The notion of a split exact sequence of diagrams and translations was defined in section (4.2). It should be noted that, by Theorem 14, if $\mathbf{X}'$ and $\mathbf{X}''$ are projective resolutions of A' and A'' respectively, then $\mathbf{X}$ will *automatically* be a projective resolution of A.

Proof. Put

$$X_n = X'_n + X''_n \quad \text{(direct sum)}$$

so that the elements of X_n are pairs (x'_n, x''_n), where x'_n belongs to X'_n and x''_n belongs to X''_n. Since $\mathbf{X}'$ and $\mathbf{X}''$ are left complexes over A' and A'', there are associated with them 0-sequences

$$\cdots \to X''_2 \xrightarrow{d''_2} X''_1 \xrightarrow{d''_1} X''_0 \xrightarrow{\epsilon''} A'' \to 0$$

and

$$\cdots \to X'_2 \xrightarrow{d'_2} X'_1 \xrightarrow{d'_1} X'_0 \xrightarrow{\epsilon'} A' \to 0.$$

Let us note that, by our hypotheses, (i) each X''_r is projective and (ii) the second sequence is exact. We now consider a family of Λ-homomorphisms

$$\sigma_0 : X''_0 \to A, \quad \sigma_n : X''_n \to X'_{n-1} \quad (n \geqslant 1),$$

which to begin with may be quite arbitrary. Using these we define Λ-homomorphisms

$$\epsilon : X_0 \to A, \quad d_n : X_n \to X_{n-1} \quad (n \geqslant 1)$$

by writing $$\epsilon(x_0', x_0'') = \phi\epsilon' x_0' + \sigma_0 x_0''$$

and $$d_n(x_n', x_n'') = (d_n' x_n' + \sigma_n x_n'', d_n'' x_n'') \quad (n \geqslant 1).$$

Here we have omitted certain brackets, as in writing $\sigma_0 x_0''$ in place of $\sigma_0(x_0'')$, in order to avoid complicating the notation. Our new mappings yield a sequence

$$\cdots \to X_n \xrightarrow{d_n} X_{n-1} \to \cdots \to X_1 \xrightarrow{d_1} X_0 \xrightarrow{\epsilon} A \to 0, \tag{5.6.12}$$

which has the property that, for every $n \geqslant 1$, the diagram

$$\begin{array}{ccccccccc} 0 & \longrightarrow & X_n' & \longrightarrow & X_n & \longrightarrow & X_n'' & \longrightarrow & 0 \\ & & \downarrow{\scriptstyle d_n'} & & \downarrow{\scriptstyle d_n} & & \downarrow{\scriptstyle d_n''} & & \\ 0 & \longrightarrow & X_{n-1}' & \longrightarrow & X_{n-1} & \longrightarrow & X_{n-1}'' & \longrightarrow & 0 \end{array}$$

is commutative, where the horizontal maps are the canonical ones associated with the direct sum structures of X_n and X_{n-1}. But we also require that

$$\begin{array}{ccccccccc} 0 & \longrightarrow & X_0' & \longrightarrow & X_0 & \longrightarrow & X_0'' & \longrightarrow & 0 \\ & & \downarrow{\scriptstyle \epsilon'} & & \downarrow{\scriptstyle \epsilon} & & \downarrow{\scriptstyle \epsilon''} & & \\ 0 & \longrightarrow & A' & \xrightarrow{\phi} & A & \xrightarrow{\psi} & A'' & \longrightarrow & 0 \end{array}$$

be commutative and this, as a simple calculation shows, will be the case if $\psi\sigma_0 = \epsilon''$. Since X_0'' is projective, we can certainly arrange that this condition is satisfied. We now see that provided we can choose σ_1, σ_2, etc., so that (5.6.12) is a 0-sequence, then

$$\mathbf{0} \to \mathbf{X}' \to \mathbf{X} \to \mathbf{X}'' \to \mathbf{0}$$

will be a split exact sequence over

$$0 \to A' \to A \to A'' \to 0,$$

and the theorem will be proved.

Now straightforward computations show that $\epsilon d_1 = 0$ is equivalent to $\phi\epsilon'\sigma_1 + \sigma_0 d_1'' = 0$ while, for $r \geqslant 2$, $d_{r-1} d_r = 0$ is equivalent to $d_{r-1}'\sigma_r + \sigma_{r-1} d_r'' = 0$. We shall therefore determine σ_1, σ_2, etc., in succession so as to satisfy these conditions. Consider the diagram

$$\begin{array}{ccccc} & & X_1'' & & \\ & & \downarrow{\scriptstyle -\sigma_0 d_1''} & & \\ X_0' & \xrightarrow{\phi\epsilon'} & A & \xrightarrow{\psi} & A'' \end{array}$$

Here the row is exact and, since $\psi\sigma_0 = \epsilon''$, we see that

$$\psi\sigma_0 d_1'' = \epsilon'' d_1'' = 0.$$

Hence, by Proposition 1, we can find $\sigma_1 : X_1'' \to X_0'$ such that

$$\phi\epsilon'\sigma_1 + \sigma_0 d_1'' = 0.$$

Next consider

$$\begin{array}{ccccc} & & X_2'' & & \\ & & \downarrow{\scriptstyle -\sigma_1 d_2''} & & \\ X_1' & \xrightarrow{d_1'} & X_0' & \xrightarrow{\phi\epsilon'} & A \end{array}$$

The row is exact and, since $\phi\epsilon'\sigma_1 + \sigma_0 d_1'' = 0$, we have

$$\phi\epsilon'\sigma_1 d_2'' = -\sigma_0 d_1'' d_2'' = 0.$$

Thus Proposition 1 shows that there exists a homomorphism

$$\sigma_2 : X_2'' \to X_1' \quad \text{such that} \quad d_1'\sigma_2 + \sigma_1 d_2'' = 0.$$

For the last step we suppose that $n \geqslant 2$ and that $\sigma_1, \sigma_2, \ldots, \sigma_n$ have all been found with the above properties. In particular, we have $d_{n-1}'\sigma_n + \sigma_{n-1}d_n'' = 0$. Then in the diagram

$$\begin{array}{ccccc} & & X_{n+1}'' & & \\ & & \downarrow{\scriptstyle -\sigma_n d_{n+1}''} & & \\ X_n' & \xrightarrow{d_n'} & X_{n-1}' & \xrightarrow{d_{n-1}'} & X_{n-2}' \end{array}$$

the row is exact and $d_{n-1}'\sigma_n d_{n+1}'' = -\sigma_{n-1}d_n'' d_{n+1}'' = 0$. Accordingly, there exists a Λ-homomorphism $\sigma_{n+1} : X_{n+1}'' \to X_n'$ such that $d_n'\sigma_{n+1} + \sigma_n d_{n+1}'' = 0$, and now the theorem follows by induction.

The dual of Theorem 16 is

Theorem 17. *Let* $$0 \to A' \to A \to A'' \to 0 \tag{5.6.13}$$

be an exact sequence, let $\mathbf{Y}'$ *be an injective right complex over* A' *and* $\mathbf{Y}''$ *an acyclic right complex over* A''. *Then it is possible to find a right complex* $\mathbf{Y}$ *over* A, *which can be embedded in a split exact sequence*

$$\mathbf{0} \to \mathbf{Y}' \to \mathbf{Y} \to \mathbf{Y}'' \to \mathbf{0}$$

over (5.6.13). *Furthermore, when* $\mathbf{Y}'$ *and* $\mathbf{Y}''$ *are injective resolutions of* A' *and* A'' *respectively, then* $\mathbf{Y}$ *will automatically be an injective resolution of* A.

5.7 Further results on resolutions of sequences

We shall now supplement the results of the last section by considering situations which are somewhat more complicated. First, however, we introduce a notational device, of a temporary nature, which will be very convenient in the discussion. Let

$$0 \to V' \to V \to V'' \to 0$$

be a split exact sequence of Λ-modules. Then, by Proposition 2 of section (1.11), there exist Λ-homomorphisms $V \to V'$ and $V'' \to V$ such that $V' \to V \to V'$; $V'' \to V \to V''$ is a complete representation of V as a direct sum. For each element v' of V' and v'' of V'' write $(v', 0)$ and $(0, v'')$ for their respective images in V. If we now write

$$(v', v'') = (v', 0) + (0, v''),$$

then each element of V has a unique representation in the form (v', v''). We shall call such a representation a *split representation* for the elements of V.

After these preliminaries we consider a commutative diagram (over Λ)

$$\begin{array}{ccccccccc} 0 & \longrightarrow & A' & \xrightarrow{\phi} & A & \xrightarrow{\psi} & A'' & \longrightarrow & 0 \\ & & \downarrow{h'} & & \downarrow{h} & & \downarrow{h''} & & \\ 0 & \longrightarrow & B' & \xrightarrow{\xi} & B & \xrightarrow{\eta} & B'' & \longrightarrow & 0 \end{array} \tag{5.7.1}$$

with exact rows, and also split exact sequences

$$0 \longrightarrow \mathbf{X}' \longrightarrow \mathbf{X} \longrightarrow \mathbf{X}'' \longrightarrow 0$$

and $$0 \longrightarrow \mathbf{W}' \longrightarrow \mathbf{W} \longrightarrow \mathbf{W}'' \longrightarrow 0$$

over the upper and lower rows respectively of (5.7.1). Here $\mathbf{X}'$, $\mathbf{X}$, $\mathbf{X}''$ and $\mathbf{W}'$, $\mathbf{W}$, $\mathbf{W}''$ are left complexes over A', A, A'' and B', B, B'' respectively.

Theorem 18. *Assume, in addition to the above suppositions, that* $\mathbf{X}''$ *is a projective left complex over* A'' *and* $\mathbf{W}'$ *an acyclic left complex over* B'*. Further, let* $\mathbf{f}' : \mathbf{X}' \to \mathbf{W}'$ *and* $\mathbf{f}'' : \mathbf{X}'' \to \mathbf{W}''$ *be translations over* h' *and* h'' *respectively. Then there exists a translation* $\mathbf{f} : \mathbf{X} \to \mathbf{W}$ *over* h *such that*

$$\begin{array}{ccccccccc} 0 & \longrightarrow & \mathbf{X}' & \longrightarrow & \mathbf{X} & \longrightarrow & \mathbf{X}'' & \longrightarrow & 0 \\ & & \downarrow{\mathbf{f}'} & & \downarrow{\mathbf{f}} & & \downarrow{\mathbf{f}''} & & \\ 0 & \longrightarrow & \mathbf{W}' & \longrightarrow & \mathbf{W} & \longrightarrow & \mathbf{W}'' & \longrightarrow & 0 \end{array}$$

is a commutative diagram of complexes and translations.

Proof. For each $n \geq 0$

$$0 \longrightarrow X'_n \longrightarrow X_n \longrightarrow X''_n \longrightarrow 0$$

and $$0 \longrightarrow W'_n \longrightarrow W_n \longrightarrow W''_n \longrightarrow 0$$

are split exact sequences and, in each case, we introduce split representations for the elements of X_n and W_n. Having done this we see that, since

$$\begin{array}{ccccccccc} 0 & \longrightarrow & X'_n & \longrightarrow & X_n & \longrightarrow & X''_n & \longrightarrow & 0 \\ & & \downarrow f'_n & & \downarrow f_n & & \downarrow f''_n & & \\ 0 & \longrightarrow & W'_n & \longrightarrow & W_n & \longrightarrow & W''_n & \longrightarrow & 0 \end{array}$$

needs to be commutative, $f_n(x'_n, x''_n)$ has to be of the form

$$f_n(x'_n, x''_n) = (f'_n x'_n + q_n x''_n, f''_n x''_n), \tag{5.7.2}$$

where $q_n : X''_n \to W'_n$ $(n = 0, 1, 2, \ldots)$ are certain Λ-homomorphisms which have still to be determined. Here, as in the last section, we have left out certain brackets in order to prevent the notation from becoming too cumbersome. In what follows we shall use d'_n, d_n, d''_n, where $n \geqslant 1$, for the boundary homomorphisms of $\mathbf{X}', \mathbf{X}, \mathbf{X}''$ respectively, and e', e, e'' will denote the augmentation homomorphisms of these same complexes. The corresponding symbols for $\mathbf{W}', \mathbf{W}, \mathbf{W}''$ will be $\delta'_n, \delta_n, \delta''_n$ and $\epsilon', \epsilon, \epsilon''$. The problem is now to show that we can choose q_0, q_1, q_2, etc., so that

$$\begin{array}{ccccccccccccc} \cdots & \longrightarrow & X_n & \xrightarrow{d_n} & X_{n-1} & \longrightarrow & \cdots & \xrightarrow{d_1} & X_0 & \xrightarrow{e} & A & \longrightarrow & 0 \\ & & \downarrow f_n & & \downarrow f_{n-1} & & & & \downarrow f_0 & & \downarrow h & & \\ \cdots & \longrightarrow & W_n & \xrightarrow{\delta_n} & W_{n-1} & \longrightarrow & \cdots & \xrightarrow{\delta_1} & W_0 & \xrightarrow{\epsilon} & B & \longrightarrow & 0 \end{array} \tag{5.7.3}$$

is a commutative diagram. However, before we attempt to do this, it will be convenient to express our data in a more tractable form.

From the commutative properties of

$$\begin{array}{ccccc} X'_0 & \longrightarrow & X_0 & \longrightarrow & X''_0 \\ \downarrow e' & & \downarrow e & & \downarrow e'' \\ A' & \xrightarrow{\phi} & A & \xrightarrow{\psi} & A'' \end{array}$$

we see that
$$e(x'_0, x''_0) = \phi e' x'_0 + \sigma_0 x''_0, \tag{5.7.4}$$

where $\sigma_0 : X''_0 \to A$ and
$$\psi\sigma_0 = e''. \tag{5.7.5}$$

In the same way, the fact that $\mathbf{0} \to \mathbf{W}' \to \mathbf{W} \to \mathbf{W}'' \to \mathbf{0}$ is over $0 \to B' \to B \to B'' \to 0$ gives

$$\epsilon(w'_0, w''_0) = \xi\epsilon' w'_0 + \tau_0 w''_0, \tag{5.7.6}$$

where $\tau_0 : W''_0 \to B$ and
$$\eta\tau_0 = \epsilon''. \tag{5.7.7}$$

Again, for $n > 1$, the diagram

$$\begin{array}{ccccc} X'_n & \longrightarrow & X_n & \longrightarrow & X''_n \\ \downarrow d'_n & & \downarrow d_n & & \downarrow d''_n \\ X'_{n-1} & \longrightarrow & X_{n-1} & \longrightarrow & X''_{n-1} \end{array}$$

is commutative and therefore

$$d_n(x'_n, x''_n) = (d'_n x'_n + \sigma_n x''_n, d''_n x''_n), \qquad (5.7.8)$$

where $\sigma_n : X''_n \to X'_{n-1}$. But $ed_1 = 0$ and $d_{n-1}d_n = 0$ $(n > 1)$ whence, using (5.7.4) and (5.7.8) to express these in a different form, we obtain, as equivalents

$$\phi e'\sigma_1 + \sigma_0 d''_1 = 0, \quad d'_{n-1}\sigma_n + \sigma_{n-1}d''_n = 0 \quad (n > 1). \qquad (5.7.9)$$

Similarly, we can show that, for $n > 1$,

$$\delta_n(w'_n, w''_n) = (\delta'_n w'_n + \tau_n w''_n, \delta''_n w''_n), \qquad (5.7.10)$$

where $\tau_n : W''_n \to W'_{n-1}$ and

$$\xi\epsilon'\tau_1 + \tau_0\delta''_1 = 0, \quad \delta'_{n-1}\tau_n + \tau_{n-1}\delta''_n = 0 \quad (n > 1). \qquad (5.7.11)$$

Finally, since $\mathbf{f}'$, $\mathbf{f}''$ are translations over h', h'' respectively, we have

$$h'e' = \epsilon' f'_0; \quad f'_{n-1}d'_n = \delta'_n f'_n \quad (n \geq 1) \qquad (5.7.12)$$

and

$$h''e'' = \epsilon'' f''_0; \quad f''_{n-1}d''_n = \delta''_n f''_n \quad (n \geq 1). \qquad (5.7.13)$$

These various relations represent different aspects of our hypotheses and, with these before us, we return to the problem of choosing q_0, q_1, q_2, etc., so as to make (5.7.3) commutative. What we require is that we should have $he = \epsilon f_0$ and also $f_{n-1}d_n = \delta_n f_n$ for $n \geq 1$. The first of these is found, after a little reduction using (5.7.12), to be equivalent to

$$\xi\epsilon' q_0 = -\tau_0 f''_0 + h\sigma_0, \quad \text{(A)}$$

while the second takes the form

$$\delta'_n q_n = f'_{n-1}\sigma_n + q_{n-1}d''_n - \tau_n f''_n \quad (n \geq 1). \quad \text{(B)}$$

We proceed to show that it is possible to find $q_0, q_1, q_2, \ldots$ in succession so that (A) and (B) are both satisfied.

The diagram

$$\begin{array}{ccccc} & & X''_0 & & \\ & & \downarrow -\tau_0 f''_0 + h\sigma_0 & & \\ W'_0 & \xrightarrow{\xi\epsilon'} & B & \xrightarrow{\eta} & B'' \end{array}$$

has its row exact and X''_0 is, of course, projective. Further, by (5.7.7), (5.7.5) and (5.7.13),

$$-\eta\tau_0 f''_0 + \eta h\sigma_0 = -\epsilon'' f''_0 + h''\psi\sigma_0 = -\epsilon'' f''_0 + h''e'' = 0,$$

consequently, by Proposition 1, there exists $q_0 : X_0'' \to W_0'$ such that $\xi\epsilon' q_0 = -\tau_0 f_0'' + h\sigma_0$. Thus (A) has been satisfied.

Next the row in

$$\begin{array}{ccccc} & & X_1'' & & \\ & & \downarrow {\scriptstyle f_0'\sigma_1 + q_0 d_1'' - \tau_1 f_1''} & & \\ W_1' & \xrightarrow{\delta_1'} & W_0' & \xrightarrow{\xi\epsilon'} & B \end{array}$$

is exact and, by (5.7.12), (A) and (5.7.11),

$$\begin{aligned} \xi\epsilon' f_0'\sigma_1 + \xi\epsilon' q_0 d_1'' - \xi\epsilon'\tau_1 f_1'' &= \xi h'e'\sigma_1 - \tau_0 f_0'' d_1'' + h\sigma_0 d_1'' + \tau_0 \delta_1'' f_1'' \\ &= (h\phi e'\sigma_1 + h\sigma_0 d_1'') + \tau_0(\delta_1'' f_1'' - f_0'' d_1''), \end{aligned}$$

and this vanishes by (5.7.9) and (5.7.13). Accordingly, by Proposition 1, there exists $q_1 : X_1'' \to W_1'$ such that $\delta_1' q_1 = f_0'\sigma_1 + q_0 d_1'' - \tau_1 f_1''$.

Finally, suppose that $n \geqslant 2$ and that $q_0, q_1, q_2, \ldots, q_{n-1}$ have been found to satisfy the requirements (A) and (B). Turning our attention to the diagram

$$\begin{array}{ccccc} & & X_n'' & & \\ & & \downarrow {\scriptstyle f_{n-1}'\sigma_n + q_{n-1} d_n'' - \tau_n f_n''} & & \\ W_n' & \xrightarrow{\delta_n'} & W_{n-1}' & \xrightarrow{\delta_{n-1}'} & W_{n-2}' \end{array}$$

we observe that, by (5.7.12), the inductive hypothesis and (5.7.13),

$$\begin{aligned} &\delta_{n-1}' f_{n-1}' \sigma_n + \delta_{n-1}' q_{n-1} d_n'' - \delta_{n-1}' \tau_n f_n'' \\ &\quad = f_{n-2}' d_{n-1}' \sigma_n + (f_{n-2}' \sigma_{n-1} + q_{n-2} d_{n-1}'' - \tau_{n-1} f_{n-1}'') d_n'' - \delta_{n-1}' \tau_n f_n'' \\ &\quad = f_{n-2}'(d_{n-1}' \sigma_n + \sigma_{n-1} d_n'') - \tau_{n-1} \delta_n'' f_n'' - \delta_{n-1}' \tau_n f_n'' \\ &\quad = f_{n-2}'(d_{n-1}' \sigma_n + \sigma_{n-1} d_n'') - (\tau_{n-1} \delta_n'' + \delta_{n-1}' \tau_n) f_n'', \end{aligned}$$

and this is zero, by (5.7.9) and (5.7.11). It follows, again by Proposition 1, that there exists $q_n : X_n'' \to W_n'$ such that

$$\delta_n' q_n = f_{n-1}' \sigma_n + q_{n-1} d_n'' - \tau_n f_n'',$$

and now the theorem follows by induction.

We conclude this chapter by stating, without proof, the dual of the last theorem. In the dual problem we are concerned with a commutative diagram

$$\begin{array}{ccccccccc} 0 & \longrightarrow & B' & \longrightarrow & B & \longrightarrow & B'' & \longrightarrow & 0 \\ & & \downarrow {\scriptstyle h'} & & \downarrow {\scriptstyle h} & & \downarrow {\scriptstyle h''} & & \\ 0 & \longrightarrow & A' & \longrightarrow & A & \longrightarrow & A'' & \longrightarrow & 0 \end{array} \qquad (5.7.14)$$

whose rows are exact, and two split exact sequences

$$\mathbf{0} \longrightarrow \mathbf{W}' \longrightarrow \mathbf{W} \longrightarrow \mathbf{W}'' \longrightarrow \mathbf{0} \qquad (5.7.15)$$

and

$$\mathbf{0} \longrightarrow \mathbf{Y}' \longrightarrow \mathbf{Y} \longrightarrow \mathbf{Y}'' \longrightarrow \mathbf{0}. \qquad (5.7.16)$$

This time $\mathbf{Y}'$, $\mathbf{Y}$, $\mathbf{Y}''$ are right complexes over A', A, A''; $\mathbf{W}'$, $\mathbf{W}$, $\mathbf{W}''$ are right complexes over B', B, B''; while (5.7.15) and (5.7.16) are over the upper and lower rows respectively of (5.7.14).

Theorem 19. *Let* $\mathbf{Y}'$ *be an injective right complex over* A' *and* $\mathbf{W}''$ *an acyclic right complex over* B''. *Further, let* $\mathbf{f}' : \mathbf{W}' \to \mathbf{Y}'$, $\mathbf{f}'' : \mathbf{W}'' \to \mathbf{Y}''$ *be translations over* h', h'' *respectively. Then there exists a translation* $\mathbf{f} : \mathbf{W} \to \mathbf{Y}$ *over* h *such that*

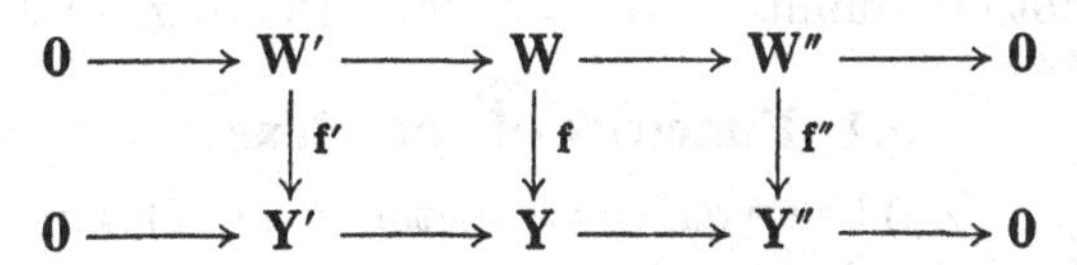

$$\begin{array}{ccccccccc} 0 & \longrightarrow & \mathbf{W}' & \longrightarrow & \mathbf{W} & \longrightarrow & \mathbf{W}'' & \longrightarrow & 0 \\ & & \downarrow{\mathbf{f}'} & & \downarrow{\mathbf{f}} & & \downarrow{\mathbf{f}''} & & \\ 0 & \longrightarrow & \mathbf{Y}' & \longrightarrow & \mathbf{Y} & \longrightarrow & \mathbf{Y}'' & \longrightarrow & 0 \end{array}$$

is commutative.

The importance of Theorems 12–19 lies in their essential role in the construction of what are called *derived functors*. The notion of a derived functor occupies a central position in Homological Algebra and, indeed, the last three chapters have been largely concerned with preparation for the introduction of this concept. In Chapter 6 we accomplish the definition and establish a number of basic properties.

6

DERIVED FUNCTORS

Notation. Λ, Λ_1, Λ_2, etc., denote rings, with identity elements, which need not be commutative. Z denotes the ring of integers.

6.1 Functors of complexes

Let $T(A_1, A_2, ..., A_k)$ be an *additive* functor of modules in the sense of section (3.9). As in that section we suppose that A_r varies in the category $\mathscr{G}_{\Lambda_r}$ and that T takes its values in $\mathscr{G}_\Lambda$. According to our usual practice, we allow some of $\mathscr{G}_{\Lambda_1}, ..., \mathscr{G}_{\Lambda_k}, \mathscr{G}_\Lambda$ to be categories of left modules while the others are categories of right modules. The situation in this respect is assumed to be known and to be fixed once and for all. On this understanding if, for example, $\mathscr{G}_{\Lambda_r}$ consists of left Λ_r-modules, then, when we speak of *complexes over* Λ_r, it is to be understood that we are referring to complexes of left Λ_r-modules. Finally, so far as the distribution of covariant and contravariant variables is concerned, we suppose that T is covariant in A_i for suffixes i in I and contravariant in A_j, when j belongs to J.

Our immediate objective is to construct, from a given functor $T(A_1, A_2, ..., A_k)$, a new functor $T(\mathbf{X}_1, \mathbf{X}_2, ..., \mathbf{X}_k)$, where $\mathbf{X}_r$ varies in the translation category of complexes over Λ_r and where

$$T(\mathbf{X}_1, \mathbf{X}_2, ..., \mathbf{X}_k)$$

is a complex in the translation category of complexes over Λ. In (i)–(v) below there are set out some of the basic properties which will be possessed by $T(\mathbf{X}_1, \mathbf{X}_2, ..., \mathbf{X}_k)$. First, however, we recall that in section (5.4) we decided that if A_r is a Λ_r-module, then we should denote by $\mathbf{A}_r$ the associated complex

$$\cdots \to 0 \to 0 \to A_r \to 0 \to 0 \to \cdots,$$

where the position of A_r is such that $\mathbf{A}_r$ is both a right and a left complex. Also, if $f_r : A_r \to A'_r$ is a Λ_r-homomorphism, then, by the *associated translation* $\mathbf{A}_r \to \mathbf{A}'_r$ *of the associated complexes*, we shall mean that translation whose particulars are set out in the following diagram:

$$\begin{array}{ccccccccccc}
\cdots & \longrightarrow & 0 & \longrightarrow & 0 & \longrightarrow & A_r & \longrightarrow & 0 & \longrightarrow & 0 & \longrightarrow & \cdots \\
 & & \downarrow & & \downarrow & & \downarrow{\scriptstyle f_r} & & \downarrow & & \downarrow & & \\
\cdots & \longrightarrow & 0 & \longrightarrow & 0 & \longrightarrow & A'_r & \longrightarrow & 0 & \longrightarrow & 0 & \longrightarrow & \cdots
\end{array}$$

We come now to the statement of some of the main properties that will be possessed by the functor which is to be constructed. In this statement $\mathbf{X}_r$ and $\mathbf{X}'_r$ denote typical complexes over Λ_r.

(i) $T(\mathbf{X}_1, \mathbf{X}_2, \ldots, \mathbf{X}_k)$ is an *additive* functor covariant in $\mathbf{X}_i$ $(i \in I)$ and contravariant in $\mathbf{X}_j$ $(j \in J)$.

(ii) If, for each r, where $1 \leqslant r \leqslant k$, $\mathbf{f}_r : \mathbf{X}_r \to \mathbf{X}'_r$ and $\mathbf{g}_r : \mathbf{X}_r \to \mathbf{X}'_r$ are homotopic translations, then $T(\mathbf{f}_1, \mathbf{f}_2, \ldots, \mathbf{f}_k)$ and $T(\mathbf{g}_1, \mathbf{g}_2, \ldots, \mathbf{g}_k)$ are also homotopic.

(iii) For each r $(1 \leqslant r \leqslant k)$, let A_r be a Λ_r-module and let $\mathbf{A}_r$ be the associated complex. Then $T(\mathbf{A}_1, \mathbf{A}_2, \ldots, \mathbf{A}_k)$ is the complex associated with the Λ-module $T(A_1, A_2, \ldots, A_k)$.

(iv) Suppose that, for each covariant suffix i, we are given a Λ_i-homomorphism $f_i : A_i \to A'_i$, and for each contravariant suffix j, a Λ_j-homomorphism $f_j : A'_j \to A_j$. Let $\mathbf{f}_i : \mathbf{A}_i \to \mathbf{A}'_i$, $\mathbf{f}_j : \mathbf{A}'_j \to \mathbf{A}_j$ be the associated translations. Then

$$T(\mathbf{f}_1, \mathbf{f}_2, \ldots, \mathbf{f}_k) : T(\mathbf{A}_1, \mathbf{A}_2, \ldots, \mathbf{A}_k) \to T(\mathbf{A}'_1, \mathbf{A}'_2, \ldots, \mathbf{A}'_k)$$

is the associated translation of the associated complexes corresponding to

$$T(f_1, f_2, \ldots, f_k) : T(A_1, A_2, \ldots, A_k) \to T(A'_1, A'_2, \ldots, A'_k).$$

(v) A common special situation is that in which $\Lambda_1, \Lambda_2, \ldots, \Lambda_k, \Lambda$ are all one and the same *commutative* ring. Suppose that we have this situation and that T possesses a further property, namely, that

$$T(f_1, \ldots, \lambda f_r, \ldots, f_k) = \lambda T(f_1, f_2, \ldots, f_k) \quad (1 \leqslant r \leqslant k)$$

for every set $f_1, f_2, \ldots, f_k$ of homomorphisms and every λ in Λ. In these circumstances, if λ belongs to Λ, then

$$T(\boldsymbol{\phi}_1, \ldots, \lambda \boldsymbol{\phi}_r, \ldots, \boldsymbol{\phi}_k) = \lambda T(\boldsymbol{\phi}_1, \boldsymbol{\phi}_2, \ldots, \boldsymbol{\phi}_k) \quad (1 \leqslant r \leqslant k),$$

where $\boldsymbol{\phi}_1, \boldsymbol{\phi}_2, \ldots, \boldsymbol{\phi}_k$ are arbitrary translations of complexes.

For our purposes, the particularly important cases are those in which the number k of variables in T is either 1 or 2, and so we shall limit most of our attention to these values of k. The case $k = 1$ will receive immediate attention while discussion of the case $k = 2$, which is considerably more complicated, will be held over until we come to section (6.2).

Let us begin by supposing that $T(A)$ is a covariant additive functor of a single variable. We have to explain what we shall mean by $T(\mathbf{X})$, where

$$(\mathbf{X}) \quad \cdots \to X^{n-1} \to X^n \to X^{n+1} \to \cdots$$

is a complex over Λ_1, and by $T(\mathbf{f})$, where $\mathbf{f}: \mathbf{X} \to \mathbf{X}'$ is a translation

$$\begin{array}{ccccccccc} \cdots \longrightarrow & X^{n-1} & \longrightarrow & X^{n} & \longrightarrow & X^{n+1} & \longrightarrow \cdots \\ & \downarrow & & \downarrow & & \downarrow & \\ \cdots \longrightarrow & X'^{n-1} & \longrightarrow & X'^{n} & \longrightarrow & X'^{n+1} & \longrightarrow \cdots \end{array}$$

of such complexes. This we do by taking $T(\mathbf{X})$ to be the complex

$$\cdots \to T(X^{n-1}) \to T(X^n) \to T(X^{n+1}) \to \cdots,$$

and by taking $T(\mathbf{f})$ to be the obvious translation

$$\begin{array}{ccccccccc} \cdots \longrightarrow & T(X^{n-1}) & \longrightarrow & T(X^{n}) & \longrightarrow & T(X^{n+1}) & \longrightarrow \cdots \\ & \downarrow & & \downarrow & & \downarrow & \\ \cdots \longrightarrow & T(X'^{n-1}) & \longrightarrow & T(X'^{n}) & \longrightarrow & T(X'^{n+1}) & \longrightarrow \cdots \end{array}$$

Since not only the order of the modules in a complex is important, but also the actual numbers attached to them, we make the above definition quite precise by saying that $T(\mathbf{X}) = \mathbf{V}$, where $V^n = T(X^n)$. It is easily verified that $T(\mathbf{X})$ is a functor of $\mathbf{X}$ and that (i)–(v) are all satisfied.

Consider now the case where $T(A)$ is a contravariant additive functor of a single variable. This time we put† $T(\mathbf{X}) = \mathbf{V}$, where $V^n = T(X_n)$, and take as boundary mapping $V^n \to V^{n+1}$ the homomorphism obtained by operating with T on $X_{n+1} \to X_n$. Thus $T(\mathbf{X})$ is the complex

$$\cdots \to T(X_{n-1}) \to T(X_n) \to T(X_{n+1}) \to \cdots,$$

where it is to be understood that $T(X_0)$ is the module which separates the right half of the complex from the left. Suppose now that $\mathbf{f}: \mathbf{X} \to \mathbf{X}'$ is a given translation, then $T(\mathbf{f}): T(\mathbf{X}') \to T(\mathbf{X})$ is the translation represented by the diagram

$$\begin{array}{ccccccccc} \cdots \longrightarrow & T(X'_{n-1}) & \longrightarrow & T(X'_{n}) & \longrightarrow & T(X'_{n+1}) & \longrightarrow \cdots \\ & \downarrow & & \downarrow & & \downarrow & \\ \cdots \longrightarrow & T(X_{n-1}) & \longrightarrow & T(X_{n}) & \longrightarrow & T(X_{n+1}) & \longrightarrow \cdots \end{array}$$

It is clear that this makes $T(\mathbf{X})$ a contravariant functor of $\mathbf{X}$, and again it is easily checked that the construction has all the properties (i)–(v).

Our first theorem plays an important auxiliary role in a number of applications.

† We use the convention, introduced in section (4.6), that $X_n = X^{-n}$.

Theorem 1. *Let $T(A)$ be an additive exact functor of a single variable module and let* $\mathbf{X}$ *be a complex.*† *Then, if T is covariant in A, we have, for each n, a canonical isomorphism*

$$H_n(T(\mathbf{X})) \approx T(H_n(\mathbf{X})),$$

while, if T is contravariant in A, this must be replaced by an isomorphism

$$H_n(T(\mathbf{X})) \approx T(H^n(\mathbf{X})).$$

In either case, the isomorphism determines a natural equivalence between the two sides regarded as functors of $\mathbf{X}$.

Proof. For definiteness we shall deal with the case in which T is contravariant in A. Put

$$A^n = \operatorname{Im}(X^{n-1} \to X^n) \quad \text{and} \quad B^n = \operatorname{Ker}(X^n \to X^{n+1}).$$

Then the homomorphism $X^{n-1} \to X^n$ can be regarded as the combined map

$$X^{n-1} \to A^n \to B^n \to X^n.$$

Now $0 \to A^n \to B^n \to H^n(\mathbf{X}) \to 0$ is an exact sequence and T is an exact functor; consequently

$$0 \to T(H^n(\mathbf{X})) \to T(B^n) \to T(A^n) \to 0 \tag{6.1.1}$$

is exact. Again, since $0 \to B^n \to X^n$ is exact, $T(X^n) \to T(B^n)$ is an epimorphism. A suitable restriction of this epimorphism will map

$$\operatorname{Ker}(T(X^n) \to T(A^n)) \quad \textit{on to} \quad \operatorname{Ker}(T(B^n) \to T(A^n)),$$

and so we obtain an epimorphism

$$\operatorname{Ker}(T(X^n) \to T(A^n)) \to \operatorname{Ker}(T(B^n) \to T(A^n)),$$

whose kernel is seen to be $\operatorname{Ker}(T(X^n) \to T(B^n))$. Accordingly, there is induced an isomorphism

$$\operatorname{Ker}(T(X^n) \to T(A^n))/\operatorname{Ker}(T(X^n) \to T(B^n)) \approx \operatorname{Ker}(T(B^n) \to T(A^n))$$

by means of which an exact sequence

$$0 \to [\operatorname{Ker}(T(X^n) \to T(A^n))/\operatorname{Ker}(T(X^n) \to T(B^n))] \to T(B^n) \to T(A^n) \to 0 \tag{6.1.2}$$

can be constructed. Now $X^{n-1} \to A^n \to 0$ is exact and therefore $T(A^n) \to T(X^{n-1})$ is a monomorphism. But $T(X^n) \to T(X^{n-1})$ can be regarded as the combined map $T(X^n) \to T(A^n) \to T(X^{n-1})$ and thus we see that

$$\operatorname{Ker}(T(X^n) \to T(X^{n-1})) = \operatorname{Ker}(T(X^n) \to T(A^n)). \tag{6.1.3}$$

† The component modules of $\mathbf{X}$ are, of course, to belong to the category in which A varies.

Next, $B^n \to X^n \to X^{n+1}$ is exact and therefore $T(X^{n+1}) \to T(X^n) \to T(B^n)$ is exact, which shows that

$$\operatorname{Im}(T(X^{n+1}) \to T(X^n)) = \operatorname{Ker}(T(X^n) \to T(B^n)). \tag{6.1.4}$$

Substituting from (6.1.3) and (6.1.4) in (6.1.2), we obtain the exact sequence

$$0 \to H_n(T(\mathbf{X})) \to T(B^n) \to T(A^n) \to 0, \tag{6.1.5}$$

and by combining this with (6.1.1) there results an isomorphism

$$T(H^n(\mathbf{X})) \approx H_n(T(\mathbf{X})). \tag{6.1.6}$$

Indeed, we can describe the isomorphism in (6.1.6) as the unique homomorphism

$$T(H^n(\mathbf{X})) \to H_n(T(\mathbf{X})), \tag{6.1.7}$$

which preserves the commutative properties of the diagram

$$\begin{array}{ccccccccc} 0 & \longrightarrow & T(H^n(\mathbf{X})) & \longrightarrow & T(B^n) & \longrightarrow & T(A^n) & \longrightarrow & 0 \\ & & & & \big\downarrow \text{identity} & & \big\downarrow \text{identity} & & \\ 0 & \longrightarrow & H_n(T(\mathbf{X})) & \longrightarrow & T(B^n) & \longrightarrow & T(A^n) & \longrightarrow & 0 \end{array} \tag{6.1.8}$$

But, as is easily verified, every translation $\mathbf{X}' \to \mathbf{X}$ leads to a translation of (6.1.8), and therefore, by the corollary to Proposition 2 of section (4.3), to a translation of (6.1.7). This proves that (6.1.6) is not only an isomorphism of modules but also an equivalence of functors.

6.2 Functors of two complexes

In this section we shall consider an additive functor $T(A, B)$ of two variables, where A belongs to the category $\mathscr{G}_{\Lambda_1}$ and B to $\mathscr{G}_{\Lambda_2}$, and we shall construct a functor $T(\mathbf{X}, \mathbf{Y})$ of complexes which, among other things, will have the properties described in the last section. The details of the construction of $T(\mathbf{X}, \mathbf{Y})$ depend on the variances of the variables A and B and, in this respect, there are four possibilities, which we shall refer to as follows:

Case (a): T is covariant in A and covariant in B;
Case (b): T is covariant in A and contravariant in B;
Case (c): T is contravariant in A and covariant in B;
Case (d): T is contravariant in A and contravariant in B.

Let $\mathbf{X}$ and $\mathbf{Y}$ be the complexes

$$\cdots \to X^{n-1} \to X^n \xrightarrow{d_X^n} X^{n+1} \to \cdots \tag{6.2.1}$$

and

$$\cdots \to Y^{n-1} \to Y^n \xrightarrow{d_Y^n} Y^{n+1} \to \cdots \tag{6.2.2}$$

over Λ_1 and Λ_2 respectively and write

in Case (*a*), $$T^{r,s}(\mathbf{X},\mathbf{Y}) = T(X^r, Y^s); \tag{6.2.3a}$$

in Case (*b*), $$T^{r,s}(\mathbf{X},\mathbf{Y}) = T(X^r, Y_s); \tag{6.2.3b}$$

in Case (*c*), $$T^{r,s}(\mathbf{X},\mathbf{Y}) = T(X_r, Y^s); \tag{6.2.3c}$$

in Case (*d*), $$T^{r,s}(\mathbf{X},\mathbf{Y}) = T(X_r, Y_s); \tag{6.2.3d}$$

and in all four cases define $T^n(\mathbf{X},\mathbf{Y})$ by the direct sum formula

$$T^n(\mathbf{X},\mathbf{Y}) = \sum_{r+s=n} T^{r,s}(\mathbf{X},\mathbf{Y}). \tag{6.2.4}$$

These modules $T^n(\mathbf{X},\mathbf{Y})$ will be the constituent modules of the complex $T(\mathbf{X},\mathbf{Y})$, but, of course, we still have to define the boundary homomorphisms. Before doing this, however, we shall need to make some auxiliary investigations. Let $\mathbf{X}'$ and $\mathbf{Y}'$ be further complexes over Λ_1 and Λ_2 and let $\boldsymbol{\phi} = [\phi^r]$, $\boldsymbol{\psi} = [\psi^s]$ be families of homomorphisms

$$\phi^r : X^r \to X'^{r+p} \quad (-\infty < r < \infty), \tag{6.2.5}$$

$$\psi^s : Y^s \to Y'^{s+q} \quad (-\infty < s < \infty). \tag{6.2.6}$$

Here p and q are independent of r and s and the homomorphisms are not assumed to commute with the boundary mappings. The integers p and q will be called the *degrees* of $\boldsymbol{\phi}$ and $\boldsymbol{\psi}$ and it will be convenient to write $\phi_r = \phi^{-r}$ and $\psi_s = \psi^{-s}$. The next step is to define a system $T^{r,s}(\boldsymbol{\phi},\boldsymbol{\psi})$ of Λ-homomorphisms. The particulars of the definition differ in the four cases, the appropriate formulae being

$$T^{r,s}(\boldsymbol{\phi},\boldsymbol{\psi}) = (-1)^{rq}\, T(\phi^r, \psi^s), \tag{6.2.7a}$$

$$T^{r,s}(\boldsymbol{\phi},\boldsymbol{\psi}) = (-1)^{rq}\, T(\phi^r, \psi_{s+q}), \tag{6.2.7b}$$

$$T^{r,s}(\boldsymbol{\phi},\boldsymbol{\psi}) = (-1)^{rq}\, T(\phi_{r+p}, \psi^s), \tag{6.2.7c}$$

$$T^{r,s}(\boldsymbol{\phi},\boldsymbol{\psi}) = (-1)^{rq}\, T(\phi_{r+p}, \psi_{s+q}). \tag{6.2.7d}$$

Here, of course, (6.2.7*a*) refers to Case (*a*) and so on for the other cases. Note the way in which these homomorphisms act. We have, in fact, for the four cases

$$T^{r,s}(\boldsymbol{\phi},\boldsymbol{\psi}) : T(X^r, Y^s) \to T(X'^{r+p}, Y'^{s+q}), \tag{6.2.8a}$$

$$T^{r,s}(\boldsymbol{\phi},\boldsymbol{\psi}) : T(X^r, Y'_s) \to T(X'^{r+p}, Y_{s+q}), \tag{6.2.8b}$$

$$T^{r,s}(\boldsymbol{\phi},\boldsymbol{\psi}) : T(X'_r, Y^s) \to T(X_{r+p}, Y'^{s+q}), \tag{6.2.8c}$$

$$T^{r,s}(\boldsymbol{\phi},\boldsymbol{\psi}) : T(X'_r, Y'_s) \to T(X_{r+p}, Y_{s+q}). \tag{6.2.8d}$$

Now define $T^n(\phi, \psi)$ so that, when $r+s = n$, it extends $T^{r,s}(\phi, \psi)$ and, in the various cases,

$$T^n(\phi, \psi) : T^n(\mathbf{X}, \mathbf{Y}) \to T^{n+p+q}(\mathbf{X}', \mathbf{Y}'), \tag{6.2.9a}$$

$$T^n(\phi, \psi) : T^n(\mathbf{X}, \mathbf{Y}') \to T^{n+p+q}(\mathbf{X}', \mathbf{Y}), \tag{6.2.9b}$$

$$T^n(\phi, \psi) : T^n(\mathbf{X}', \mathbf{Y}) \to T^{n+p+q}(\mathbf{X}, \mathbf{Y}'), \tag{6.2.9c}$$

$$T^n(\phi, \psi) : T^n(\mathbf{X}', \mathbf{Y}') \to T^{n+p+q}(\mathbf{X}, \mathbf{Y}). \tag{6.2.9d}$$

In all cases we write $T(\phi, \psi)$ for the family $[T^n(\phi, \psi)]$ so that $T(\phi, \psi)$ has degree $p+q$.

If ϕ_1 and ϕ_2 are two families of homomorphisms $[X^r \to X'^{r+p}]$ of the same degree p and we define $\phi_1 + \phi_2$ in the obvious manner, then, from the formulae (6.2.7), we see at once that

$$T(\phi_1 + \phi_2, \psi) = T(\phi_1, \psi) + T(\phi_2, \psi), \tag{6.2.10}$$

and, with a self explanatory notation, we have a complementary relation, namely,

$$T(\phi, \psi_1 + \psi_2) = T(\phi, \psi_1) + T(\phi, \psi_2). \tag{6.2.11}$$

Let us now suppose that, in addition to the families ϕ and ψ of (6.2.5) and (6.2.6), we also have families ϕ', ψ' of degrees p', q', where

$$\phi'^r : X'^r \to X''^{r+p'} \quad (-\infty < r < \infty), \tag{6.2.12}$$

$$\psi'^s : Y'^s \to Y''^{s+q'} \quad (-\infty < s < \infty). \tag{6.2.13}$$

Then, with a natural rule of composition, $\phi'\phi$ and $\psi'\psi$ are families $[X^r \to X''^{r+p+p'}]$ and $[Y^s \to Y''^{s+q+q'}]$ of degrees $p+p'$ and $q+q'$ respectively. Further, in Case (a),

$$\begin{aligned} T^{r,s}(\phi'\phi, \psi'\psi) &= (-1)^{r(q+q')} T(\phi'^{r+p}\phi^r, \psi'^{s+q}\psi^s) \\ &= (-1)^{r(q+q')} T(\phi'^{r+p}, \psi'^{s+q})\, T(\phi^r, \psi^s) \\ &= (-1)^{pq'} T^{r+p,s+q}(\phi', \psi')\, T^{r,s}(\phi, \psi), \end{aligned}$$

which gives $T(\phi'\phi, \psi'\psi) = (-1)^{pq'} T(\phi', \psi')\, T(\phi, \psi)$. There are, of course, similar results in the other cases and, in fact, the complete system of relations is as follows:

$$T(\phi'\phi, \psi'\psi) = (-1)^{pq'} T(\phi', \psi')\, T(\phi, \psi), \tag{6.2.14a}$$

$$T(\phi'\phi, \psi'\psi) = (-1)^{pq} T(\phi', \psi)\, T(\phi, \psi'), \tag{6.2.14b}$$

$$T(\phi'\phi, \psi'\psi) = (-1)^{p'q'} T(\phi, \psi')\, T(\phi', \psi), \tag{6.2.14c}$$

$$T(\phi'\phi, \psi'\psi) = (-1)^{p'q} T(\phi, \psi)\, T(\phi', \psi'). \tag{6.2.14d}$$

The numerical factors which occur are of fundamental importance. They are readily remembered, for in each case the power to which -1

is raised is the product of the degrees of the two middle families on the right-hand side.

We are now ready to complete the definition of the complex $T(\mathbf{X}, \mathbf{Y})$ by defining suitable boundary homomorphisms

$$T^n(\mathbf{X}, \mathbf{Y}) \to T^{n+1}(\mathbf{X}, \mathbf{Y}).$$

The fact that $\mathbf{X}$ and $\mathbf{Y}$ are *complexes* involves the existence of families

$$[X^r \xrightarrow{d_X^r} X^{r+1}] \quad \text{and} \quad [Y^s \xrightarrow{d_Y^s} Y^{s+1}]$$

of boundary homomorphisms. These we denote by $\mathbf{d}_X$ and $\mathbf{d}_Y$ respectively and observe that they are families of degree unity. Further, we let $\mathbf{i}_X$ and $\mathbf{i}_Y$ denote the families $[X^r \to X^r]$ and $[Y^s \to Y^s]$ of identity maps so that $\mathbf{i}_X$ and $\mathbf{i}_Y$ are of degree zero. Put

$$T(\mathbf{d}_X, \mathbf{i}_Y) = \boldsymbol{\delta}_1 \quad \text{and} \quad T(\mathbf{i}_X, \mathbf{d}_Y) = \boldsymbol{\delta}_2, \tag{6.2.15}$$

then $\boldsymbol{\delta}_1$ and $\boldsymbol{\delta}_2$ are families

$$[\delta_1^n : T^n(\mathbf{X}, \mathbf{Y}) \to T^{n+1}(\mathbf{X}, \mathbf{Y})], \quad [\delta_2^n : T^n(\mathbf{X}, \mathbf{Y}) \to T^{n+1}(\mathbf{X}, \mathbf{Y})] \tag{6.2.16}$$

of degree unity. Furthermore, it follows, from (6.2.14), that (whichever case we are in)

$$\boldsymbol{\delta}_1\boldsymbol{\delta}_1 = \mathbf{0}, \quad \boldsymbol{\delta}_2\boldsymbol{\delta}_2 = \mathbf{0}, \quad \boldsymbol{\delta}_1\boldsymbol{\delta}_2 + \boldsymbol{\delta}_2\boldsymbol{\delta}_1 = \mathbf{0}, \tag{6.2.17}$$

where, in this context, $\mathbf{0}$ denotes the family consisting of all the null maps $T^n(\mathbf{X}, \mathbf{Y}) \to T^{n+2}(\mathbf{X}, \mathbf{Y})$. If therefore we write

$$\mathbf{d} = \boldsymbol{\delta}_1 + \boldsymbol{\delta}_2, \tag{6.2.18}$$

then $\mathbf{d}$ is a family of maps $[d^n : T^n(\mathbf{X}, \mathbf{Y}) \to T^{n+1}(\mathbf{X}, \mathbf{Y})]$ and, by (6.2.17), $\mathbf{dd} = \mathbf{0}$. Accordingly

$$\cdots \to T^{n-1}(\mathbf{X}, \mathbf{Y}) \xrightarrow{d^{n-1}} T^n(\mathbf{X}, \mathbf{Y}) \xrightarrow{d^n} T^{n+1}(\mathbf{X}, \mathbf{Y}) \to \cdots \tag{6.2.19}$$

is a complex, and it is this complex that we denote by $T(\mathbf{X}, \mathbf{Y})$.

We now turn to the functor properties of $T(\mathbf{X}, \mathbf{Y})$. Suppose that $\boldsymbol{\phi}$ and $\boldsymbol{\psi}$ are respectively *translations* $\mathbf{X} \to \mathbf{X}'$ and $\mathbf{Y} \to \mathbf{Y}'$ of complexes, then they are families of degree zero and, with an obvious notation,

$$\boldsymbol{\phi}\mathbf{d}_X = \mathbf{d}_{X'}\boldsymbol{\phi}, \quad \boldsymbol{\psi}\mathbf{d}_Y = \mathbf{d}_{Y'}\boldsymbol{\psi}.$$

Accordingly, if we are in Case (*a*) and $\mathbf{d}'$ denotes the family of boundary homomorphisms of $T(\mathbf{X}', \mathbf{Y}')$, then

$$\begin{aligned} T(\boldsymbol{\phi}, \boldsymbol{\psi})\,\mathbf{d} &= T(\boldsymbol{\phi}, \boldsymbol{\psi})\,T(\mathbf{d}_X, \mathbf{i}_Y) + T(\boldsymbol{\phi}, \boldsymbol{\psi})\,T(\mathbf{i}_X, \mathbf{d}_Y) \\ &= T(\boldsymbol{\phi}\mathbf{d}_X, \boldsymbol{\psi}\mathbf{i}_Y) + T(\boldsymbol{\phi}\mathbf{i}_X, \boldsymbol{\psi}\mathbf{d}_Y) \\ &= T(\mathbf{d}_{X'}\boldsymbol{\phi}, \mathbf{i}_{Y'}\boldsymbol{\psi}) + T(\mathbf{i}_{X'}\boldsymbol{\phi}, \mathbf{d}_{Y'}\boldsymbol{\psi}) \\ &= T(\mathbf{d}_{X'}, \mathbf{i}_{Y'})\,T(\boldsymbol{\phi}, \boldsymbol{\psi}) + T(\mathbf{i}_{X'}, \mathbf{d}_{Y'})\,T(\boldsymbol{\phi}, \boldsymbol{\psi}) \\ &= \mathbf{d}'T(\boldsymbol{\phi}, \boldsymbol{\psi}), \end{aligned}$$

which shows that $T(\boldsymbol{\phi}, \boldsymbol{\psi})$ is a translation of complexes. Very similar arguments will now show that the same conclusion holds in the remaining three cases.

This having been established, it follows, from (6.2.7), (6.2.9), (6.2.10), (6.2.11) and (6.2.14), that $T(\mathbf{X}, \mathbf{Y})$ may be regarded as an additive functor of $\mathbf{X}$ and $\mathbf{Y}$ having the same pattern of variances as the original functor $T(A, B)$. It is now a simple matter to verify that, of the five properties listed in section (6.1), (i), (iii) and (iv) are all satisfied by our new functor and that, under the appropriate conditions, (v) is satisfied as well. It remains for us to consider (ii) of section (6.1), which, it will be recalled, was concerned with homotopic translations. Suppose therefore that $\boldsymbol{\phi}_1$, $\boldsymbol{\phi}_2$ are homotopic translations $\mathbf{X} \to \mathbf{X}'$ and that $\boldsymbol{\psi}_1$, $\boldsymbol{\psi}_2$ are homotopic translations $\mathbf{Y} \to \mathbf{Y}'$. What needs to be shown is that $T(\boldsymbol{\phi}_1, \boldsymbol{\psi}_1)$ and $T(\boldsymbol{\phi}_2, \boldsymbol{\psi}_2)$ are also homotopic. But

$$T(\boldsymbol{\phi}_1, \boldsymbol{\psi}_1) - T(\boldsymbol{\phi}_2, \boldsymbol{\psi}_2) = T(\boldsymbol{\phi}_1 - \boldsymbol{\phi}_2, \boldsymbol{\psi}_1) + T(\boldsymbol{\phi}_2, \boldsymbol{\psi}_1 - \boldsymbol{\psi}_2),$$

consequently the required result will be established if we prove the following lemma:

Lemma 1. *If $\boldsymbol{\phi} : \mathbf{X} \to \mathbf{X}'$ and $\boldsymbol{\psi} : \mathbf{Y} \to \mathbf{Y}'$ are translations of com plexes and either is null homotopic, then $T(\boldsymbol{\phi}, \boldsymbol{\psi})$ is null homotopic.*

Proof. Assume that $\boldsymbol{\phi}$ is null homotopic. Then there exists a family $\mathbf{s}$ of homomorphisms $[X^n \xrightarrow{s^n} X'^{n-1}]$ of degree -1 such that

$$\mathbf{d}_{X'}\mathbf{s} + \mathbf{s}\mathbf{d}_X = \boldsymbol{\phi}.$$

Assuming for definiteness, that we are dealing with Case (a), it follows that

$$\begin{aligned} T(\boldsymbol{\phi}, \boldsymbol{\psi}) &= T(\mathbf{d}_{X'}\mathbf{s} + \mathbf{s}\mathbf{d}_X, \boldsymbol{\psi}) = T(\mathbf{d}_{X'}\mathbf{s}, \mathbf{i}_{Y'}\boldsymbol{\psi}) + T(\mathbf{s}\mathbf{d}_X, \boldsymbol{\psi}\mathbf{i}_Y) \\ &= T(\mathbf{d}_{X'}, \mathbf{i}_{Y'})\, T(\mathbf{s}, \boldsymbol{\psi}) + T(\mathbf{s}, \boldsymbol{\psi})\, T(\mathbf{d}_X, \mathbf{i}_Y) \\ &= \mathbf{d}' T(\mathbf{s}, \boldsymbol{\psi}) + T(\mathbf{s}, \boldsymbol{\psi})\,\mathbf{d} - T(\mathbf{i}_{X'}, \mathbf{d}_{Y'})\, T(\mathbf{s}, \boldsymbol{\psi}) \\ &\qquad - T(\mathbf{s}, \boldsymbol{\psi})\, T(\mathbf{i}_X, \mathbf{d}_Y), \end{aligned}$$

where, as before, $\mathbf{d}'$ denotes the family of boundary homomorphisms of the complex $T(\mathbf{X}', \mathbf{Y}')$. But

$$T(\mathbf{i}_{X'}, \mathbf{d}_{Y'})\, T(\mathbf{s}, \boldsymbol{\psi}) + T(\mathbf{s}, \boldsymbol{\psi})\, T(\mathbf{i}_X, \mathbf{d}_Y) = -T(\mathbf{s}, \mathbf{d}_{Y'}\boldsymbol{\psi}) + T(\mathbf{s}, \boldsymbol{\psi}\mathbf{d}_Y),$$

which consists entirely of null maps since $\boldsymbol{\psi}\mathbf{d}_Y = \mathbf{d}_{Y'}\boldsymbol{\psi}$. Thus we have proved that

$$T(\boldsymbol{\phi}, \boldsymbol{\psi}) = \mathbf{d}' T(\mathbf{s}, \boldsymbol{\psi}) + T(\mathbf{s}, \boldsymbol{\psi})\,\mathbf{d},$$

which shows that $T(\phi, \psi)$ is null homotopic. The proof for the case in which ψ is null homotopic is entirely similar.

There is an additional observation concerning the functor $T(\mathbf{X}, \mathbf{Y})$ which it is convenient to make at this point in order to have it available for future reference. Let us keep A fixed and write $U(B) = T(A, B)$; then U will be an additive functor of the single variable B. We can therefore use the method of section (6.1) to construct a functor $U(\mathbf{Y})$, and this we may designate by $T(A, \mathbf{Y})$. The component modules of $T(A, \mathbf{Y})$ we shall denote by $T^n(A, \mathbf{Y})$. On the other hand, if $\mathbf{A}$ denotes the complex associated with A, we can use the present section to construct a complex $T(\mathbf{A}, \mathbf{Y})$. Let us compare $T(A, \mathbf{Y})$ with $T(\mathbf{A}, \mathbf{Y})$.

It is clear that $T^n(A, \mathbf{Y}) = T^n(\mathbf{A}, \mathbf{Y})$ and that the boundary homomorphisms $T^n(A, \mathbf{Y}) \to T^{n+1}(A, \mathbf{Y})$ and $T^n(\mathbf{A}, \mathbf{Y}) \to T^{n+1}(\mathbf{A}, \mathbf{Y})$ are the same except for a possible difference in sign. However, using (6.2.7), we find on inspection that the boundary homomorphisms have the *same* sign, and therefore we may write

$$T(\mathbf{A}, \mathbf{Y}) = T(A, \mathbf{Y}). \tag{6.2.20}$$

In a similar way, and with an obvious extension of our notation, we can also show that

$$T(\mathbf{X}, \mathbf{B}) = T(\mathbf{X}, B), \tag{6.2.21}$$

but it should be noted that the reasons why the signs of the boundary homomorphisms agree in (6.2.20) and again in (6.2.21) are somewhat different in the two cases.

6.3 Right-derived functors

The present section will be devoted to developing a method by means of which we can derive new functors of modules from a given additive functor $T(A_1, A_2, \ldots, A_k)$. The cases $k = 1$ and $k = 2$ are those which concern us most, and therefore, as the actual number of variables makes, in fact, but little difference, we shall work throughout with an additive functor $T(A, B)$, which is covariant in A and contravariant in B. It will be clear how our results are to be modified to make them applicable to all other situations which are of interest to us. The modules A, B will be assumed to vary in $\mathscr{G}_{\Lambda_1}$, $\mathscr{G}_{\Lambda_2}$ respectively and the values of $T(A, B)$ are to be in $\mathscr{G}_{\Lambda}$.

Let $f : A \to A'$, $f' : A' \to A''$, $g : B' \to B$ and $g' : B'' \to B'$ be given homomorphisms, and suppose that $\mathbf{Q}$, $\mathbf{Q}'$, $\mathbf{Q}''$ are *injective* resolutions of A, A', A'' while $\mathbf{P}$, $\mathbf{P}'$, $\mathbf{P}''$ are *projective* resolutions of B, B', B''.

Consider the complex $T(\mathbf{Q}, \mathbf{P})$, which is to be constructed in the manner described in the last section. By (6.2.3*b*) and (6.2.4)

$$T^n(\mathbf{Q}, \mathbf{P}) = \sum_{r+s=n} T(Q^r, P_s) \quad \text{(direct sum)},$$

hence $T^n(\mathbf{Q}, \mathbf{P}) = 0$, if $n < 0$, and therefore $T(\mathbf{Q}, \mathbf{P})$ is a right complex. Again, the fact that $\mathbf{Q}$ is a right complex over A means that we have an augmentation translation $\mathbf{A} \to \mathbf{Q}$, where, as usual, $\mathbf{A}$ denotes the complex associated with A. Similarly, since $\mathbf{P}$ is a left complex over B we have an augmentation translation $\mathbf{P} \to \mathbf{B}$. The two translations $\mathbf{A} \to \mathbf{Q}$ and $\mathbf{P} \to \mathbf{B}$ now determine (by the last section) a translation

$$T(\mathbf{A}, \mathbf{B}) \to T(\mathbf{Q}, \mathbf{P}).$$

This means, by (iii) of section (6.1), that $T(\mathbf{Q}, \mathbf{P})$ is a right complex over $T(A, B)$.

Next, by Theorems 12 and 13 of section (5.5), there exist translations $\phi : \mathbf{Q} \to \mathbf{Q}'$ and $\psi : \mathbf{P}' \to \mathbf{P}$ over f and g respectively, and each of them is fully determined to within a homotopy. Accordingly we have commutative diagrams

$$\begin{array}{ccc} \mathbf{A} & \longrightarrow & \mathbf{Q} \\ \downarrow & & \downarrow \phi \\ \mathbf{A}' & \longrightarrow & \mathbf{Q}' \end{array} \quad \text{and} \quad \begin{array}{ccc} \mathbf{P}' & \longrightarrow & \mathbf{B}' \\ \psi \downarrow & & \downarrow \\ \mathbf{P} & \longrightarrow & \mathbf{B} \end{array}$$

and these determine a further commutative diagram, namely,

$$\begin{array}{ccc} T(\mathbf{A}, \mathbf{B}) & \longrightarrow & T(\mathbf{Q}, \mathbf{P}) \\ \downarrow & & \downarrow T(\phi, \psi) \\ T(\mathbf{A}', \mathbf{B}') & \longrightarrow & T(\mathbf{Q}', \mathbf{P}') \end{array} \tag{6.3.1}$$

which shows, by virtue of (iv) of section (6.1), that $T(\phi, \psi)$ is a translation $T(\mathbf{Q}, \mathbf{P}) \to T(\mathbf{Q}', \mathbf{P}')$ over $T(A, B) \to T(A', B')$. We can also say, this time by (ii) of section (6.1), that this translation is determined by f and g to within a homotopy. It follows, using Theorem 7 of section (4.7), that, for each n, the homomorphism

$$H^n(T(\mathbf{Q}, \mathbf{P})) \to H^n(T(\mathbf{Q}', \mathbf{P}')) \tag{6.3.2}$$

of homology modules, that results from the translation, is determined by f and g *without any ambiguity at all*.

In order to analyse what is happening in greater detail we shall take advantage of the fact that T is a fixed functor and introduce a convenient auxiliary notation by writing

$$H^n(T(\mathbf{Q}, \mathbf{P})) = H^n(A\text{—}\mathbf{Q}, \mathbf{P}\text{—}B). \tag{6.3.3}$$

A similar procedure is to be adopted in regard to $H^n(T(\mathbf{Q}', \mathbf{P}'))$ and $H^n(T(\mathbf{Q}'', \mathbf{P}''))$. In this notation, homomorphisms $A \to A'$ and $B' \to B$ determine a uniquely defined homomorphism

$$H^n(A\text{—}\mathbf{Q}, \mathbf{P}\text{—}B) \to H^n(A'\text{—}\mathbf{Q}', \mathbf{P}'\text{—}B'), \tag{6.3.4}$$

some of whose more obvious properties are described in the remarks below.

Remarks. (*a*) If

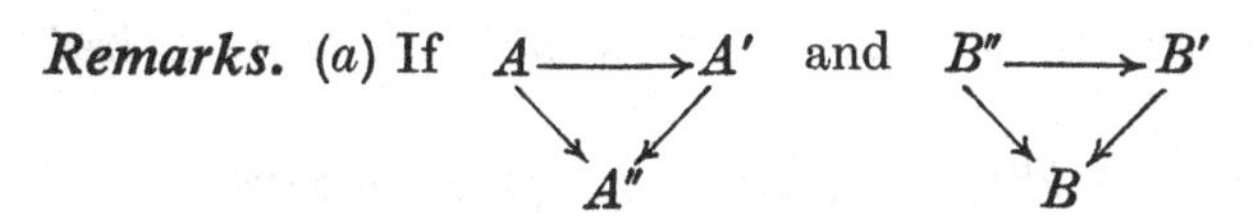

are commutative diagrams then

$$\begin{array}{ccc} H^n(A\text{—}\mathbf{Q}, \mathbf{P}\text{—}B) & \longrightarrow & H^n(A'\text{—}\mathbf{Q}', \mathbf{P}'\text{—}B') \\ & \searrow \quad \swarrow & \\ & H^n(A''\text{—}\mathbf{Q}'', \mathbf{P}''\text{—}B'') & \end{array}$$

is also commutative.

(*b*) Corresponding to identity maps $A \to A$ and $B \to B$, the homomorphism $H^n(A\text{—}\mathbf{Q}, \mathbf{P}\text{—}B) \to H^n(A\text{—}\mathbf{Q}, \mathbf{P}\text{—}B)$ is also an identity map.

(*c*) It follows from (*a*) and (*b*) that if $A \to A'$ and $B' \to B$ are isomorphisms and $A' \to A$ and $B \to B'$ are the inverse isomorphisms, then

$$H^n(A\text{—}\mathbf{Q}, \mathbf{P}\text{—}B) \to H^n(A'\text{—}\mathbf{Q}', \mathbf{P}'\text{—}B')$$

and

$$H^n(A'\text{—}\mathbf{Q}', \mathbf{P}'\text{—}B') \to H^n(A\text{—}\mathbf{Q}, \mathbf{P}\text{—}B)$$

are inverse isomorphisms.

(*d*) As an important special case of (*c*) we may note that if $\mathbf{Q}$, $^*\mathbf{Q}$ are injective resolutions of A while $\mathbf{P}$, $^*\mathbf{P}$ are projective resolutions of B, then, corresponding to the identity maps $A \to A$ and $B \to B$, we have homomorphisms

$$H^n(A\text{—}\mathbf{Q}, \mathbf{P}\text{—}B) \to H^n(A\text{—}{}^*\mathbf{Q}, {}^*\mathbf{P}\text{—}B)$$

$$H^n(A\text{—}{}^*\mathbf{Q}, {}^*\mathbf{P}\text{—}B) \to H^n(A\text{—}\mathbf{Q}, \mathbf{P}\text{—}B),$$

and these are inverse isomorphisms. Accordingly, we have a canonical isomorphism

$$H^n(A\text{—}\mathbf{Q}, \mathbf{P}\text{—}B) \approx H^n(A\text{—}{}^*\mathbf{Q}, {}^*\mathbf{P}\text{—}B). \tag{6.3.5}$$

Suppose now that we select a definite injective resolution $\mathbf{Q}$ for each module A of $\mathscr{G}_{\Lambda_1}$, and a definite projective resolution $\mathbf{P}$ for each module B of $\mathscr{G}_{\Lambda_2}$. On this understanding we shall write

$$R^nT(A, B) = H^n(T(\mathbf{Q}, \mathbf{P})), \tag{6.3.6}$$

and then it follows, from (6.3.4), that homomorphisms $A \to A'$ and $B' \to B$ give rise to a homomorphism

$$R^nT(A,B) \to R^nT(A',B').$$

Indeed, Remarks (*a*) and (*b*) above show that, for each n, $R^nT(A,B)$ is a functor, which is covariant in A and contravariant in B, and it is a simple matter to verify that these functors are additive.

Definition. The functors $R^nT(A,B)$ are said to be *right-derived* from $T(A,B)$.

Let us note that, since $T(\mathbf{Q},\mathbf{P})$ is a right complex, we have

$$R^nT(A,B) = 0 \quad \text{for all } n < 0. \tag{6.3.7}$$

The defining system. The functors R^nT, as defined above, depend on the resolutions chosen for the individual modules of the two categories $\mathscr{G}_{\Lambda_1}$ and $\mathscr{G}_{\Lambda_2}$. These chosen resolutions will be referred to collectively as the *defining system* for the right-derived functors. If we change the defining system, then we obtain, in place of the functors $R^nT(A,B)$, another system, say $^*R^nT(A,B)$, of right-derived functors. But, by (6.3.5), we see that there exists a well-defined module isomorphism

$$R^nT(A,B) \approx {}^*R^nT(A,B), \tag{6.3.8}$$

and then Remark (*a*) shows that (6.3.8) is not only a module isomorphism but also a natural equivalence of functors in the sense of section (3.8). We incorporate these observations in the following theorem:

Theorem 2. *Let $T(A,B)$ be an additive functor, covariant in A and contravariant in B, then each of the right-derived functors $R^nT(A,B)$ also has these same properties. The $R^nT(A,B)$ are uniquely defined to within natural equivalences, depending on the defining system, and, for all $n < 0$, we have*

$$R^nT(A,B) = 0.$$

Remark. It is convenient to take this opportunity of observing that if $\Lambda_1 = \Lambda_2 = \Lambda$ and all are commutative, and if further $T(A,B)$ is Λ-linear in the sense of section (3.7), then each of the functors $R^nT(A,B)$ will also be Λ-linear. This follows at once from (v) of section (6.1) in view of the method of construction of derived functors.

The connecting homomorphisms. There are certain connexions between the various right-derived functors R^nT which must now be established. For this purpose we consider an exact sequence

$$0 \to A' \to A \to A'' \to 0 \tag{6.3.9}$$

of Λ_1-modules and a translation

$$\begin{array}{ccccccccc} 0 & \longrightarrow & A' & \longrightarrow & A & \longrightarrow & A'' & \longrightarrow & 0 \\ & & \downarrow & & \downarrow & & \downarrow & & \\ 0 & \longrightarrow & \tilde{A}' & \longrightarrow & \tilde{A} & \longrightarrow & \tilde{A}'' & \longrightarrow & 0 \end{array} \tag{6.3.10}$$

of (6.3.9) into another exact sequence of the same kind. In addition, we suppose ourselves given a homomorphism

$$\tilde{B} \to B \tag{6.3.11}$$

of Λ_2-modules.

Suppose now that we construct *arbitrary* injective resolutions $\mathbf{Q}'$, $\mathbf{Q}''$, $\tilde{\mathbf{Q}}'$, $\tilde{\mathbf{Q}}''$ for A', A'', $\tilde{A}'$, $\tilde{A}''$ respectively, and also *arbitrary* projective resolutions $\mathbf{P}$, $\tilde{\mathbf{P}}$ for B, $\tilde{B}$ respectively. By Theorem 17 of section (5.6), we can find injective resolutions $\mathbf{Q}$, $\tilde{\mathbf{Q}}$ for A, $\tilde{A}$ respectively, which can be embedded in *split exact* sequences

$$\mathbf{0} \to \mathbf{Q}' \to \mathbf{Q} \to \mathbf{Q}'' \to \mathbf{0} \tag{6.3.12}$$

and

$$\mathbf{0} \to \tilde{\mathbf{Q}}' \to \tilde{\mathbf{Q}} \to \tilde{\mathbf{Q}}'' \to \mathbf{0}, \tag{6.3.13}$$

where (6.3.12) is over the upper row of (6.3.10) and (6.3.13) is over the lower row of the same diagram. From (6.3.12) we obtain a sequence

$$\mathbf{0} \to T(\mathbf{Q}', \mathbf{P}) \to T(\mathbf{Q}, \mathbf{P}) \to T(\mathbf{Q}'', \mathbf{P}) \to \mathbf{0}, \tag{6.3.14}$$

and this, we contend, is exact. To establish this fact, let us observe that the component sequences of (6.3.14) are

$$0 \to T^n(\mathbf{Q}', \mathbf{P}) \to T^n(\mathbf{Q}, \mathbf{P}) \to T^n(\mathbf{Q}'', \mathbf{P}) \to 0, \tag{6.3.15}$$

where, by (6.2.3*b*) and (6.2.4), the various modules are given by the direct-sum formulae

$$T^n(\mathbf{Q}', \mathbf{P}) = \sum_{r+s=n} T(Q'^r, P_s),$$

$$T^n(\mathbf{Q}, \mathbf{P}) = \sum_{r+s=n} T(Q^r, P_s),$$

$$T^n(\mathbf{Q}'', \mathbf{P}) = \sum_{r+s=n} T(Q''^r, P_s).$$

But we know that the sequences

$$0 \to Q'^r \to Q^r \to Q''^r \to 0$$

are split exact sequences; hence, by Theorem 2 of section (3.9),

$$0 \to T(Q'^r, P_s) \to T(Q^r, P_s) \to T(Q''^r, P_s) \to 0$$

is an exact sequence, for every pair r, s of integers. Accordingly, by combining these results for all r, s for which $r+s=n$, we establish the exactness of (6.3.15) and hence that of (6.3.14) as well.

Since (6.3.14) is an exact sequence, there exists, by Theorem 5 of section (4.6), an exact homology sequence

$$\cdots \to H^n(T(\mathbf{Q}', \mathbf{P})) \to H^n(T(\mathbf{Q}, \mathbf{P})) \to H^n(T(\mathbf{Q}'', \mathbf{P})) \xrightarrow{\Delta} H^{n+1}(T(\mathbf{Q}', \mathbf{P})) \to H^{n+1}(T(\mathbf{Q}, \mathbf{P})) \to \cdots. \qquad (6.3.16)$$

It is important to observe that, at this stage, we do not know that the connecting homomorphism

$$H^n(T(\mathbf{Q}'', \mathbf{P})) \xrightarrow{\Delta} H^{n+1}(T(\mathbf{Q}', \mathbf{P})) \qquad (6.3.17)$$

is independent of the freedom of choice which we had for $\mathbf{Q}$, but this will emerge presently.

In order to obtain more detailed information about the homomorphism (6.3.17) let us construct translations

$$\boldsymbol{\phi}' : \mathbf{Q}' \to \tilde{\mathbf{Q}}', \quad \boldsymbol{\phi}'' : \mathbf{Q}'' \to \tilde{\mathbf{Q}}''$$

over $A' \to \tilde{A}'$ and $A'' \to \tilde{A}''$ respectively and also a translation $\tilde{\mathbf{P}} \to \mathbf{P}$ over $\tilde{B} \to B$. If now we apply Theorem 19 of section (5.7) we see that there can be found a translation $\boldsymbol{\phi} : \mathbf{Q} \to \tilde{\mathbf{Q}}$ over $A \to \tilde{A}$ for which

$$\begin{array}{ccccccccc} 0 & \longrightarrow & \mathbf{Q}' & \longrightarrow & \mathbf{Q} & \longrightarrow & \mathbf{Q}'' & \longrightarrow & 0 \\ & & \downarrow{\scriptstyle \boldsymbol{\phi}'} & & \downarrow{\scriptstyle \boldsymbol{\phi}} & & \downarrow{\scriptstyle \boldsymbol{\phi}''} & & \\ 0 & \longrightarrow & \tilde{\mathbf{Q}}' & \longrightarrow & \tilde{\mathbf{Q}} & \longrightarrow & \tilde{\mathbf{Q}}'' & \longrightarrow & 0 \end{array}$$

is commutative, and from this we derive a further commutative system, namely

$$\begin{array}{ccccccccc} 0 & \longrightarrow & T(\mathbf{Q}', \mathbf{P}) & \longrightarrow & T(\mathbf{Q}, \mathbf{P}) & \longrightarrow & T(\mathbf{Q}'', \mathbf{P}) & \longrightarrow & 0 \\ & & \downarrow & & \downarrow & & \downarrow & & \\ 0 & \longrightarrow & T(\tilde{\mathbf{Q}}', \tilde{\mathbf{P}}) & \longrightarrow & T(\tilde{\mathbf{Q}}, \tilde{\mathbf{P}}) & \longrightarrow & T(\tilde{\mathbf{Q}}'', \tilde{\mathbf{P}}) & \longrightarrow & 0 \end{array}$$

in which we know the rows are exact. Let us again apply Theorem 5 of section (4.6). This yields commutative diagrams

$$\begin{array}{ccc} H^n(T(\mathbf{Q}'', \mathbf{P})) & \longrightarrow & H^{n+1}(T(\mathbf{Q}', \mathbf{P})) \\ \downarrow & & \downarrow \\ H^n(T(\tilde{\mathbf{Q}}'', \tilde{\mathbf{P}})) & \longrightarrow & H^{n+1}(T(\tilde{\mathbf{Q}}', \tilde{\mathbf{P}})) \end{array} \qquad (6.3.18)$$

which, in the notation of (6.3.3), can also be written as

$$\begin{array}{ccc} H^n(A''\text{—}\mathbf{Q}'', \mathbf{P}\text{—}B) & \longrightarrow & H^{n+1}(A'\text{—}\mathbf{Q}', \mathbf{P}\text{—}B) \\ \downarrow & & \downarrow \\ H^n(\tilde{A}''\text{—}\tilde{\mathbf{Q}}'', \tilde{\mathbf{P}}\text{—}\tilde{B}) & \longrightarrow & H^{n+1}(\tilde{A}'\text{—}\tilde{\mathbf{Q}}', \tilde{\mathbf{P}}\text{—}\tilde{B}) \end{array} \qquad (6.3.19)$$

We are now in a position to draw a number of important conclusions. First let us take $0 \to \tilde{A}' \to \tilde{A} \to \tilde{A}'' \to 0$ to be the same sequence as $0 \to A' \to A \to A'' \to 0$, the maps $A' \to \tilde{A}'$, $A \to \tilde{A}$ and $A'' \to \tilde{A}''$ all being identity maps. In addition, let $\tilde{B} = B$ and let $\tilde{B} \to B$ also be an identity map. In these special circumstances (6.3.19) becomes a commutative diagram

$$\begin{array}{ccc} H^n(A''\text{—}\mathbf{Q}'', \mathbf{P}\text{—}B) & \longrightarrow & H^{n+1}(A'\text{—}\mathbf{Q}', \mathbf{P}\text{—}B) \\ \downarrow & & \downarrow \\ H^n(A''\text{—}\tilde{\mathbf{Q}}'', \tilde{\mathbf{P}}\text{—}B) & \longrightarrow & H^{n+1}(A'\text{—}\tilde{\mathbf{Q}}', \tilde{\mathbf{P}}\text{—}B) \end{array} \qquad (6.3.20)$$

in which, to take one instance, $\mathbf{Q}''$ and $\tilde{\mathbf{Q}}''$ are both injective resolutions of A''. If now we specialize still further and take $\tilde{\mathbf{Q}}'' = \mathbf{Q}''$, $\tilde{\mathbf{Q}}' = \mathbf{Q}'$ and $\tilde{\mathbf{P}} = \mathbf{P}$ but without *necessarily supposing that* $\mathbf{Q}$ *and* $\tilde{\mathbf{Q}}$ *are the same*, then the vertical maps in (6.3.20) become identity maps (Remark (*b*)), and this shows that the *connecting homomorphism* (6.3.17) *is indeed independent of the choice of* $\mathbf{Q}$. Bearing this in mind and returning to (6.3.6) and (6.3.16) we see that, for the derived functors $R^nT(A, B)$, there is, for each n, a well-defined connecting homomorphism

$$R^nT(A'', B) \xrightarrow{\Delta} R^{n+1}T(A', B), \qquad (6.3.21)$$

and that these connecting homomorphisms make

$$\cdots \to R^nT(A', B) \to R^nT(A, B) \to R^nT(A'', B) \\ \xrightarrow{\Delta} R^{n+1}T(A', B) \to R^{n+1}T(A, B) \to \cdots \qquad (6.3.22)$$

an exact sequence. But we can say more, for the commutative properties of (6.3.20) show that, if we change the defining system, the connecting homomorphisms

$$*R^nT(A'', B) \to *R^{n+1}T(A', B),$$

for the new functors, agree with (6.3.21) when we identify the two systems of functors by means of the canonical equivalences

$$*R^nT(A, B) \approx R^nT(A, B),$$
$$*R^{n+1}T(A, B) \approx R^{n+1}T(A, B)$$

of (6.3.8). Again, from (6.3.18),

$$\begin{array}{ccc} R^nT(A'', B) & \longrightarrow & R^{n+1}T(A', B) \\ \downarrow & & \downarrow \\ R^nT(\tilde{A}'', \tilde{B}) & \longrightarrow & R^{n+1}T(\tilde{A}', \tilde{B}) \end{array}$$

is commutative and therefore we may say that (6.3.22), regarded as belonging to the translation category of infinite exact sequences, depends covariantly on the exact sequence $0 \to A' \to A \to A'' \to 0$ and contravariantly on the module B.

So far we have dealt only with exact sequences of the form

$$0 \to A' \to A \to A'' \to 0.$$

If, instead, we consider an exact sequence

$$0 \to B' \to B \to B'' \to 0,$$

then a similar discussion leads to an infinite exact sequence

$$\cdots \to R^nT(A,B'') \to R^nT(A,B) \to R^nT(A,B') \xrightarrow{\Delta} R^{n+1}T(A,B'') \to R^{n+1}T(A,B) \to \cdots$$

with properties similar to those of (6.3.22). No new ideas are needed, so we need not go through the details again. The complete body of results is set out in the next theorem.

Theorem 3. *Let $T(A,B)$ be an additive functor, covariant in A and contravariant in B, and let $R^nT(A,B)$, $n = 0, \pm 1, \pm 2, \ldots$, be its right-derived functors. Further let*

$$0 \to A' \to A \to A'' \to 0, \quad 0 \to B' \to B \to B'' \to 0$$

be exact sequences. Then, for each n, there exist well-defined connecting homomorphisms

$$R^nT(A'',B) \to R^{n+1}T(A',B), \quad R^nT(A,B') \to R^{n+1}T(A,B''),$$

and these connecting homomorphisms make the sequences

$$\cdots \to R^nT(A',B) \to R^nT(A,B) \to R^nT(A'',B) \to R^{n+1}T(A',B) \to R^{n+1}T(A,B) \to \cdots \quad (6.3.23)$$

and

$$\cdots \to R^nT(A,B'') \to R^nT(A,B) \to R^nT(A,B') \to R^{n+1}T(A,B'') \to R^{n+1}T(A,B) \to \cdots \quad (6.3.24)$$

exact. Moreover, if (6.3.23) *and* (6.3.24) *are regarded as belonging to the translation category of (infinite) exact sequences, then* (6.3.23) *is a covariant functor of the exact sequence $0 \to A' \to A \to A'' \to 0$ and a contravariant functor of the module B. On the other hand* (6.3.24) *is a covariant functor of the module A and a contravariant functor of the exact sequence $0 \to B' \to B \to B'' \to 0$. Furthermore, the natural equivalences associated with a change of the defining system for the right-derived functors (see* (6.3.8)*) commute with both sets of connecting homomorphisms.*

The functor R^0T. Of the right-derived functors R^nT $(n = 0, 1, 2, \ldots)$ the first of these, namely, R^0T, is the one most closely related to the original functor T. The theorems about to be established are all concerned with the connexions between these two functors and we shall show, presently, that we can regard T and R^0T as being essentially the same in those situations where T is left exact.

Theorem 4. *The derived functor R^0T is left exact.*

Proof. Let $0 \to A' \to A \to A'' \to 0$ and $0 \to B' \to B \to B'' \to 0$ be exact sequences, then, since R^nT takes only null values when $n < 0$, it follows, from (6.3.23) and (6.3.24), that the sequences

$$0 \to R^0T(A', B) \to R^0T(A, B) \to R^0T(A'', B)$$

and $$0 \to R^0T(A, B'') \to R^0T(A, B) \to R^0T(A, B')$$

are both exact. This completes the proof.

Now let $A \to A'$ and $B' \to B$ be homomorphisms and let $\mathbf{Q}$, $\mathbf{Q}'$ be arbitrary injective resolutions of A, A', while $\mathbf{P}$, $\mathbf{P}'$ are arbitrary projective resolutions of B, B'. Suppose, too, that $\mathbf{Q} \to \mathbf{Q}'$ and $\mathbf{P}' \to \mathbf{P}$ are translations, which are over $A \to A'$ and $B' \to B$ respectively. The various augmentation translations $\mathbf{A} \to \mathbf{Q}$, $\mathbf{A}' \to \mathbf{Q}'$, $\mathbf{P} \to \mathbf{B}$, $\mathbf{P}' \to \mathbf{B}'$ then lead to a commutative system

$$\begin{array}{ccc} T(\mathbf{A}, \mathbf{B}) & \longrightarrow & T(\mathbf{Q}, \mathbf{P}) \\ \downarrow & & \downarrow \\ T(\mathbf{A}', \mathbf{B}') & \longrightarrow & T(\mathbf{Q}', \mathbf{P}') \end{array}$$

of complexes and translations, and hence to a commutative diagram

$$\begin{array}{ccc} H^0(T(\mathbf{A}, \mathbf{B})) & \longrightarrow & H^0(T(\mathbf{Q}, \mathbf{P})) \\ \downarrow & & \downarrow \\ H^0(T(\mathbf{A}', \mathbf{B}')) & \longrightarrow & H^0(T(\mathbf{Q}', \mathbf{P}')) \end{array} \tag{6.3.25}$$

of modules and homomorphisms. But $T(\mathbf{A}, \mathbf{B})$ is the complex associated with $T(A, B)$ and therefore $H^0(T(\mathbf{A}, \mathbf{B})) = T(A, B)$. Accordingly, the upper row of (6.3.25) yields a homomorphism

$$T(A, B) \to R^0T(A, B), \tag{6.3.26}$$

which, by the commutative properties of (6.3.25), defines a natural transformation $T \to R^0T$. We shall refer to this as the *canonical transformation* $T \to R^0T$.

Theorem 5. *The canonical transformation $T \to R^0T$ commutes with the equivalence* (6.3.8) *associated with a change in the defining system. Further $T \to R^0T$ is an equivalence of functors if and only if T is left exact.*

Proof. The first assertion is an immediate consequence of the commutative properties of (6.3.25) applied to the case in which $A' = A$, $B' = B$ and $A \to A'$, $B' \to B$ are identity maps. Next, if $T \to R^0T$ is an equivalence, then Theorem 4 shows at once that T is a left exact functor. We have now to prove the converse and so, for the rest of the proof, we assume that T is a left exact functor.

Let the notation be as in the paragraph immediately preceding the statement of Theorem 5. Then the translation $T(\mathbf{A}, \mathbf{B}) \to T(\mathbf{Q}, \mathbf{P})$, arising from the augmentation translations, shows that $T(\mathbf{Q}, \mathbf{P})$ is a right complex over $T(A, B)$ and therefore we have a 0-sequence

$$0 \to T(A, B) \to T^0(\mathbf{Q}, \mathbf{P}) \xrightarrow{d^0} T^1(\mathbf{Q}, \mathbf{P}). \qquad (6.3.27)$$

What we wish to show is that the homomorphism $T(A, B) \to R^0T(A, B)$ of (6.3.26) is an isomorphism and this is equivalent to showing that (6.3.27) is exact. Now

$$T^0(\mathbf{Q}, \mathbf{P}) = T(Q^0, P_0), \quad T^1(\mathbf{Q}, \mathbf{P}) = T(Q^1, P_0) + T(Q^0, P_1) \text{ (direct sum)}$$

and so (6.3.27) takes the form

$$0 \to T(A, B) \to T(Q^0, P_0) \to T(Q^1, P_0) + T(Q^0, P_1), \qquad (6.3.28)$$

where the homomorphism on the right of (6.3.28) is compounded from

$$T(Q^0, P_0) \to T(Q^1, P_0) \quad \text{and} \quad T(Q^0, P_0) \to T(Q^0, P_1),$$

possibly after some changes of sign have been made. But we may, in fact, proceed as though no changes in sign are involved because this will not affect the exactness of (6.3.28). However, T is a left exact functor and both $0 \to A \to Q^0 \to Q^1$ and $P_1 \to P_0 \to B \to 0$ are exact sequences; hence the exactness of (6.3.28) follows from Theorem 5 of section (3.12). This completes the proof.

Concluding remarks. Throughout the present section we have considered the right-derived functors of $T(A, B)$, where T is covariant in A and contravariant in B. The rule for obtaining right-derived functors in other cases is to replace each covariant variable by an injective resolution and each contravariant variable by a projective resolution. Apart from trivial modifications, the discussion given then goes through without alteration.

6.4 Left-derived functors

As before, let $T(A, B)$ be an additive functor, which is covariant in A and contravariant in B. We shall now examine another process for obtaining new functors, which is akin to that considered in the last section. The basic ideas are very similar to those already described in detail and therefore we shall only give a broad outline. As in section (6.3), the case considered is typical. In other words, the results obtained (when suitably modified) hold for additive functors of one, two, or more variables, which exhibit all possible situations in respect of the covariance and contravariance of the variables.

Suppose that $f: A \to A'$, $g: B' \to B$ are homomorphisms and let $\mathbf{P}$, $\mathbf{P}'$ be projective resolutions of A, A', while $\mathbf{Q}$, $\mathbf{Q}'$ are injective resolutions of B, B'. By (6.2.3*b*) and (6.2.4), $T_n(\mathbf{P}, \mathbf{Q})$ is given by the direct sum formula

$$T_n(\mathbf{P}, \mathbf{Q}) = T^{-n}(\mathbf{P}, \mathbf{Q}) = \sum_{r+s=n} T(P_r, Q^s), \tag{6.4.1}$$

which shows that $T(\mathbf{P}, \mathbf{Q})$ is a left complex. Moreover, we can regard this complex as being over the module $T(A, B)$, by virtue of the translation

$$T(\mathbf{P}, \mathbf{Q}) \to T(\mathbf{A}, \mathbf{B}), \tag{6.4.2}$$

which results from the augmentation maps $\mathbf{P} \to \mathbf{A}$ and $\mathbf{B} \to \mathbf{Q}$.

By Theorems 12 and 13 of section (5.5), there exist translations $\phi: \mathbf{P} \to \mathbf{P}'$ and $\psi: \mathbf{Q}' \to \mathbf{Q}$ over $A \to A'$ and $B' \to B$ respectively, and from these we obtain a translation

$$T(\phi, \psi): T(\mathbf{P}, \mathbf{Q}) \to T(\mathbf{P}', \mathbf{Q}'), \tag{6.4.3}$$

which is over $T(A, B) \to T(A', B')$. Further, ϕ and ψ are determined by $A \to A'$ and $B' \to B$ to within homotopies and therefore $T(\phi, \psi)$ is also determined up to a homotopy. Hence, passing to the homology modules, there results a well-defined homomorphism

$$H_n(T(\mathbf{P}, \mathbf{Q})) \to H_n(T(\mathbf{P}', \mathbf{Q}')). \tag{6.4.4}$$

Assume now that we select a definite projective resolution $\mathbf{P}$ for each module A of $\mathscr{G}_{\Lambda_1}$ and a definite injective resolution $\mathbf{Q}$ for each module B of $\mathscr{G}_{\Lambda_2}$. We can then write

$$L_n T(A, B) = H_n(T(\mathbf{P}, \mathbf{Q})), \tag{6.4.5}$$

and it follows, by (6.4.4), that we have a well-defined homomorphism

$$L_n T(A, B) \to L_n T(A', B'). \tag{6.4.6}$$

This enables us to regard the $L_n T$ as a family of additive functors.

Definition. The functors $L_n T(A,B)$ are said to be *left-derived* from $T(A,B)$.

It should be observed that, since $T(\mathbf{P},\mathbf{Q})$ is a left complex, we have

$$L_n T(A,B) = 0 \quad \text{if} \quad n < 0. \tag{6.4.7}$$

The defining system. The functors $L_n T$ depend on the choice of the resolutions of the modules of $\mathscr{G}_{\Lambda_1}$ and $\mathscr{G}_{\Lambda_2}$, so we shall refer to the selected resolutions as the *defining system* for the left-derived functors. If we change the defining system then we obtain new functors, $*L_n T$ say, but these are related to the original ones by means of well-defined functor equivalences

$$L_n T(A,B) \approx *L_n T(A,B). \tag{6.4.8}$$

These equivalences arise from the homomorphisms (6.4.4) by consideration of the identity maps $A \to A$, $B \to B$, the situation being similar to that described in (6.3.8) in the case of right-derived functors. Our next theorem corresponds to Theorem 2.

Theorem 6. *Let $T(A,B)$ be an additive functor, covariant in A and contravariant in B, then the left-derived functors $L_n T(A,B)$ also have these same properties. The $L_n T(A,B)$ are uniquely defined to within natural equivalences depending on the defining system, and*

$$L_n T(A,B) = 0, \quad \textit{for all } n < 0.$$

As usual, when $\Lambda_1 = \Lambda_2 = \Lambda$ and all are commutative, then we can say a little more; namely, if T is not only additive but also Λ-linear, then the same will be true for the left-derived functors.

The connecting homomorphisms. Let

$$0 \to A' \to A \to A'' \to 0 \tag{6.4.9}$$

be an exact sequence of Λ_1-modules and let B be a Λ_2-module. Suppose now that we are given projective resolutions $\mathbf{P}'$, $\mathbf{P}''$ for A', A'' and an injective resolution $\mathbf{Q}$ for B. In these circumstances it is possible, by Theorem 16 of section (5.6), to find a split exact sequence

$$\mathbf{0} \to \mathbf{P}' \to \mathbf{P} \to \mathbf{P}'' \to \mathbf{0} \tag{6.4.10}$$

over (6.4.9), where $\mathbf{P}$ is a suitable projective resolution of A. If we do this then it turns out that

$$\mathbf{0} \to T(\mathbf{P}',\mathbf{Q}) \to T(\mathbf{P},\mathbf{Q}) \to T(\mathbf{P}'',\mathbf{Q}) \to \mathbf{0}$$

is an exact sequence and as such possesses, by virtue of Theorem 5 of section (4.6), an exact homology sequence. In particular, we obtain connecting homomorphisms

$$H_n(T(\mathbf{P}'',\mathbf{Q})) \to H_{n-1}(T(\mathbf{P}',\mathbf{Q})),$$

which, on further investigation, prove to be independent of the choice of **P**, within the limits prescribed by the construction. This implies, therefore, that we have well-defined connecting homomorphisms

$$L_n T(A'', B) \to L_{n-1} T(A', B), \tag{6.4.11}$$

which make

$$\cdots \to L_n T(A', B) \to L_n T(A, B) \to L_n T(A'', B)$$
$$\to L_{n-1} T(A', B) \to L_{n-1} T(A, B) \to \cdots \tag{6.4.12}$$

an exact sequence. One discovers that every translation

$$\begin{array}{ccccccccc} 0 & \longrightarrow & A' & \longrightarrow & A & \longrightarrow & A'' & \longrightarrow & 0 \\ & & \downarrow & & \downarrow & & \downarrow & & \\ 0 & \longrightarrow & \tilde{A}' & \longrightarrow & \tilde{A} & \longrightarrow & \tilde{A}'' & \longrightarrow & 0 \end{array}$$

of (6.4.9) into a similar exact sequence, produces a translation of (6.4.12) and the same is true for every Λ_2-homomorphism $\tilde{B} \to B$. Thus we may say briefly that (6.4.12) is a covariant functor of the exact sequence $0 \to A' \to A \to A'' \to 0$ and a contravariant functor of B.

Of course, there also exist connecting homomorphisms

$$L_n T(A, B') \to L_{n-1} T(A, B'')$$

corresponding to exact sequences $0 \to B' \to B \to B'' \to 0$ and these have similar properties. To make the whole position quite clear we set out the relevant facts in the next theorem.

Theorem 7. *Let $T(A, B)$ be an additive functor, which is covariant in A and contravariant in B, and let $L_n T(A, B)$, $n = 0, \pm 1, \pm 2, \ldots$, be its left-derived functors. Further let*

$$0 \to A' \to A \to A'' \to 0, \quad 0 \to B' \to B \to B'' \to 0$$

be exact sequences. Then, for each n, there exist well-defined connecting homomorphisms

$$L_n T(A'', B) \to L_{n-1} T(A', B), \quad L_n T(A, B') \to L_{n-1} T(A, B'')$$

which make the sequences

$$\cdots \to L_n T(A', B) \to L_n T(A, B) \to L_n T(A'', B)$$
$$\to L_{n-1} T(A', B) \to L_{n-1} T(A, B) \to \cdots \tag{6.4.13}$$

and

$$\cdots \to L_n T(A, B'') \to L_n T(A, B) \to L_n T(A, B')$$
$$\to L_{n-1} T(A, B'') \to L_{n-1} T(A, B) \to \cdots \tag{6.4.14}$$

exact. Moreover, if (6.4.13) *and* (6.4.14) *are regarded as belonging to the translation category of infinite exact sequences, then* (6.4.13) *is a covariant functor of the exact sequence* $0 \to A' \to A \to A'' \to 0$ *and a contravariant functor of the module* B. *On the other hand,* (6.4.14) *is a covariant functor of the module* A *and a contravariant functor of the exact sequence* $0 \to B' \to B \to B'' \to 0$. *Finally, if the defining system for the left-derived functors is changed, then the canonical equivalences, relating the new functors* (*see* (6.4.8)) *to the old, commute with both sets of connecting homomorphisms.*

The functor $L_0 T$. Let us now examine the relations which connect the left-derived functor $L_0 T$ with the original functor T. To begin with we have

Theorem 8. $L_0 T$ *is right exact.*

This follows from the exactness of the sequences (6.4.13) and (6.4.14) taken in conjunction with the fact that $L_n T(A, B) = 0$ whenever $n < 0$.

Now suppose that $\mathbf{P}$ is a projective resolution of a Λ_1-module A and that $\mathbf{Q}$ is an injective resolution of a Λ_2-module B. Then the augmentation translations determine a translation

$$T(\mathbf{P}, \mathbf{Q}) \to T(\mathbf{A}, \mathbf{B}), \tag{6.4.15}$$

which in turn gives rise to a homomorphism

$$H_0(T(\mathbf{P}, \mathbf{Q})) \to H_0(T(\mathbf{A}, \mathbf{B})). \tag{6.4.16}$$

But $H_0(T(\mathbf{A}, \mathbf{B})) = T(A, B)$, and so (6.4.16) yields a homomorphism

$$L_0 T(A, B) \to T(A, B). \tag{6.4.17}$$

Further, by consideration of the translation properties of (6.4.16), we see that (6.4.17) is more than a module homomorphism for it defines a functor transformation $L_0 T \to T$. We shall refer to this as the *canonical transformation* $L_0 T \to T$.

Theorem 9. *The canonical transformation* $L_0 T \to T$ *commutes with the equivalence* $L_0 T \approx {}^*L_0 T$ (*see* (6.4.8)) *associated with a change in the defining system for the left-derived functors. Furthermore,* $L_0 T \to T$ *is a natural equivalence of functors if and only if* T *is right exact.*

Here it is only the last assertion that presents any difficulty, and this can be demonstrated in a manner analogous to that used in the discussion of the corresponding section of Theorem 5.

Concluding remarks. Throughout the whole of section (6.4) we have concerned ourselves with a functor of two variables, covariant in the first and contravariant in the second. If one wants to consider left-derived functors in other situations then the rule is to replace each covariant variable by a projective resolution and each contravariant variable by an injective resolution. The preceding discussion then goes through with only trivial modifications.

6.5 Connected sequences of functors

In sections (6.3) and (6.4) we saw how, starting from an additive functor of modules, one can derive the sequence of so-called right-derived functors and also the sequence of left-derived functors. In each case we obtain a *sequence of functors*, and in each case the functors of the sequence are connected to one another by means of certain homomorphisms that are associated with exact sequences of the type $0 \to A' \to A \to A'' \to 0$.

The idea of a connected sequence of functors is of such importance that we shall formulate it explicitly and then establish results which enable us to say that, under suitable circumstances, two connected sequences are essentially the same. In other words, we propose to prove certain isomorphism theorems for connected sequences of functors. Later these will be used to justify methods of computing derived functors, which differ from those on which the definitions are based. Since, in our applications, it will be possible to consider separately the different variables which occur, it turns out to be sufficient for the sequel if we confine our attention to connected sequences of functors of a single variable.

After these preliminaries let us consider an infinite sequence

$$T^0(A),\ T^1(A),\ T^2(A),\ \dots \qquad (6.5.1)$$

of covariant functors from $\mathscr{G}_{\Lambda_1}$ to $\mathscr{G}_{\Lambda}$. It will be supposed that, for each exact sequence

$$0 \to A' \to A \to A'' \to 0 \qquad (6.5.2)$$

and each integer $n \geqslant 0$, there is defined a Λ-homomorphism

$$T^n(A'') \to T^{n+1}(A'), \qquad (6.5.3\,(\text{i}))$$

and it will also be supposed that

$$T^0(A') \to T^0(A) \to T^0(A'') \to T^1(A') \to \cdots$$
$$\to T^n(A') \to T^n(A) \to T^n(A'') \to T^{n+1}(A') \to \cdots \qquad (6.5.4\,(\text{i}))$$

is a 0-sequence.

Definition. If (6.5.4 (i)), when regarded as belonging to the translation category of right semi-infinite 0-sequences, is a covariant functor of the exact sequence (6.5.2), then we shall say that $[T^n]_{n\geq 0}$ is a *connected right sequence of covariant functors*, by virtue of the connecting homomorphisms (6.5.3 (i)).

It follows from the definition that, if $[T^n]_{n\geq 0}$ is such a connected sequence and

$$\begin{array}{ccccccccc} 0 & \longrightarrow & A' & \longrightarrow & A & \longrightarrow & A'' & \longrightarrow & 0 \\ & & \downarrow & & \downarrow & & \downarrow & & \\ 0 & \longrightarrow & \tilde{A}' & \longrightarrow & \tilde{A} & \longrightarrow & \tilde{A}'' & \longrightarrow & 0 \end{array} \tag{6.5.5}$$

is a commutative diagram with exact rows, then, for each $n\geq 0$, the diagram

$$\begin{array}{ccc} T^n(A'') & \longrightarrow & T^{n+1}(A') \\ \downarrow & & \downarrow \\ T^n(\tilde{A}'') & \longrightarrow & T^{n+1}(\tilde{A}') \end{array} \tag{6.5.6 (i)}$$

is also commutative. Indeed, this simply expresses the condition that (6.5.4 (i)) be a covariant functor of (6.5.2).

By a minor modification of the definition we arrive at the notion of a *connected right sequence of contravariant functors*. Quite precisely the functors T^n, when $n\geq 0$, must all be contravariant and, for each exact sequence (6.5.2), we require connecting homomorphisms

$$T^n(A')\to T^{n+1}(A'') \quad (n\geq 0). \tag{6.5.3 (ii)}$$

Further,

$$\begin{aligned} &T^0(A'')\to T^0(A)\to T^0(A')\to T^1(A'')\to\cdots \\ &\quad\to T^n(A'')\to T^n(A)\to T^n(A')\to T^{n+1}(A'')\to\cdots \end{aligned} \tag{6.5.4 (ii)}$$

must be a 0-sequence, and the 0-sequence itself has to be a contravariant functor of the exact sequence $0\to A'\to A\to A''\to 0$. Note that this time (6.5.5) will give rise to a commutative diagram

$$\begin{array}{ccc} T^n(\tilde{A}') & \longrightarrow & T^{n+1}(\tilde{A}'') \\ \downarrow & & \downarrow \\ T^n(A') & \longrightarrow & T^{n+1}(A'') \end{array} \tag{6.5.6 (ii)}$$

To take an example, let $T(A_1, A_2, \ldots, A_k)$ be an arbitrary additive functor of modules and let $R^nT(A_1, A_2, \ldots, A_k)$, for $n\geq 0$, be its right-derived functors. If now we keep all but one of $A_1, A_2, \ldots, A_k$ fixed then, by Theorem 3,

$$R^0T(A_1, \ldots, A_k),\ R^1T(A_1, \ldots, A_k),\ R^2T(A_1, \ldots, A_k),\ \ldots$$

is a connected right sequence for the remaining variable.

Now suppose that we have two connected right sequences $[T^n]_{n\geq 0}$ and $[U^n]_{n\geq 0}$ of covariant functors from $\mathscr{G}_{\Lambda_1}$ to $\mathscr{G}_{\Lambda}$ and suppose also that we have a family $\Phi = [\phi^n]_{n\geq 0}$ of natural transformations

$$\phi^n : T^n \to U^n. \tag{6.5.7}$$

Definition. (Covariant case) Φ is said to be a *homomorphism* of $[T^n]_{n\geq 0}$ into $[U^n]_{n\geq 0}$ if, for every exact sequence

$$0 \to A' \to A \to A'' \to 0$$

and every $n \geq 0$, the diagram

$$\begin{array}{ccc} T^n(A'') & \longrightarrow & T^{n+1}(A') \\ \downarrow{\scriptstyle \phi^n} & & \downarrow{\scriptstyle \phi^{n+1}} \\ U^n(A'') & \longrightarrow & U^{n+1}(A') \end{array} \tag{6.5.8 (i)}$$

is commutative. If Φ is such a homomorphism then it is called an *isomorphism* of $[T^n]_{n\geq 0}$ on to $[U^n]_{n\geq 0}$ if each $\phi^n : T^n \to U^n$ is a natural equivalence of functors.

It is clear that, in the case of an isomorphism, the equivalences inverse to the ϕ^n define an isomorphism of $[U^n]_{n\geq 0}$ on to $[T^n]_{n\geq 0}$.

Similar definitions apply to connected right sequences $[T^n]_{n\geq 0}$ and $[U^n]_{n\geq 0}$ of contravariant functors, the only difference being that (6.5.8 (i)) has to be replaced by

$$\begin{array}{ccc} T^n(A') & \longrightarrow & T^{n+1}(A'') \\ \downarrow{\scriptstyle \phi^n} & & \downarrow{\scriptstyle \phi^{n+1}} \\ U^n(A') & \longrightarrow & U^{n+1}(A'') \end{array} \tag{6.5.8 (ii)}$$

Theorem 10. *Let $[T^n]_{n\geq 0}$ and $[U^n]_{n\geq 0}$ be connected right sequences of covariant functors (from $\mathscr{G}_{\Lambda_1}$ to $\mathscr{G}_{\Lambda}$) and let $\phi^0 : T^0 \to U^0$ be a given natural transformation. Suppose further that whenever $0 \to A \to Q \to L \to 0$ is an exact sequence with Q a Λ_1- injective module, then*

$$T^n(Q) \to T^n(L) \to T^{n+1}(A) \to 0$$

is an exact sequence for every $n \geq 0$. In these circumstances there exist unique natural transformations $\phi^n : T^n \to U^n$ $(n \geq 1)$ such that $\Phi = [\phi^n]_{n\geq 0}$ is a homomorphism $[T_n]_{n\geq 0} \to [U^n]_{n\geq 0}$ of connected sequences.

Remark. It is convenient to describe Φ as an extension of

$$\phi^0 : T^0 \to U^0$$

to a homomorphism of $[T^n]_{n\geq 0}$ into $[U^n]_{n\geq 0}$.

Proof. We shall construct ϕ^0, ϕ^1, ϕ^2, etc., by an inductive process and therefore we assume that natural transformations $\phi^0, \phi^1, \ldots, \phi^n$, which commute with the connecting homomorphisms, have already been found. Let A and A' be two given Λ_1-modules, then it is possible to set up exact sequences

$$0 \to A \to Q \to L \to 0 \quad \text{and} \quad 0 \to A' \to Q' \to L' \to 0$$

in which Q and Q' are Λ_1 injective. Consider the commutative diagram

$$\begin{array}{ccccccc} T^n(Q) & \longrightarrow & T^n(L) & \longrightarrow & T^{n+1}(A) & \longrightarrow & 0 \\ \downarrow{\scriptstyle \phi^n} & & \downarrow{\scriptstyle \phi^n} & & & & \\ U^n(Q) & \longrightarrow & U^n(L) & \longrightarrow & U^{n+1}(A) & & \end{array} \tag{6.5.9}$$

in which we know that the upper row is exact and the lower row is a 0-sequence. It follows, by the corollary to Proposition 1 of section (4.3), that there is a *unique* Λ-homomorphism

$$T^{n+1}(A) \to U^{n+1}(A),$$

which preserves the commutative property of (6.5.9), and this proves, in particular, that ϕ^{n+1} (if it exists) is unique.

Now let $f: A \to A'$ be a given Λ_1-homomorphism. Since Q' is injective, we can find a homomorphism $Q \to Q'$ such that the combined maps $A \to Q \to Q'$ and $A \to A' \to Q'$ coincide and then, again using the corollary to Proposition 1 of section (4.3), there exists a further homomorphism $L \to L'$ for which

$$\begin{array}{ccccccccc} 0 & \longrightarrow & A & \longrightarrow & Q & \longrightarrow & L & \longrightarrow & 0 \\ & & \downarrow{\scriptstyle f} & & \downarrow & & \downarrow & & \\ 0 & \longrightarrow & A' & \longrightarrow & Q' & \longrightarrow & L' & \longrightarrow & 0 \end{array} \tag{6.5.10}$$

is a commutative diagram. Furthermore, from $0 \to A' \to Q' \to L' \to 0$ is obtained a diagram

$$\begin{array}{ccccccc} T^n(Q') & \longrightarrow & T^n(L') & \longrightarrow & T^{n+1}(A') & \longrightarrow & 0 \\ \downarrow{\scriptstyle \phi^n} & & \downarrow{\scriptstyle \phi^n} & & & & \\ U^n(Q') & \longrightarrow & U^n(L') & \longrightarrow & U^{n+1}(A') & & \end{array} \tag{6.5.11}$$

similar in all respects to (6.5.9) and now, by virtue of (6.5.10) and the fact that ϕ^n is a natural transformation, we have a translation of (6.5.9) into (6.5.11). But, once again using the corollary to Proposition 1 of section (4.3), this translation will remain a translation when the diagrams are supplemented by

$$T^{n+1}(A) \to U^{n+1}(A) \quad \text{and} \quad T^{n+1}(A') \to U^{n+1}(A'),$$

which implies that

$$\begin{array}{ccc} T^{n+1}(A) & \longrightarrow & U^{n+1}(A) \\ \downarrow & & \downarrow \\ T^{n+1}(A') & \longrightarrow & U^{n+1}(A') \end{array} \tag{6.5.12}$$

is commutative. Let us, for the moment, take $A' = A$ and allow f to be the identity map. In (6.5.12) the vertical maps are then also identity maps and this shows that $T^{n+1}(A) \to U^{n+1}(A)$ *does not depend on the choice of the variable modules* Q, L *in the sequence* $0 \to A \to Q \to L \to 0$. We are now justified in denoting this homomorphism by $\phi^{n+1}(A)$. Returning to (6.5.12), as it applies to the general case, we see that the $\phi^{n+1}(A)$ constitute a natural transformation

$$\phi^{n+1} : T^{n+1} \to U^{n+1}.$$

Finally, let $0 \to A_1' \to A_1 \to A_1'' \to 0$ be an exact sequence and let us form an exact sequence $0 \to A_1' \to Q_1' \to L_1' \to 0$, where Q_1' is injective. If now $A_1' \to A_1'$ denotes the identity mapping, we can construct a commutative diagram

$$\begin{array}{ccccccccc} 0 & \longrightarrow & A_1' & \longrightarrow & A_1 & \longrightarrow & A_1'' & \longrightarrow & 0 \\ & & \downarrow & & \downarrow & & \downarrow & & \\ 0 & \longrightarrow & A_1' & \longrightarrow & Q_1' & \longrightarrow & L_1' & \longrightarrow & 0 \end{array} \tag{6.5.13}$$

and this, by the translation properties of connected sequences, implies that

$$\begin{array}{ccc} T^n(A_1'') & & \\ \downarrow & \searrow & \\ T^n(L_1') & \to & T^{n+1}(A_1'), \end{array} \qquad \begin{array}{ccc} U^n(A_1'') & & \\ \downarrow & \searrow & \\ U^n(L_1') & \to & U^{n+1}(A_1') \end{array} \tag{6.5.14}$$

are also commutative. Again, by the construction of ϕ^{n+1},

$$\begin{array}{ccccc} T^n(Q_1') & \longrightarrow & T^n(L_1') & \longrightarrow & T^{n+1}(A_1') \\ \downarrow{\scriptstyle \phi^n} & & \downarrow{\scriptstyle \phi^n} & & \downarrow{\scriptstyle \phi^{n+1}} \\ U^n(Q_1') & \longrightarrow & U^n(L_1') & \longrightarrow & U^{n+1}(A_1') \end{array} \tag{6.5.15}$$

is commutative, and if we observe that the same is true of

$$\begin{array}{ccc} T^n(A_1'') & \xrightarrow{\phi^n} & U^n(A_1'') \\ \downarrow & & \downarrow \\ T^n(L_1') & \xrightarrow{\phi^n} & U^n(L_1') \end{array} \tag{6.5.16}$$

then, by combining (6.5.14), (6.5.15) and (6.5.16), it emerges that

$$\begin{array}{ccc} T^n(A_1'') & \longrightarrow & T^{n+1}(A_1') \\ \downarrow{\scriptstyle \phi^n} & & \downarrow{\scriptstyle \phi^{n+1}} \\ U^n(A_1'') & \longrightarrow & U^{n+1}(A_1') \end{array}$$

is also commutative. Thus ϕ^{n+1} has all the required properties and now the proof is complete.

Corollary. *Let $\phi^0 : T^0 \to U^0$ be a natural equivalence and suppose that, whenever $0 \to A \to Q \to L \to 0$ is exact and Q is injective, then both the sequences*

$$T^n(Q) \to T^n(L) \to T^{n+1}(A) \to 0$$

and

$$U^n(Q) \to U^n(L) \to U^{n+1}(A) \to 0$$

are exact for every $n \geqslant 0$. If now $\Phi = [\phi^n]_{n\geqslant 0}$ is the unique extension of ϕ^0 to a homomorphism $[T^n]_{n\geqslant 0} \to [U^n]_{n\geqslant 0}$, then Φ is an isomorphism and therefore ϕ^n is an equivalence for every value of n.

Proof. Let $\psi^0 : U^0 \to T^0$ be the equivalence inverse to ϕ^0 and let Ψ be the extension of ψ^0 to a homomorphism $[U^n]_{n\geqslant 0} \to [T^n]_{n\geqslant 0}$. Then, with an obvious notation, $\Psi\Phi$ is a homomorphism $[T^n]_{n\geqslant 0} \to [T^n]_{n\geqslant 0}$ and it extends the identity equivalence $T^0 \to T^0$. Consequently, by the uniqueness of the extension, $\psi^n\phi^n$ is the identity equivalence $T^n \to T^n$ for each value of n. Similarly, $\phi^n\psi^n$ is the identity equivalence $U^n \to U^n$, and it therefore follows that ϕ^n and ψ^n are inverse equivalences.

There are, of course, analogues of Theorem 10 and its corollary for connected right sequences of contravariant functors. These will now be stated, without proofs, since no essentially new ideas are needed in the demonstrations.

Theorem 11. *Let $[T^n]_{n\geqslant 0}$ and $[U^n]_{n\geqslant 0}$ be connected right sequences of contravariant functors (from $\mathscr{G}_{\Lambda_1}$ to $\mathscr{G}_{\Lambda}$) and let $\phi^0 : T^0 \to U^0$ be a given natural transformation. Suppose further that, whenever*

$$0 \to L \to P \to A \to 0$$

is an exact sequence with P a Λ_1-projective module, then

$$T^n(P) \to T^n(L) \to T^{n+1}(A) \to 0$$

is an exact sequence for every $n \geqslant 0$. In these circumstances there exists a unique extension of ϕ^0 to a homomorphism

$$\Phi : [T^n]_{n\geqslant 0} \to [U^n]_{n\geqslant 0}$$

of connected sequences.

Corollary. *Let $\phi^0 : T^0 \to U^0$ be a natural equivalence and suppose that, for every exact sequence $0 \to L \to P \to A \to 0$, with P projective, both sequences*

$$T^n(P) \to T^n(L) \to T^{n+1}(A) \to 0$$

and

$$U^n(P) \to U^n(L) \to U^{n+1}(A) \to 0$$

are exact for all $n \geqslant 0$. Then the unique extension of ϕ^0 to a homomorphism $[T^n]_{n\geqslant 0} \to [U^n]_{n\geqslant 0}$ is, in fact, an isomorphism of connected sequences.

Connected left sequences. In the previous discussion we have considered what have been called connected *right sequences*. We could, however, have started equally well with a sequence

$$\ldots, T_2(A), T_1(A), T_0(A) \tag{6.5.17}$$

of covariant functors from $\mathscr{G}_{\Lambda_1}$ to $\mathscr{G}_{\Lambda}$ and assumed that, for each exact sequence $0 \to A' \to A \to A'' \to 0$ and each $n \geqslant 0$, there was defined a connecting homomorphism

$$T_{n+1}(A'') \to T_n(A'). \tag{6.5.18}$$

This would have given rise to the semi-infinite sequence

$$\cdots \to T_{n+1}(A') \to T_{n+1}(A) \to T_{n+1}(A'') \to T_n(A') \to \cdots$$
$$\to T_1(A'') \to T_0(A') \to T_0(A) \to T_0(A''). \tag{6.5.19}$$

Definition. If, for every exact sequence $0 \to A' \to A \to A'' \to 0$, (6.5.19) is a 0-sequence, and if, as a 0-sequence, it is a covariant functor of $0 \to A' \to A \to A'' \to 0$, then $[T_n]_{n\geqslant 0}$ will be said to be a *connected left sequence of covariant functors*.

For contravariant functors there is a similar definition. To obtain this we have only to interchange A' and A'' in (6.5.18) and (6.5.19) and then, in the above definition, substitute 'contravariant' for 'covariant' wherever it occurs.

An important example of a connected left sequence arises as follows: if $T(A_1, A_2, \ldots, A_k)$ is an additive functor of modules and we keep all but one of the variables $A_1, A_2, \ldots, A_k$ fixed, then (with the usual notation for left-derived functors)

$$\ldots, L_2T(A_1, \ldots, A_k), L_1T(A_1, \ldots, A_k), L_0T(A_1, \ldots, A_k)$$

is a connected left sequence for the remaining variable.

Again if $[T_n]_{n\geqslant 0}$ and $[U_n]_{n\geqslant 0}$ are connected left sequences of covariant (or contravariant) functors, then by a *homomorphism*

$$[T_n]_{n\geqslant 0} \to [U_n]_{n\geqslant 0}$$

we mean a family $\Phi = [\phi_n]_{n\geq 0}$ of natural transformations $\phi_n : T_n \to U_n$ which commute with the connecting homomorphisms. If, in addition, each ϕ_n is a natural equivalence, then Φ is said to be an *isomorphism* of $[T_n]_{n\geq 0}$ on to $[U_n]_{n\geq 0}$. Corresponding to Theorem 10 and its corollary we have

Theorem 12. *Let $[T_n]_{n\geq 0}$ and $[U_n]_{n\geq 0}$ be connected left sequences of covariant functors from $\mathscr{G}_{\Lambda_1}$ to $\mathscr{G}_{\Lambda}$ and let $\phi_0 : T_0 \to U_0$ be a given natural transformation. Suppose further that whenever $0 \to L \to P \to A \to 0$ is an exact sequence with P a Λ_1-projective module, then*

$$0 \to U_{n+1}(A) \to U_n(L) \to U_n(P)$$

is an exact sequence for every $n \geq 0$. In these circumstances, ϕ_0 has a unique extension to a homomorphism

$$\Phi : [T_n]_{n\geq 0} \to [U_n]_{n\geq 0}$$

of connected sequences.

Corollary. *Let $\phi_0 : T_0 \to U_0$ be a natural equivalence and suppose that, for every exact sequence $0 \to L \to P \to A \to 0$ with P projective, both*

$$0 \to U_{n+1}(A) \to U_n(L) \to U_n(P)$$

and

$$0 \to T_{n+1}(A) \to T_n(L) \to T_n(P)$$

are exact sequences for every $n \geq 0$. Then the unique extension of ϕ_0 to a homomorphism $[T_n]_{n\geq 0} \to [U_n]_{n\geq 0}$ is, in fact, an isomorphism.

There is, of course, a form of Theorem 12 which is applicable to connected left sequences of contravariant functors, but, since we shall not have occasion to use this result, we shall not formulate it.

One final remark. The notion of a connected sequence of functors, as defined in this section, can be adapted, in an obvious manner, so as to apply to sequences

$$\ldots, T_2(A), T_1(A), T_0(A), T_{-1}(A), T_{-2}(A), \ldots$$

which are infinite in both directions. Later we shall have occasion to consider sequences of this type. For the moment, however, we shall merely take the opportunity to observe that, for such sequences, one can define homomorphisms and isomorphisms just as in the case of connected sequences of functors, which are infinite in only one direction.

7

TORSION AND EXTENSION FUNCTORS

Notation. As usual Λ denotes a ring (not necessarily commutative) with an identity element and Z denotes the ring of integers. In accordance with the terminology introduced in section (3.5), $\mathscr{G}_\Lambda^R$ and $\mathscr{G}_\Lambda^L$ will be used to signify the categories of right Λ-modules and left Λ-modules respectively, these two categories being regarded as the same when Λ is commutative. Finally, as in section (5.4), if A is a Λ-module then $\mathbf{A}$ will be used to denote the associated complex.

7.1 Torsion functors

The constructions described in the last chapter, when applied to $\otimes$ and Hom, yield some of the most important non-elementary functors in the theory of modules, namely, the so-called torsion and extension functors. It is our immediate intention to give an account of their fundamental properties and then to go on and show how they may be used to develop a 'dimension theory' for modules and rings. When we come, in a later chapter, to consider the particular case of commutative Noetherian rings, it will be found that the new concept of homological dimension has connexions with Hilbert's systems of 'syzygies' in the classical theory of polynomial rings.†

Let us start by recording the most obvious facts concerning $A \otimes_\Lambda B$. To begin with, it is an additive functor from $\mathscr{G}_\Lambda^R \times \mathscr{G}_\Lambda^L$ to $\mathscr{G}_Z$, which is covariant in both variables. However, should Λ happen to be commutative, the functor can be regarded as taking values in $\mathscr{G}_\Lambda$ and then it has the further property that, for arbitrary Λ-homomorphisms $f: A \to A'$ and $g: B \to B'$,

$$\lambda f \otimes g = \lambda(f \otimes g) = f \otimes \lambda g$$

for all λ in Λ.

Returning now to the non-commutative case, we propose to denote the *left-derived* functors of $A \otimes_\Lambda B$ by $\mathrm{Tor}_n^\Lambda(A, B)$, where n takes all integral values. By the general theory of Chapter 6, we can say at

† This is discussed in the notes on Chapter 9.

once that these are all additive functors (taking values in $\mathscr{G}_Z$) and they are all covariant in both variables. These functors will be called *torsion functors*. The origin of this name is to be found in an application to the theory of torsion modules over integral domains and, although we shall not be discussing this particular application, it is convenient to retain the name.

Now let $0 \to A' \to A \to A'' \to 0$ and $0 \to B' \to B \to B'' \to 0$ be exact sequences of Λ-modules and Λ-homomorphisms. Then, by Theorem 7 of section (6.4), these give rise to exact sequences

$$\cdots \to \mathrm{Tor}_n^\Lambda(A', B) \to \mathrm{Tor}_n^\Lambda(A, B) \to \mathrm{Tor}_n^\Lambda(A'', B) \to \mathrm{Tor}_{n-1}^\Lambda(A', B)$$
$$\cdots \to \mathrm{Tor}_1^\Lambda(A'', B) \to \mathrm{Tor}_0^\Lambda(A', B) \to \mathrm{Tor}_0^\Lambda(A, B) \to \mathrm{Tor}_0^\Lambda(A'', B) \to 0 \tag{7.1.1}$$

and

$$\cdots \to \mathrm{Tor}_n^\Lambda(A, B') \to \mathrm{Tor}_n^\Lambda(A, B) \to \mathrm{Tor}_n^\Lambda(A, B'') \to \mathrm{Tor}_{n-1}^\Lambda(A, B')$$
$$\cdots \to \mathrm{Tor}_1^\Lambda(A, B'') \to \mathrm{Tor}_0^\Lambda(A, B') \to \mathrm{Tor}_0^\Lambda(A, B) \to \mathrm{Tor}_0^\Lambda(A, B'') \to 0. \tag{7.1.2}$$

In addition, these sequences have certain translation properties so that, if

$$\begin{array}{ccccccccc} 0 & \longrightarrow & A' & \longrightarrow & A & \longrightarrow & A'' & \longrightarrow & 0 \\ & & \downarrow & & \downarrow & & \downarrow & & \\ 0 & \longrightarrow & A_1' & \longrightarrow & A_1 & \longrightarrow & A_1'' & \longrightarrow & 0 \end{array}$$

and

$$\begin{array}{ccccccccc} 0 & \longrightarrow & B' & \longrightarrow & B & \longrightarrow & B'' & \longrightarrow & 0 \\ & & \downarrow & & \downarrow & & \downarrow & & \\ 0 & \longrightarrow & B_1' & \longrightarrow & B_1 & \longrightarrow & B_1'' & \longrightarrow & 0 \end{array}$$

are commutative diagrams over Λ, with exact rows, then, for every value of n, all the diagrams

$$\begin{array}{ccc} \mathrm{Tor}_n^\Lambda(A'', B) & \longrightarrow & \mathrm{Tor}_{n-1}^\Lambda(A', B) \\ \downarrow & & \downarrow \\ \mathrm{Tor}_n^\Lambda(A_1'', B) & \longrightarrow & \mathrm{Tor}_{n-1}^\Lambda(A_1', B), \end{array}$$

$$\begin{array}{ccc} \mathrm{Tor}_n^\Lambda(A'', B) & \longrightarrow & \mathrm{Tor}_{n-1}^\Lambda(A', B) \\ \downarrow & & \downarrow \\ \mathrm{Tor}_n^\Lambda(A'', B_1) & \longrightarrow & \mathrm{Tor}_{n-1}^\Lambda(A', B_1), \end{array}$$

$$\begin{array}{ccc} \operatorname{Tor}_n^\Lambda(A, B'') & \longrightarrow & \operatorname{Tor}_{n-1}^\Lambda(A, B') \\ \downarrow & & \downarrow \\ \operatorname{Tor}_n^\Lambda(A_1, B'') & \longrightarrow & \operatorname{Tor}_{n-1}^\Lambda(A_1, B') \end{array}$$

and
$$\begin{array}{ccc} \operatorname{Tor}_n^\Lambda(A, B'') & \longrightarrow & \operatorname{Tor}_{n-1}^\Lambda(A, B') \\ \downarrow & & \downarrow \\ \operatorname{Tor}_n^\Lambda(A, B_1'') & \longrightarrow & \operatorname{Tor}_{n-1}^\Lambda(A, B_1') \end{array}$$

are commutative.

Consider, for a moment, the special case in which Λ is commutative. The values of $\operatorname{Tor}_n^\Lambda(A, B)$ are then all in the category of Λ-modules and the mappings in (7.1.1) and (7.1.2) are all Λ-homomorphisms. Moreover,† for every n and every λ in Λ,

$$\operatorname{Tor}_n^\Lambda(\lambda f, g) = \lambda \operatorname{Tor}_n^\Lambda(f, g) = \operatorname{Tor}_n^\Lambda(f, \lambda g), \qquad (7.1.3)$$

where f and g are arbitrary Λ-homomorphisms.

7.2 Basic properties of torsion functors

We now continue the general discussion so that Λ is no longer assumed to be commutative. Unless otherwise stated, throughout this section A will denote a typical right Λ-module and B a typical left Λ-module.

Theorem 1. *There exists a canonical natural (functor) equivalence* $\operatorname{Tor}_0^\Lambda(A, B) \approx A \otimes_\Lambda B$.

Proof. By Theorem 4 of section (2.4), $A \otimes_\Lambda B$ is a right exact functor and now Theorem 1 is seen to be a special case of Theorem 9 of section (6.4).

It is convenient to recall here, in preparation for the proof of Theorem 2, just how the module isomorphism $\operatorname{Tor}_0^\Lambda(A, B) \approx A \otimes_\Lambda B$ works. Accordingly, let $\mathbf{X}$ and $\mathbf{Y}$ be projective resolutions of A and B respectively and let $\mathbf{X} \to \mathbf{A}$ and $\mathbf{Y} \to \mathbf{B}$ be the augmentation translations. These give rise to a translation $\mathbf{X} \otimes_\Lambda \mathbf{Y} \to \mathbf{A} \otimes_\Lambda \mathbf{B}$ and hence to a homomorphism

$$H_0(\mathbf{X} \otimes_\Lambda \mathbf{Y}) \to H_0(\mathbf{A} \otimes_\Lambda \mathbf{B}), \qquad (7.2.1)$$

and it is essentially (7.2.1) which turns out to be,‡ with a different notation, the isomorphism

$$\operatorname{Tor}_0^\Lambda(A, B) \approx A \otimes_\Lambda B.$$

† See the remark following Theorem 6 of section (6.4).
‡ See (6.4.16).

Theorem 1 shows that we can replace $\mathrm{Tor}_0^\Lambda(A, B)$ by $A \otimes_\Lambda B$ whenever this is convenient. In particular, if $0 \to A' \to A \to A'' \to 0$ and $0 \to B' \to B \to B'' \to 0$ are exact sequences, then we can substitute for (7.1.1) and (7.1.2) the equivalent exact sequences

$$\cdots \to \mathrm{Tor}_1^\Lambda(A', B) \to \mathrm{Tor}_1^\Lambda(A, B) \to \mathrm{Tor}_1^\Lambda(A'', B) \to A' \otimes_\Lambda B \to A \otimes_\Lambda B \to A'' \otimes_\Lambda B \to 0 \quad (7.2.2)$$

and

$$\cdots \to \mathrm{Tor}_1^\Lambda(A, B') \to \mathrm{Tor}_1^\Lambda(A, B) \to \mathrm{Tor}_1^\Lambda(A, B'') \to A \otimes_\Lambda B' \to A \otimes_\Lambda B \to A \otimes_\Lambda B'' \to 0. \quad (7.2.3)$$

Proposition 1. *If either A or B is Λ-projective then* $\mathrm{Tor}_n^\Lambda(A, B) = 0$ *for all* $n \geq 1$.

Proof. We shall only consider the case in which A is Λ-projective, since the other case can be treated similarly. Assume, therefore, that A is Λ-projective and let $\mathbf{A}$ be the associated complex, then this complex, together with the identity translation $\mathbf{A} \to \mathbf{A}$ is a projective resolution of A. Consequently, if $\mathbf{P}$ is a projective resolution of B, $\mathrm{Tor}_n^\Lambda(A, B) \approx H_n(\mathbf{A} \otimes_\Lambda \mathbf{P})$. But, by (6.2.20), $\mathbf{A} \otimes_\Lambda \mathbf{P}$ is the same as $A \otimes_\Lambda \mathbf{P}$ and therefore it is none other than

$$\cdots \to A \otimes_\Lambda P_n \to A \otimes_\Lambda P_{n-1} \to \cdots \to A \otimes_\Lambda P_0 \to 0 \to \cdots.$$

Moreover, if $n \geq 1$, $A \otimes_\Lambda P_{n+1} \to A \otimes_\Lambda P_n \to A \otimes_\Lambda P_{n-1}$ is exact because, by Theorem 4 of section (5.1), $A \otimes_\Lambda B$ is exact as a functor of B. Thus $H_n(\mathbf{A} \otimes_\Lambda \mathbf{P}) = 0$ provided that $n \geq 1$, and this completes the proof.

Theorem 2. *Let $S_n(A) = \mathrm{Tor}_n^\Lambda(A, B)$ and let $U(A) = A \otimes_\Lambda B$, where B is regarded as a fixed module. Then (with the usual notation for left-derived functors) $[S_n]_{n \geq 0}$ and $[L_n U]_{n \geq 0}$ are isomorphic connected left sequences of covariant functors.*

Remarks. Besides showing that the two connected sequences are isomorphic, we shall actually exhibit (see (7.2.5)) an explicit family $[\phi_n : S_n \approx L_n U]_{n \geq 0}$ of natural equivalences, which gives an isomorphism. Moreover, the same equivalences ϕ_n will be used again in Theorem 5 when the $\mathrm{Tor}_n^\Lambda(A, B)$ are considered from the point of view of their behaviour in the second variable. It follows from this that, by analysing the proofs of Theorems 2 and 5, one can arrive at a different way of defining torsion functors. However, although this other method is simpler in that it uses only the rather obvious con structions of section (6.1), as compared with the more sophisticated ones of section (6.2), it suffers from the disadvantage of putting the two variables on different footings.

Proof. Let $\mathbf{X}$ and $\mathbf{Y}$ be projective resolutions of A and B respectively, then, from the augmentation map $\mathbf{Y} \to \mathbf{B}$, is obtained a translation

$$\mathbf{X} \otimes_\Lambda \mathbf{Y} \to \mathbf{X} \otimes_\Lambda \mathbf{B} = \mathbf{X} \otimes_\Lambda B = U(\mathbf{X}). \tag{7.2.4}$$

Passing now to the homology modules there exists, for each n, a homomorphism

$$\phi_n(A) : S_n(A) = H_n(\mathbf{X} \otimes_\Lambda \mathbf{Y}) \to H_n(U(\mathbf{X})) = L_n U(A). \tag{7.2.5}$$

We shall show that these mappings set up the required isomorphism between the connected sequences.

First suppose that $A \to A^*$ is a Λ-homomorphism, that $\mathbf{X}^*$ is a projective resolution of A^* and that $\mathbf{X} \to \mathbf{X}^*$ is a translation over $A \to A^*$. Since the combined mappings

$$\mathbf{X} \otimes_\Lambda \mathbf{Y} \to \mathbf{X} \otimes_\Lambda \mathbf{B} \to \mathbf{X}^* \otimes_\Lambda \mathbf{B}$$

and

$$\mathbf{X} \otimes_\Lambda \mathbf{Y} \to \mathbf{X}^* \otimes_\Lambda \mathbf{Y} \to \mathbf{X}^* \otimes_\Lambda \mathbf{B}$$

are the same, it follows that, for each n, the diagram

$$\begin{array}{ccc} S_n(A) & \xrightarrow{\phi_n(A)} & L_n U(A) \\ \downarrow & & \downarrow \\ S_n(A^*) & \xrightarrow{\phi_n(A^*)} & L_n U(A^*) \end{array}$$

is commutative. In other words, $\phi_n : S_n \to L_n U$ is a natural transformation of functors.

We next suppose that $0 \to A' \to A \to A'' \to 0$ is an exact sequence and construct† over this a split exact sequence $\mathbf{0} \to \mathbf{X}' \to \mathbf{X} \to \mathbf{X}'' \to \mathbf{0}$, where $\mathbf{X}'$, $\mathbf{X}$, $\mathbf{X}''$ are projective resolutions of A', A, A'' respectively. Then in the commutative diagram

$$\begin{array}{ccccccccc} \mathbf{0} & \longrightarrow & \mathbf{X}' \otimes_\Lambda \mathbf{Y} & \longrightarrow & \mathbf{X} \otimes_\Lambda \mathbf{Y} & \longrightarrow & \mathbf{X}'' \otimes_\Lambda \mathbf{Y} & \longrightarrow & \mathbf{0} \\ & & \downarrow & & \downarrow & & \downarrow & & \\ \mathbf{0} & \longrightarrow & \mathbf{X}' \otimes_\Lambda \mathbf{B} & \longrightarrow & \mathbf{X} \otimes_\Lambda \mathbf{B} & \longrightarrow & \mathbf{X}'' \otimes_\Lambda \mathbf{B} & \longrightarrow & \mathbf{0} \end{array}$$

the rows are exact, the lower row being none other than

$$\mathbf{0} \to U(\mathbf{X}') \to U(\mathbf{X}) \to U(\mathbf{X}'') \to \mathbf{0}.$$

Theorem 5 of section (4.6) now shows that

$$\begin{array}{ccc} S_n(A'') & \longrightarrow & S_{n-1}(A') \\ \downarrow{\scriptstyle \phi_n(A'')} & & \downarrow{\scriptstyle \phi_{n-1}(A')} \\ L_n U(A'') & \longrightarrow & L_{n-1} U(A') \end{array}$$

† See the remarks following Theorem 16 of section (5.6).

is commutative and therefore $[\phi_n]_{n\geq 0}$ is a homomorphism

$$[S_n]_{n\geq 0} \to [L_n U]_{n\geq 0}$$

of connected sequences. It will now be shown that this homomorphism is an isomorphism.

Consider the diagram

$$\begin{array}{ccc} \mathbf{X}\otimes_\Lambda \mathbf{Y} & \to & \mathbf{A}\otimes_\Lambda \mathbf{B} \\ \downarrow & & \downarrow \\ \mathbf{X}\otimes_\Lambda \mathbf{B} & \to & \mathbf{A}\otimes_\Lambda \mathbf{B} \end{array}$$

This is commutative and, since $H_0(\mathbf{A}\otimes_\Lambda \mathbf{B}) = A\otimes_\Lambda B = U(A)$, it gives rise to a further commutative diagram, namely,

$$\begin{array}{ccc} \mathrm{Tor}_0^\Lambda(A,B) & \longrightarrow & A\otimes_\Lambda B \\ \downarrow{\scriptstyle \phi_0(A)} & & \downarrow \\ L_0 U(A) & \longrightarrow & U(A) \end{array}$$

where the right vertical map is an identity. The upper horizontal map is the isomorphism of Theorem 1 and, since $U(A)$ is right exact, the lower horizontal map is, by Theorem 9 of section (6.4), also an isomorphism. Accordingly, $\phi_0(A)$ is a module isomorphism and therefore $\phi_0 : S_0 \to L_0 U$ is a natural equivalence.

Finally, let $0\to M\to P\to A\to 0$ be an exact sequence with P a Λ-projective module. Then, for each $n\geq 0$,

$$\mathrm{Tor}_{n+1}^\Lambda(P,B)\to \mathrm{Tor}_{n+1}^\Lambda(A,B)\to \mathrm{Tor}_n^\Lambda(M,B)\to \mathrm{Tor}_n^\Lambda(P,B)$$

is exact, and therefore, by Proposition 1,

$$0\to S_{n+1}(A)\to S_n(M)\to S_n(P)$$

is also exact. Again

$$L_{n+1}U(P)\to L_{n+1}U(A)\to L_n U(M)\to L_n U(P)$$

is exact, and now the required result will follow (by the corollary to Theorem 12 of section (6.5)) if we show that $L_{n+1}U(P) = 0$. But $L_{n+1}U(P) \approx H_{n+1}(U(\mathbf{P}))$, and this is null† because of the degenerate character of $U(\mathbf{P})$.

The next theorem, which is concerned with the way in which $\mathrm{Tor}_n^\Lambda(A,B)$ $(n=0,1,2,\ldots)$ depends on B, instead of on A, can be established by the same methods.

Theorem 3. *If $T_n(B) = \mathrm{Tor}_n^\Lambda(A,B)$ and $V(B) = A\otimes_\Lambda B$, then, when A is kept fixed, $[T_n]_{n\geq 0}$ and $[L_n V]_{n\geq 0}$ are isomorphic as connected left sequences of covariant functors.*

† Observe that $\mathbf{P}$ is a projective resolution of P.

Theorem 4. *Let* $\mathbf{X}$ *be a projective resolution of* A *and* $\mathbf{Y}$ *a projective resolution of* B. *Then there exist well-defined isomorphisms*

$$H_n(\mathbf{X} \otimes_\Lambda B) \approx \mathrm{Tor}_n^\Lambda(A, B) \approx H_n(A \otimes_\Lambda \mathbf{Y}) \quad (n \geqslant 0),$$

which are Λ*-isomorphisms if* Λ *is commutative.*

Remark. It should be noted, because we shall use this fact later on a number of occasions, that the theorem shows that $\mathrm{Tor}_n^\Lambda(A, B)$ can be computed as the nth homology module of either of the complexes

$$\cdots \to X_{n+1} \otimes_\Lambda B \to X_n \otimes_\Lambda B \to X_{n-1} \otimes_\Lambda B \to \cdots$$

and $$\cdots \to A \otimes_\Lambda Y_{n+1} \to A \otimes_\Lambda Y_n \to A \otimes_\Lambda Y_{n-1} \to \cdots.$$

Proof. With the notation of Theorem 2 we have

$$H_n(\mathbf{X} \otimes_\Lambda B) \approx L_n U(A) \approx S_n(A) = \mathrm{Tor}_n^\Lambda(A, B),$$

and, in a similar manner, Theorem 3 yields an isomorphism

$$\mathrm{Tor}_n^\Lambda(A, B) \approx H_n(A \otimes_\Lambda \mathbf{Y}).$$

Theorem 3 gives an alternative description of the functors $\mathrm{Tor}_n^\Lambda(A, B)$ when they are regarded as forming a connected sequence in the variable B. But the same connected sequence can be thought of in yet another way.

Let $\mathbf{X}$ be a *fixed* projective resolution of A, then every Λ-homomorphism $B \to B^*$ (say) determines a translation $\mathbf{X} \otimes_\Lambda B \to \mathbf{X} \otimes_\Lambda B^*$ and hence a homomorphism $H_n(\mathbf{X} \otimes_\Lambda B) \to H_n(\mathbf{X} \otimes_\Lambda B^*)$. In this way we can consider $H_n(\mathbf{X} \otimes_\Lambda B)$ as a functor of B. Further, if

$$0 \to B' \to B \to B'' \to 0$$

is an exact sequence, then, by Theorem 4 of section (5.1),

$$\mathbf{0} \to \mathbf{X} \otimes_\Lambda B' \to \mathbf{X} \otimes_\Lambda B \to \mathbf{X} \otimes_\Lambda B'' \to \mathbf{0}$$

is also exact and therefore it can be used to define connecting homomorphisms $H_n(\mathbf{X} \otimes_\Lambda B'') \to H_{n-1}(\mathbf{X} \otimes_\Lambda B')$. It follows that *we can regard* $[H_n(\mathbf{X} \otimes_\Lambda B)]_{n \geqslant 0}$ *as a connected sequence of covariant functors in the variable* B.

Theorem 5. *Let* $T_n(B) = \mathrm{Tor}_n^\Lambda(A, B)$ *and let* $\mathbf{X}$ *be a fixed projective resolution of* A. *Then the connected sequences (in the variable* B*) formed by the* $T_n(B)$ *and the* $H_n(\mathbf{X} \otimes_\Lambda B)$ *are isomorphic.*

Remarks. There is, of course, a similar result in which the roles of the modules A and B are interchanged, but we do not need to state it separately. A less obvious point is that the *same* isomorphisms $\mathrm{Tor}_n^\Lambda(A,B) \approx H_n(\mathbf{X} \otimes_\Lambda B)$, as were used in Theorem 2 (see (7.2.5)) to establish an isomorphism of connected sequences in the first variable, give the required mappings in the present instance.

Proof. Let $\mathbf{Y}$ be a projective resolution of B, then the augmentation translation $\mathbf{Y} \to \mathbf{B}$ determines a translation $\mathbf{X} \otimes_\Lambda \mathbf{Y} \to \mathbf{X} \otimes_\Lambda \mathbf{B}$ and thence a homomorphism

$$\mathrm{Tor}_n^\Lambda(A,B) \to H_n(\mathbf{X} \otimes_\Lambda \mathbf{B}) = H_n(\mathbf{X} \otimes_\Lambda B), \tag{7.2.6}$$

this being the homomorphism which was denoted by $\phi_n(A)$ in the proof of Theorem 2. It is clear that (7.2.6) is a natural transformation of the functors of B under consideration and we know, from Theorem 2, that the mapping is an isomorphism for each value of n. To complete the proof we need only show that these isomorphisms commute with the connecting homomorphisms associated with an exact sequence $0 \to B' \to B \to B'' \to 0$.

Construct†, over this sequence, a split exact sequence

$$\mathbf{0} \to \mathbf{Y}' \to \mathbf{Y} \to \mathbf{Y}'' \to \mathbf{0},$$

where $\mathbf{Y}'$, $\mathbf{Y}$, $\mathbf{Y}''$ are projective resolutions of B', B, B'' respectively. Then

$$\begin{array}{ccccccccc}
\mathbf{0} & \longrightarrow & \mathbf{X} \otimes_\Lambda \mathbf{Y}' & \longrightarrow & \mathbf{X} \otimes_\Lambda \mathbf{Y} & \longrightarrow & \mathbf{X} \otimes_\Lambda \mathbf{Y}'' & \longrightarrow & \mathbf{0} \\
 & & \downarrow & & \downarrow & & \downarrow & & \\
\mathbf{0} & \longrightarrow & \mathbf{X} \otimes_\Lambda \mathbf{B}' & \longrightarrow & \mathbf{X} \otimes_\Lambda \mathbf{B} & \longrightarrow & \mathbf{X} \otimes_\Lambda \mathbf{B}'' & \longrightarrow & \mathbf{0}
\end{array}$$

is a commutative diagram with exact rows, whence, passing to the homology modules, we see that the diagram

$$\begin{array}{ccc}
\mathrm{Tor}_n^\Lambda(A,B'') & \longrightarrow & \mathrm{Tor}_{n-1}^\Lambda(A,B') \\
\downarrow & & \downarrow \\
H_n(\mathbf{X} \otimes_\Lambda B'') & \longrightarrow & H_{n-1}(\mathbf{X} \otimes_\Lambda B')
\end{array}$$

is commutative for each value of n. Thus our isomorphisms do commute with the connecting homomorphisms and so the proof is complete.

7.3 Extension functors

Throughout the present section, unless there is an explicit statement to the contrary, A and B will both denote *left* Λ-modules. In this case $\mathrm{Hom}_\Lambda(A,B)$ is an additive functor and, as such, gives rise to new functors when we apply the methods and results of the last chapter.

† See the remarks following Theorem 16 of section (5.6).

Of particular importance are its right-derived functors, which are denoted by $\operatorname{Ext}^n_\Lambda(A,B)$ $(n=0,1,2,\ldots)$ and which it is customary to call *extension functors*. The name is, in fact, used because $\operatorname{Ext}^1_\Lambda(A,B)$ provides a model of the set of solutions of the so-called 'extension problem' in which one seeks to find all essentially distinct exact sequences of the form $0\to B\to X\to A\to 0$. It is not our intention to give an account of this theory, and we mention it merely to indicate the origin of what is now standard terminology.

In order to familiarize ourselves with the new notation, we shall restate some of the results of Chapter 6 for the special case which concerns us now. To begin with, for each n, $\operatorname{Ext}^n_\Lambda(A,B)$ is an additive functor contravariant in A and covariant in B. To describe the connexions between the functors which correspond to different values of n, let

$$\begin{array}{ccccccccc} 0 & \longrightarrow & A' & \longrightarrow & A & \longrightarrow & A'' & \longrightarrow & 0 \\ & & \downarrow & & \downarrow & & \downarrow & & \\ 0 & \longrightarrow & A_1' & \longrightarrow & A_1 & \longrightarrow & A_1'' & \longrightarrow & 0 \end{array} \qquad (7.3.1)$$

and

$$\begin{array}{ccccccccc} 0 & \longrightarrow & B' & \longrightarrow & B & \longrightarrow & B'' & \longrightarrow & 0 \\ & & \downarrow & & \downarrow & & \downarrow & & \\ 0 & \longrightarrow & B_1' & \longrightarrow & B_1 & \longrightarrow & B_1'' & \longrightarrow & 0 \end{array}$$

be commutative diagrams (over Λ) with exact rows. The upper rows in the diagrams determine exact sequences

$$\begin{gathered} 0\to \operatorname{Ext}^0_\Lambda(A'',B)\to \operatorname{Ext}^0_\Lambda(A,B)\to \operatorname{Ext}^0_\Lambda(A',B)\to \operatorname{Ext}^1_\Lambda(A'',B)\to\cdots \\ \to \operatorname{Ext}^n_\Lambda(A'',B)\to \operatorname{Ext}^n_\Lambda(A,B)\to \operatorname{Ext}^n_\Lambda(A',B)\to \operatorname{Ext}^{n+1}_\Lambda(A'',B)\to\cdots \end{gathered} \qquad (7.3.2)$$

and

$$\begin{gathered} 0\to \operatorname{Ext}^0_\Lambda(A,B')\to \operatorname{Ext}^0_\Lambda(A,B)\to \operatorname{Ext}^0_\Lambda(A,B'')\to \operatorname{Ext}^1_\Lambda(A,B')\to\cdots \\ \to \operatorname{Ext}^n_\Lambda(A,B')\to \operatorname{Ext}^n_\Lambda(A,B)\to \operatorname{Ext}^n_\Lambda(A,B'')\to \operatorname{Ext}^{n+1}_\Lambda(A,B')\to\cdots, \end{gathered} \qquad (7.3.3)$$

while the translations (7.3.1) give rise to commutative diagrams

$$\begin{array}{ccc} \operatorname{Ext}^n_\Lambda(A_1',B) & \longrightarrow & \operatorname{Ext}^{n+1}_\Lambda(A_1'',B) \\ \downarrow & & \downarrow \\ \operatorname{Ext}^n_\Lambda(A',B) & \longrightarrow & \operatorname{Ext}^{n+1}_\Lambda(A'',B), \end{array}$$

$$\begin{array}{ccc} \operatorname{Ext}^n_\Lambda(A',B) & \longrightarrow & \operatorname{Ext}^{n+1}_\Lambda(A'',B) \\ \downarrow & & \downarrow \\ \operatorname{Ext}^n_\Lambda(A',B_1) & \longrightarrow & \operatorname{Ext}^{n+1}_\Lambda(A'',B_1), \end{array}$$

$$\begin{array}{ccc}\mathrm{Ext}^n_\Lambda(A_1,B'') & \longrightarrow & \mathrm{Ext}^{n+1}_\Lambda(A_1,B') \\ \downarrow & & \downarrow \\ \mathrm{Ext}^n_\Lambda(A,B'') & \longrightarrow & \mathrm{Ext}^{n+1}_\Lambda(A,B')\end{array}$$

and

$$\begin{array}{ccc}\mathrm{Ext}^n_\Lambda(A,B'') & \longrightarrow & \mathrm{Ext}^{n+1}_\Lambda(A,B') \\ \downarrow & & \downarrow \\ \mathrm{Ext}^n_\Lambda(A,B''_1) & \longrightarrow & \mathrm{Ext}^{n+1}_\Lambda(A,B'_1).\end{array}$$

So far we have spoken only of the case where A and B are both left Λ-modules. If we consider instead right Λ-modules then $\mathrm{Hom}_\Lambda(A,B)$ is still an additive functor and its typical right-derived functor is again denoted by $\mathrm{Ext}^n_\Lambda(A,B)$. The new Ext^n_Λ are functors of right Λ-modules and enjoy properties entirely similar to those just described. In other words, *we have one system of extension functors for left Λ-modules and a second system for right Λ-modules.* Usually the context makes it quite clear which system we are talking about, but there are situations in which it is important to remember that Ext^n_Λ has two essentially different meanings.

If Λ is commutative then, of course, we do not have to distinguish between extension functors for *left* Λ-modules and extension functors for *right* Λ-modules. Let us consider the commutative case in a little more detail. In this situation, $\mathrm{Hom}_\Lambda(A,B)$ takes values in $\mathscr{G}_\Lambda$ and, besides being additive, is Λ-linear. These properties, as we know, will be transmitted to its right-derived functors. More explicitly $\mathrm{Ext}^n_\Lambda(A,B)$ will be a Λ-module and the mappings in (7.3.2) and (7.3.3) will be Λ-homomorphisms. Again, if f and g are Λ-homomorphisms and λ belongs to Λ, then, for every value of n, the relations

$$\mathrm{Ext}^n_\Lambda(\lambda f,g) = \lambda\,\mathrm{Ext}^n_\Lambda(f,g) = \mathrm{Ext}^n_\Lambda(f,\lambda g) \qquad (7.3.4)$$

are satisfied.

7.4 Basic properties of extension functors

Throughout the present section we return to the non-commutative case and suppose that A and B are *left* Λ-modules. We are therefore concerned with properties of extension functors of left modules, but precisely similar results will, of course, hold for the extension functors of right modules.

The first of our results is an immediate consequence of Theorem 5 of section (6.3) when it is recalled that $\mathrm{Hom}_\Lambda(A,B)$ is a left exact functor.

Theorem 6. *The functors* $\mathrm{Hom}_\Lambda(A, B)$ *and* $\mathrm{Ext}^0_\Lambda(A, B)$ *are naturally equivalent.*

Indeed, to be more explicit, if $\mathbf{X}$ is a projective resolution of A and $\mathbf{Y}$ an injective resolution of B, then the augmentation maps $\mathbf{X} \to \mathbf{A}$ and $\mathbf{B} \to \mathbf{Y}$ give rise to a translation

$$\mathrm{Hom}_\Lambda(\mathbf{A}, \mathbf{B}) \to \mathrm{Hom}_\Lambda(\mathbf{X}, \mathbf{Y}).$$

This, in turn, produces homomorphisms of the homology modules and, in particular, a homomorphism $H_0(\mathrm{Hom}_\Lambda(\mathbf{A}, \mathbf{B})) \to H_0(\mathrm{Hom}_\Lambda(\mathbf{X}, \mathbf{Y}))$. But $H_0(\mathrm{Hom}_\Lambda(\mathbf{A}, \mathbf{B})) = \mathrm{Hom}_\Lambda(A, B)$, while

$$H_0(\mathrm{Hom}_\Lambda(\mathbf{X}, \mathbf{Y})) = \mathrm{Ext}^0_\Lambda(A, B),$$

so that, in fact, we have a mapping $\mathrm{Hom}_\Lambda(A, B) \to \mathrm{Ext}^0_\Lambda(A, B)$. Theorem 5 of section (6.3) allows us to infer that this mapping is an isomorphism, and it is this isomorphism which constitutes the natural equivalence of functors described in the theorem.

We shall often take advantage of Theorem 6 to replace the exact sequences (7.3.2) and (7.3.3) by the equivalent exact sequences

$$0 \to \mathrm{Hom}_\Lambda(A'', B) \to \mathrm{Hom}_\Lambda(A, B) \to \mathrm{Hom}_\Lambda(A', B) \to \mathrm{Ext}^1_\Lambda(A'', B) \to \cdots \tag{7.4.1}$$

and

$$0 \to \mathrm{Hom}_\Lambda(A, B') \to \mathrm{Hom}_\Lambda(A, B) \to \mathrm{Hom}_\Lambda(A, B'') \to \mathrm{Ext}^1_\Lambda(A, B') \to \cdots \tag{7.4.2}$$

Proposition 2. *If either A is Λ-projective or B is Λ-injective then, for all $n \geqslant 1$,* $\mathrm{Ext}^n_\Lambda(A, B) = 0$.

Proof. The proof of this result is very similar to that of Proposition 1 and therefore we shall not give details. The reader will find no difficulty in adapting the original proof if he bears in mind that, by Theorem 3 of section (5.1) and Theorem 7 of section (5.2), $\mathrm{Hom}_\Lambda(A, B)$ is an exact functor of B when A is Λ-projective and an exact functor of A when B is Λ-injective.

Theorem 7. *If* $S^n(A) = \mathrm{Ext}^n_\Lambda(A, B)$ *and* $U(A) = \mathrm{Hom}_\Lambda(A, B)$, *then, with the usual notation for right-derived functors,* $[S^n]_{n\geqslant 0}$ *and* $[R^n U]_{n\geqslant 0}$ *are isomorphic as connected right sequences of contravariant functors of A.*

Remarks. It is convenient to make a note of the way in which the required isomorphism is obtained. Let $\mathbf{X}$ be a projective resolution of

A and $\mathbf{Y}$ an injective resolution of B. Then, from the augmentation translation $\mathbf{B} \to \mathbf{Y}$, is obtained a further translation

$$\mathrm{Hom}_\Lambda(\mathbf{X}, B) = \mathrm{Hom}_\Lambda(\mathbf{X}, \mathbf{B}) \to \mathrm{Hom}_\Lambda(\mathbf{X}, \mathbf{Y})$$

which, on passing to the homology modules, gives homomorphisms $H^n(\mathrm{Hom}_\Lambda(\mathbf{X}, B)) \to \mathrm{Ext}^n_\Lambda(A, B)$. These, using the notation of the theorem, we write as $\phi^n(A) : R^n U(A) \to S^n(A)$, and it will be shown that the $\phi^n(A)$ are isomorphisms of modules which determine an isomorphism of connected sequences.

Proof. The proof is similar in many ways to that of Theorem 2, and therefore we shall dwell at length only on those points at which there are differences. We begin by verifying (using the same methods as were explained in detail in the discussion of Theorem 2) that the homomorphisms $\phi^n(A)$ constitute a homomorphism $\Phi : [R^n U]_{n \geq 0} \to [S^n]_{n \geq 0}$ of connected sequences. Next we observe that the commutative diagram

$$\begin{array}{ccc} \mathrm{Hom}_\Lambda(\mathbf{A}, \mathbf{B}) & \longrightarrow & \mathrm{Hom}_\Lambda(\mathbf{X}, \mathbf{B}) \\ \downarrow & & \downarrow \\ \mathrm{Hom}_\Lambda(\mathbf{A}, \mathbf{B}) & \longrightarrow & \mathrm{Hom}_\Lambda(\mathbf{X}, \mathbf{Y}) \end{array}$$

yields a further commutative diagram, namely,

$$\begin{array}{ccc} U(A) & \longrightarrow & R^0 U(A) \\ \downarrow & & \downarrow \\ \mathrm{Hom}_\Lambda(A, B) & \longrightarrow & \mathrm{Ext}^0_\Lambda(A, B) \end{array}$$

in which the horizontal mappings are isomorphisms (because $\mathrm{Hom}_\Lambda(A, B)$ is left exact) and the left vertical map is an identity. It follows that $\phi^0(A)$ is an isomorphism and therefore $\phi^0 : R^0 U \to S^0$ is a natural equivalence.

Finally, suppose that $0 \to M \to P \to A \to 0$ is an exact sequence, where P is Λ-projective, then, if $n \geq 0$, the sequences

$$S^n(P) \to S^n(M) \to S^{n+1}(A) \to S^{n+1}(P)$$

and
$$R^n U(P) \to R^n U(M) \to R^{n+1} U(A) \to R^{n+1} U(P)$$

are exact. But $S^{n+1}(P) = \mathrm{Ext}^{n+1}_\Lambda(P, B) = 0$ (by Proposition 2) while $R^{n+1} U(P) \approx H^{n+1}(U(\mathbf{P})) = 0$ by the degenerate character of the complex $U(\mathbf{P})$. The corollary to Theorem 11 of section (6.5) now shows that the homomorphism $[R^n U]_{n \geq 0} \to [S^n]_{n \geq 0}$ is in fact an isomorphism.

Theorem 8. *If $T^n(B) = \mathrm{Ext}^n_\Lambda(A, B)$ and $V(B) = \mathrm{Hom}_\Lambda(A, B)$, then, with the usual notation for right-derived functors, $[T^n]_{n \geq 0}$ and $[R^n V]_{n \geq 0}$ are isomorphic as connected right sequences of covariant functors of B.*

Proof. This is a minor variation of the proofs of Theorems 2 and 7 and so we omit details. The isomorphism of the connected sequences is established, on this occasion, by using the corollary to Theorem 10 of section (6.5).

Theorem 9. *If* $\mathbf{X}$ *is a projective resolution of* A *and* $\mathbf{Y}$ *an injective resolution of* B*, then there exist well-defined isomorphisms*

$$H^n(\mathrm{Hom}_\Lambda(\mathbf{X}, B)) \approx \mathrm{Ext}^n_\Lambda(A, B) \approx H^n(\mathrm{Hom}_\Lambda(A, \mathbf{Y}))$$

for each value of n. *When* Λ *is commutative these isomorphisms are* Λ*-isomorphisms.*

Remarks. The theorem shows, of course, that $\mathrm{Ext}^n_\Lambda(A, B)$ can be computed as the nth homology module of either of the complexes

$$\cdots \to \mathrm{Hom}_\Lambda(X_{n-1}, B) \to \mathrm{Hom}_\Lambda(X_n, B) \to \mathrm{Hom}_\Lambda(X_{n+1}, B) \to \cdots$$

and

$$\cdots \to \mathrm{Hom}_\Lambda(A, Y^{n-1}) \to \mathrm{Hom}_\Lambda(A, Y^n) \to \mathrm{Hom}_\Lambda(A, Y^{n+1}) \to \cdots.$$

Proof. With the notation of Theorem 7,

$$H^n(\mathrm{Hom}_\Lambda(\mathbf{X}, B)) \approx R^n U(A) \approx S^n(A) = \mathrm{Ext}^n_\Lambda(A, B),$$

and the isomorphism $\mathrm{Ext}^n_\Lambda(A, B) \approx H^n(\mathrm{Hom}_\Lambda(A, \mathbf{Y}))$ follows similarly from Theorem 8.

For extension functors there are two results which correspond to Theorem 5 and, of these, one will be of more interest to us than the other. To describe this result, we observe that if $\mathbf{X}$ is a *fixed* projective resolution of A, then every homomorphism $B \to B^*$ produces a translation $\mathrm{Hom}_\Lambda(\mathbf{X}, B) \to \mathrm{Hom}_\Lambda(\mathbf{X}, B^*)$ and so enables us to regard the homology modules $H^n(\mathrm{Hom}_\Lambda(\mathbf{X}, B))$ as functors of B. Further, by Theorem 3 of section (5.1), for every exact sequence

$$0 \to B' \to B \to B'' \to 0$$

the corresponding sequence

$$\mathbf{0} \to \mathrm{Hom}_\Lambda(\mathbf{X}, B') \to \mathrm{Hom}_\Lambda(\mathbf{X}, B) \to \mathrm{Hom}_\Lambda(\mathbf{X}, B'') \to \mathbf{0}$$

is an exact sequence of complexes and translations; and this makes it possible to define connecting homomorphisms

$$H^n(\mathrm{Hom}_\Lambda(\mathbf{X}, B'')) \to H^{n+1}(\mathrm{Hom}_\Lambda(\mathbf{X}, B')),$$

which turn the $H^n(\mathrm{Hom}_\Lambda(\mathbf{X}, B))$ into a connected sequence of covariant functors in the variable B.

Theorem 10. *If* $\mathbf{X}$ *is a fixed projective resolution of* A *and*

$$T^n(B) = \mathrm{Ext}^n_\Lambda(A,B),$$

then the $T^n(B)$ *and the* $H^n(\mathrm{Hom}_\Lambda(\mathbf{X},B))$ *form isomorphic connected sequences of functors in the variable* B.

Proof. Let $\mathbf{Y}$ be an injective resolution of B, then, as we saw in the proof of Theorem 7, the translation $\mathrm{Hom}_\Lambda(\mathbf{X},\mathbf{B}) \to \mathrm{Hom}_\Lambda(\mathbf{X},\mathbf{Y})$ induces isomorphisms

$$H^n(\mathrm{Hom}_\Lambda(\mathbf{X},B)) \approx \mathrm{Ext}^n_\Lambda(A,B) = T^n(B).$$

It is clear that these isomorphisms constitute functor equivalences, and to complete the proof we need only show that they commute with the connecting homomorphisms that result from exact sequences $0 \to B' \to B \to B'' \to 0$. This, however, can be verified in a straightforward manner, the details being similar to those encountered in the proof of Theorem 5 at the corresponding stage.

There is a result, analogous to Theorem 10, which deals with the behaviour, as A varies, of $\mathrm{Hom}_\Lambda(A,\mathbf{Q})$, where $\mathbf{Q}$ is a fixed injective resolution of B, but, as has already been remarked, this will not be needed, and therefore we leave the reader to examine the details for himself if he so wishes.

7.5 The homological dimension of a module

Let $A \neq 0$ be a left Λ-module and $n \geqslant 0$ an integer.

Definition. The left Λ-module A is said to have *projective dimension* equal to n or to have *homological dimension* equal to n, if

(i) there exists an exact sequence of the form

$$0 \to P_n \to P_{n-1} \to \cdots \to P_0 \to A \to 0, \tag{7.5.1}$$

where each P_i is Λ-projective,

and (ii) there is no exact sequence of this type with fewer terms.

If A has homological dimension n, then we shall write $\mathrm{l.dh}_\Lambda(A) = n$, the letter l being a reminder that we are concerned with left Λ-modules. If no such exact sequences as (7.5.1) exist, then we shall write $\mathrm{l.dh}_\Lambda(A) = \infty$. Finally, as a convention, we put $\mathrm{l.dh}_\Lambda(0) = -1$. In this way the homological dimension $\mathrm{l.dh}_\Lambda(M)$ is defined for every left Λ-module M.

Remarks. (*a*) Let M be a left Λ-module and let $n \geqslant 0$ be an integer, then we have $\mathrm{l.dh}_\Lambda(M) < n$ if and only if M has a projective resolution $\mathbf{P}$ in which $P_s = 0$ for all $s \geqslant n$.

(*b*) A left Λ-module M is Λ-projective if and only if $\mathrm{l.dh}_\Lambda(M) \leqslant 0$.

(*c*) It is, of course, possible to define the homological dimension of a right Λ-module and, indeed, if N is a typical right Λ-module then its homological dimension will be written as $\mathrm{r.dh}_\Lambda(N)$. As usual, results proved for left Λ-modules can be adapted so that they apply to right Λ-modules by making the obvious formal changes.

(*d*) In place of (7.5.1) one can also consider exact sequences of the form

$$0 \to A \to Q^0 \to Q^1 \to \cdots \to Q^n \to 0,$$

where the Q^i are Λ-injective. This leads to the notion of *injective dimension*, but we shall not be concerned with this concept and will make no further reference to it.

Although the proposition which follows is a special case of Theorem 12 (below), it is convenient to have it stated separately.

Proposition 3. *The left Λ-module A is Λ-projective if and only if $\mathrm{Ext}^1_\Lambda(A, C) = 0$ for all left Λ-modules C.*

Proof. We already know, by Proposition 2, that if A is Λ-projective, then $\mathrm{Ext}^1_\Lambda(A, C) = 0$ whatever the Λ-module C. Assume now that $\mathrm{Ext}^1_\Lambda(A, C)$ vanishes for all left Λ-modules C and let

$$0 \to C' \to C \to C'' \to 0$$

be an exact sequence. Then, since

$$0 \to \mathrm{Hom}_\Lambda(A, C') \to \mathrm{Hom}_\Lambda(A, C) \to \mathrm{Hom}_\Lambda(A, C'') \to \mathrm{Ext}^1_\Lambda(A, C')$$

is exact and $\mathrm{Ext}^1_\Lambda(A, C') = 0$, we see that $\mathrm{Hom}_\Lambda(A, C)$ is an exact functor of C and therefore, by Theorem 3 of section (5.1), A is a Λ-projective module.

Lemma 1. *Let A be a left Λ-module such that $\mathrm{l.dh}_\Lambda(A) < n$ $(n \geqslant 0)$, then, if $s \geqslant n$, we have $\mathrm{Ext}^s_\Lambda(A, C) = 0$ for all left Λ-modules C.*

Proof. We can find a projective resolution $\mathbf{P}$ of A such that $P_s = 0$ for all $s \geqslant n$. If now $s \geqslant n$ and C is a left Λ-module, then, by Theorem 9,

$$\mathrm{Ext}^s_\Lambda(A, C) \approx H^s(\mathrm{Hom}_\Lambda(\mathbf{P}, C)) = 0$$

because of the semi-degenerate character of $\mathbf{P}$.

Theorem 11. *Let A be a left Λ-module, let $n \geqslant 0$ be an integer and finally let*

$$0 \to X \to P_{n-1} \to P_{n-2} \to \ldots \to P_0 \to A \to 0$$

be an exact sequence, where each P_i is Λ-projective. If now $\mathrm{l.dh}_\Lambda(A) \leqslant n$, then X is also Λ-projective.

Proof. We shall suppose that $n \geqslant 1$ since otherwise the assertion is trivial. Put $A_s = \operatorname{Im}(P_s \to P_{s-1})$, then we have exact sequences

$$0 \to A_1 \to P_0 \to A \to 0$$
$$0 \to A_2 \to P_1 \to A_1 \to 0$$
$$\cdots\cdots\cdots\cdots\cdots\cdots$$
$$0 \to X \to P_{n-1} \to A_{n-1} \to 0.$$

Now suppose that C is an arbitrary left Λ-module then, since $\operatorname{Ext}^s_\Lambda(P, C) = 0$ if $s \geqslant 1$ and P is Λ-projective, the sequences

$$0 \to \operatorname{Ext}^1_\Lambda(X, C) \to \operatorname{Ext}^2_\Lambda(A_{n-1}, C) \to 0$$
$$0 \to \operatorname{Ext}^2_\Lambda(A_{n-1}, C) \to \operatorname{Ext}^3_\Lambda(A_{n-2}, C) \to 0$$
$$\cdots\cdots\cdots\cdots\cdots\cdots\cdots\cdots\cdots\cdots$$
$$0 \to \operatorname{Ext}^n_\Lambda(A_1, C) \to \operatorname{Ext}^{n+1}_\Lambda(A, C) \to 0$$

are also exact. By combining, we obtain from these an isomorphism $\operatorname{Ext}^1_\Lambda(X, C) \approx \operatorname{Ext}^{n+1}_\Lambda(A, C)$. However, Lemma 1 shows that

$$\operatorname{Ext}^{n+1}_\Lambda(A, C) = 0,$$

consequently $\operatorname{Ext}^1_\Lambda(X, C) = 0$ and now the required result follows from Proposition 3.

Lemma 1 is now superseded by the following theorem:

Theorem 12. *Let A be a left Λ-module and n $(n \geqslant 0)$ an integer. Then* $\text{l.dh}_\Lambda(A) < n$ *if and only if* $\operatorname{Ext}^n_\Lambda(A, C) = 0$ *for all left Λ-modules C.*

Proof. We shall assume that $\operatorname{Ext}^n_\Lambda(A, C) = 0$ for all Λ-modules C and show that $\text{l.dh}_\Lambda(A) < n$. This will complete the proof since the converse follows by Lemma 1.

If $n = 0$ then, by Theorem 6, $\operatorname{Hom}_\Lambda(A, C) = 0$ for all C and therefore $\operatorname{Hom}_\Lambda(A, A) = 0$. However, this implies that $A = 0$ and therefore the result that is required has been established for the case $n = 0$. On the other hand, if $n = 1$ then what we need has already been established in Proposition 3.

From now on it will be supposed that $n \geqslant 2$. Let us construct an exact sequence

$$0 \to X \to P_{n-2} \to P_{n-3} \to \cdots \to P_0 \to A \to 0,$$

where the P_i are Λ-projective, then (as in the proof of Theorem 11 but with $n-1$ in the place of n) we arrive at an isomorphism

$$\operatorname{Ext}^1_\Lambda(X, C) \approx \operatorname{Ext}^n_\Lambda(A, C) = 0,$$

C being an arbitrary left Λ-module. But, by Proposition 3, this implies that X is Λ-projective and we therefore have $\text{l.dh}_\Lambda(A) \leqslant n-1$. This establishes the theorem.

Corollary. *Let m ($m \geqslant 0$) be an integer, then $\text{l.dh}_\Lambda(A) \leqslant m$ if and only if $\text{Ext}^m_\Lambda(A, C)$ is a right exact functor of C.*

Proof. Assume first that $\text{l.dh}_\Lambda(A) \leqslant m$ and let $0 \to C' \to C \to C'' \to 0$ be an exact sequence. Then

$$\text{Ext}^m_\Lambda(A, C') \to \text{Ext}^m_\Lambda(A, C) \to \text{Ext}^m_\Lambda(A, C'') \to \text{Ext}^{m+1}_\Lambda(A, C')$$

is exact, and since, by the theorem, $\text{Ext}^{m+1}_\Lambda(A, C') = 0$, the right exactness of $\text{Ext}^m_\Lambda(A, C)$ as a functor of C is proved.

Now assume that $\text{Ext}^m_\Lambda(A, C)$ is a right exact functor of the second variable and, for a given module C, let us construct an exact sequence $0 \to C \to Q \to M \to 0$, where Q is Λ-injective. From this we obtain the exact sequence

$$\text{Ext}^m_\Lambda(A, Q) \to \text{Ext}^m_\Lambda(A, M) \to \text{Ext}^{m+1}_\Lambda(A, C) \to \text{Ext}^{m+1}_\Lambda(A, Q).$$

But, by the assumption of right exactness, $\text{Ext}^m_\Lambda(A, Q) \to \text{Ext}^m_\Lambda(A, M)$ is an epimorphism, consequently $\text{Ext}^{m+1}_\Lambda(A, C) \to \text{Ext}^{m+1}_\Lambda(A, Q)$ is a monomorphism. However, by Proposition 2, $\text{Ext}^{m+1}_\Lambda(A, Q) = 0$ and therefore $\text{Ext}^{m+1}_\Lambda(A, C) = 0$. That $\text{l.dh}_\Lambda(A) \leqslant m$ now follows from the theorem.

Theorem 13. *Let $0 \to A' \to P \to A \to 0$ be an exact sequence of Λ-modules, where P is Λ-projective. If A is not Λ-projective then*

$$\text{l.dh}_\Lambda(A) = 1 + \text{l.dh}_\Lambda(A'),$$

where, of course, both sides may be infinite. If, on the other hand, A is Λ-projective then so is A'. In either case $\text{l.dh}_\Lambda(A) \leqslant 1 + \text{l.dh}_\Lambda(A')$.

Proof. Assume first that A is not Λ-projective. For any left Λ-module C and any integer $n \geqslant 1$ the sequence

$$0 \to \text{Ext}^n_\Lambda(A', C) \to \text{Ext}^{n+1}_\Lambda(A, C) \to 0$$

is exact and therefore $\text{Ext}^n_\Lambda(A', C) \approx \text{Ext}^{n+1}_\Lambda(A, C)$. The first assertion now follows readily from Theorem 12.

Next suppose that A is Λ-projective. Then, by Theorem 1 of section (5.1), the sequence $0 \to A' \to P \to A \to 0$ splits and therefore A' is isomorphic to a direct summand of P. But this, by Proposition 2 of section (5.1), implies that A' is Λ-projective and now the proof is complete.

7.6 Global dimension

In the last section we defined the homological dimension of a Λ-module. We shall now use this concept to define the left and right *global dimensions* of the ring Λ itself. Let M denote a variable left Λ-module, N a variable right Λ-module and let us write

$$\text{l.gl.dim}\,\Lambda = \sup_M \text{l.dh}_\Lambda(M), \tag{7.6.1}$$

$$\text{r.gl.dim}\,\Lambda = \sup_N \text{r.dh}_\Lambda(N). \tag{7.6.2}$$

Thus l.gl.dim Λ, which is called the *left global dimension* of Λ, is the upper bound of the homological dimensions of all left Λ-modules and it is therefore either a non-negative integer or 'plus infinity'. Further, by Theorem 12, we have the following result:

Theorem 14. *If m ($m \geq 0$) is an integer, then we have* l.gl.dim $\Lambda \leq m$ *if and only if* Ext_Λ^{m+1}, *regarded as a functor of left Λ-modules, takes only null values.*

Analogously r.gl.dim Λ is used as an abbreviation for the *right global dimension* of Λ. It should be noted that for the right global dimension we have a result corresponding to Theorem 14 but so closely related to it that it is not necessary to state it separately.

When Λ is commutative the distinction between right and left Λ-modules disappears and therefore right and left global dimension are one and the same. However, when Λ is not commutative the connexion between the two is not at all apparent. We shall, indeed, prove later that, for what are called Noetherian rings, the dimensions are equal even in the non-commutative case, but this will be achieved only after some substantial preliminary investigations.

Returning to the definition (7.6.1) we see that the left global dimension is defined as an upper bound in which the left Λ-module M varies quite freely. However, we shall find that we can greatly reduce the range of M without affecting the upper bound. To this end we make the following

Definition. A Λ-module which can be generated by a single element will be called a *cyclic* Λ-module.

The next theorem will be to the effect that the left global dimension of Λ is the upper bound of $\text{l.dh}_\Lambda(A)$, where A ranges over all (left) cyclic Λ-modules. Suppose, for the moment, that A is a cyclic left Λ-module generated (say) by the element x, then we obtain an epimorphism $\Lambda \to A$ by mapping the element λ of Λ into λx. The kernel L of this epimorphism is a submodule of Λ regarded as a left Λ-module, that is to say (using the language of ring theory), L is a *left ideal* of Λ. Since $A \approx \Lambda/L$, this shows that (to within isomorphisms) all the cyclic left Λ-modules can be obtained as residue modules Λ/L, where L is a typical left ideal.

Lemma 2. *Let $[A_i]_{i \in I}$ and $[B_i]_{i \in I}$ be two families of submodules of a Λ-module M, with the same system I of parameters. Let I be well ordered with $\leqslant$ denoting the ordering relation, and suppose the following conditions are satisfied:*

(i) *$B_i \subseteq A_i$, for every i in I;*
(ii) *whenever $i < j$ then $A_i \subseteq B_j$;*
(iii) *if $x \in B_j$ and $x \neq 0$, then $x \in A_i$ for some $i < j$;*
(iv) *$\text{l.dh}_\Lambda(A_i/B_i) \leqslant n$ for† all i in I.*

In these circumstances the smallest submodule of M which contains all the A_i has homological dimension at most equal to n.

Proof. Let us observe, before we come to the less obvious details, that, if $i \leqslant j$, then $A_i \subseteq A_j$ and $B_i \subseteq B_j$. Further, we may suppose that M itself is the smallest submodule containing all the A_i and therefore is their set-theoretic union. Also we may note that, by (iii), if i_0 is the first element of I, then $B_{i_0} = 0$.

We now proceed by induction on the integer n. Suppose first that $n = 0$, then A_i/B_i is projective for each value of i. But

$$0 \to B_i \to A_i \to A_i/B_i \to 0$$

is exact, consequently, by Theorem 1 of section (5.1), the sequence splits and therefore $A_i = B_i + C_i$ (direct sum), where C_i is a suitably chosen submodule of A_i. Further, since C_i is isomorphic to A_i/B_i, we see that C_i is projective. To complete the proof for the case $n = 0$ we shall show that M is the direct sum of the modules C_i. To begin with, let us prove that every element of M can be written in the form $\sum_{j \in I} c_j$, where c_j belongs to C_j and $c_j = 0$ for almost all j. Indeed, if this were not so we could choose the first i such that A_i contains an element x (say), which is not expressible in this form. Then, since $A_i = B_i + C_i$, we could write $x = y + c_i$, where y belongs to B_i and c_i to C_i. Clearly y cannot be zero, and so, by (iii), y belongs to A_j, for some $j < i$. Accordingly, y must be expressible in the required form and this implies that the same is true of x. Thus we have a contradiction.

† The symbol $\leqslant$ is here used in two different senses but there is, in fact, no danger of confusion.

Next suppose that $\sum_{j \in I} c_j = 0$, where c_j belongs to C_j and $c_j = 0$ for almost all j. We wish to show that $c_j = 0$ for *all* j. Now if this were not true we should have $c_{j_1} + c_{j_2} + \ldots + c_{j_s} = 0$, where $j_1 < j_2 < \ldots < j_s$, and, for each ν, $c_{j_\nu} \neq 0$. But, by (ii), $c_{j_1} + c_{j_2} + \ldots + c_{j_{s-1}}$ belongs to B_{j_s}, whence, since $(c_{j_1} + c_{j_2} + \ldots + c_{j_{s-1}}) + c_{j_s} = 0$ and the sum $B_{j_s} + C_{j_s}$ is direct, we see that $c_{j_s} = 0$. Thus again we have a contradiction and the lemma is proved for the case $n = 0$.

For the inductive step we suppose that n is at least unity and that the lemma has been proved for all smaller values of the inductive variable. Let Φ be the free-module generated by the non-zero elements of M (regarded, for this purpose, as independent symbols), let F_i be the free module generated by the non-zero elements of A_i, and G_i the free module generated by the non-zero elements of B_i. (Here it is understood that a free module generated by an empty set is to be taken to be a zero module.) By means of certain obvious identifications, we regard F_i and G_i as submodules of Φ so that we have $G_i \subseteq F_i \subseteq \Phi$. We next construct an epimorphism $\Phi \to M$ by mapping each element of the base of Φ into itself regarded as an element of M. Put $N = \operatorname{Ker}(\Phi \to M)$, then $\Phi \to M$ determines, by restriction, epimorphisms $F_i \to A_i$ and $G_i \to B_i$ whose kernels are $K_i = N \cap F_i$ and $L_i = N \cap G_i$ respectively. From these various mappings and inclusion relations we obtain a commutative diagram

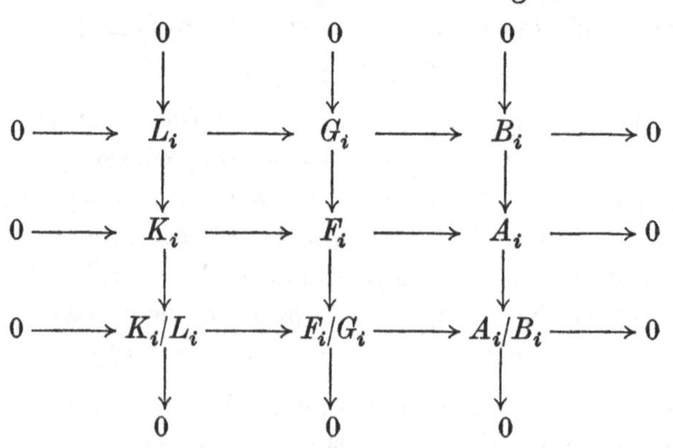

in which the columns are exact, the two upper rows are exact and the bottom row is a 0-sequence. But, by Theorem 6 of section (4.6), this implies that

$$0 \to K_i/L_i \to F_i/G_i \to A_i/B_i \to 0$$

is exact. Furthermore, by the construction, F_i/G_i is free, hence (Theorem 13) $\operatorname{l.dh}_\Lambda(K_i/L_i) \leqslant n-1$. We now contend that we can apply the inductive hypothesis to the module N, with K_i replacing A_i and L_i replacing B_i. For it is obvious that conditions (i) and (ii) are satisfied while the correctness of (iv) has just been established with the value of the inductive variable reduced by unity. Furthermore, if x belongs to L_j and $x \neq 0$, then the non-zero coefficients in the expression for x as an element of G_j are associated with a finite number of non-zero elements of B_j considered as basis elements of G_j. We can therefore find $i < j$ so that all these elements belong to A_i and this shows that x belongs to $N \cap F_i = K_i$. The application of the inductive hypothesis is therefore justified and we can thus conclude that

$$\bigcup_i K_i = \bigcup_i (F_i \cap N) = (\bigcup_i F_i) \cap N = \Phi \cap N = N$$

has homological dimension at most equal to $n-1$. But

$$0 \to N \to \Phi \to M \to 0$$

is exact and Φ is free. Hence, by Theorem 13,

$$\text{l.dh}_\Lambda(M) \leqslant \text{l.dh}_\Lambda(N) + 1 \leqslant n,$$

and with this the lemma is proved.

Theorem 15. *The left global dimension of* Λ *is equal to* $\sup_A \text{l.dh}_\Lambda(A)$, *where* A *ranges over all cyclic left* Λ*-modules.*

Proof. We assume that, for every cyclic module A, $\text{l.dh}_\Lambda(A) \leqslant n$, where n ($n \geqslant 0$) is a finite integer, and deduce that if M is an arbitrary Λ-module then $\text{l.dh}_\Lambda(M) \leqslant n$. This will prove that

$$\text{l.gl.dim}\,\Lambda \leqslant \sup_A \text{l.dh}_\Lambda(A),$$

and the opposite inequality is obvious.

Let $[x_i]_{i \in I}$ be a family of elements which generate M and let us well order the set I of parameters. For each i in I, denote by A_i the submodule of M generated by all the x_j with $j \leqslant i$, and by B_i the submodule generated by all x_j with $j < i$. The union of the A_i is the whole module M and A_i/B_i is a cyclic module generated by the natural image of x_i, consequently $\text{l.dh}_\Lambda(A_i/B_i) \leqslant n$. It is clear that conditions (i), (ii) and (iii) of Lemma 2 are satisfied, hence, by that result, we see that $\text{l.dh}_\Lambda(M) \leqslant n$. This completes the proof.

Since every cyclic left Λ-module is isomorphic to a residue module Λ/L, where L is a left Λ-ideal, the left global dimension of Λ can be described in terms of its left ideals. For the purposes of illustration we shall develop this idea for rings of global dimension zero, that is to say, for rings all of whose modules are projective. Note that, in this case, the modules are also all injective. For, if a module A is given, we can construct an exact sequence $0 \to A \to Q \to B \to 0$ with Q injective. But, by hypothesis, B is projective and so the sequence splits. Hence A, being a direct summand of an injective module, is itself injective by Proposition 5 of section (5.2).

We recall the following familiar definition:

Definition. A left Λ-module A $(A \neq 0)$ is said to be *simple* if 0 and A are its only submodules.

Proposition 4. *Let* l.gl.dim $\Lambda = 0$ *and let A be a left Λ-module. Then A is a direct sum of simple left Λ-modules.*

Proof. It is convenient to divide the proof into three parts.

(i) Let $N \subseteq M$ be submodules of A, then *N is a direct summand of M*. This follows at once from the observation that

$$0 \to N \to M \to M/N \to 0$$

is an exact sequence which splits because M/N is projective.

(ii) *Every non-zero submodule B of A contains a simple submodule.* For let b be a non-zero element of B and consider the family Ω of those submodules of B which do not contain b, these submodules being partially ordered by inclusion. Suppose now that we have a subfamily Ω' (of the family Ω) such that each pair of its modules has the property that one of the two contains the other. The set-theoretic union of the modules belonging to Ω' is a submodule of B not containing b. This shows that Ω is an inductive system to which, therefore, Zorn's lemma is applicable. Accordingly, let M be maximal in Ω. We can now find, by (i), a submodule N $(N \neq 0)$ of B such that $B = M + N$ (direct sum). *This module N is simple.* For, if $0 \subseteq N_1 \subset N$, where N_1 is a submodule, then $N = N_1 + N_2$ (direct sum) for a suitable module $N_2 \neq 0$. Now b cannot belong to both $M + N_1$ and $M + N_2$, because $(M + N_1) \cap (M + N_2) = M$, and therefore one at least of $M + N_1$ and $M + N_2$ belongs to Ω. But M is maximal in Ω and $M + N_2 \supset M$, hence $M + N_1 = M$, which implies that $N_1 = 0$. This proves that N is simple.

(iii) Finally, let $[S_i]_{i \in I}$ be the family composed of all the simple submodules of A. For the purpose of proof we may suppose that $A \neq 0$, which secures the existence of at least one member of the family. We shall denote by $\sum_{i \in I'} S_i$, where $I' \subseteq I$, the smallest submodule of A containing all the S_i with $i \in I'$; and by $\mathfrak{S}$ the family of all those subsets J of I for which the sum $\sum_{j \in J} S_j$ is direct, these subsets being partially ordered by inclusion. Let $[J_\gamma]_{\gamma \in G}$ be a subfamily of $\mathfrak{S}$ with the property that, if γ_1, γ_2 belong to G, then either $J_{\gamma_1} \subseteq J_{\gamma_2}$ or $J_{\gamma_2} \subseteq J_{\gamma_1}$. Put $J = \bigcup_{\gamma \in G} J_\gamma$ and suppose that $s_{j_1} + s_{j_2} + \ldots + s_{j_n} = 0$, where $j_1, j_2, \ldots, j_n$ are distinct elements of J and s_{j_r} belongs to S_{j_r}. Then we can find γ in G such that all of $j_1, j_2, \ldots, j_n$ belong to J_γ. But $\sum_{i \in J_\gamma} S_i$ is direct, hence $s_{j_1} = s_{j_2} = \ldots = s_{j_n} = 0$, and therefore $\sum_{j \in J} S_j$ is also direct. Accordingly J belongs to $\mathfrak{S}$. It has now been shown that the partially ordered set $\mathfrak{S}$ is one to which Zorn's lemma can be applied. Let J^* be maximal in $\mathfrak{S}$. *We claim that* $A = \sum_{j \in J^*} S_j$, and, since the sum is direct, this will complete the proof. Indeed, if $A \neq \sum_{j \in J^*} S_j$, then $A = (\sum_{j \in J^*} S_j) + D$ (direct sum), where $D \neq 0$. But, by (ii), D contains a simple submodule S_{i_0} (say), where, because $(\sum_{j \in J^*} S_j) \cap S_{i_0} = 0$, i_0 does not belong to J^*. This shows that J^*, augmented by i_0, again belongs to $\mathfrak{S}$, contradicting the maximal property of J^*. Accordingly, A is the direct sum of the modules S_j $(j \in J^*)$ and the proposition is proved.

Theorem 16. *In order that* $\text{l.gl.dim}\,\Lambda = 0$ *it is necessary and sufficient that* Λ *be expressible as a finite direct sum of simple left ideals.*

Proof. Suppose first that $\text{l.gl.dim}\,\Lambda = 0$, then, by Proposition 4, $\Lambda = \sum_{i \in I} S_i$ (say), where the sum is direct and the S_i are simple left ideals. In particular, $1 = e_{i_1} + e_{i_2} + \ldots + e_{i_n}$, where $i_1, i_2, \ldots, i_n$ are distinct elements of I and e_{i_r} belongs to S_{i_r}. Now, if λ is an arbitrary element of Λ, then $\lambda = \lambda e_{i_1} + \lambda e_{i_2} + \ldots + \lambda e_{i_n}$, and this belongs to

$$S_{i_1} + S_{i_2} + \ldots + S_{i_n}.$$

Hence $\Lambda = S_{i_1} + S_{i_2} + \ldots + S_{i_n}$ (direct sum).

To prove the converse, suppose that $\Lambda = S_1 + S_2 + \ldots + S_n$ (direct sum), where $S_1, S_2, \ldots, S_n$ are simple left ideals, and let L be an arbitrary left ideal. Choose distinct integers $m_1, m_2, \ldots, m_p$, between 1 and n, so that

$$L \cap (S_{m_1} + S_{m_2} + \ldots + S_{m_p}) = 0, \tag{7.6.3}$$

and, subject to this condition, let p be maximal. If now $1 \leqslant k \leqslant n$ and k is different from $m_1, m_2, \ldots, m_p$, then

$$L \cap (S_{m_1} + S_{m_2} + \ldots + S_{m_p} + S_k) \neq 0,$$

and so we can find an element x of $S_k \cap (L+S_{m_1}+\ldots+S_{m_p})$, where $x \neq 0$. But S_k is simple, hence $\Lambda x = S_k$ and therefore

$$S_k \subseteq L+S_{m_1}+\ldots+S_{m_p}.$$

Thus $S_1, S_2, \ldots, S_n$ are *all* contained in $L+S_{m_1}+\ldots+S_{m_p}$; accordingly, $\Lambda = L+S_{m_1}+S_{m_2}+\ldots+S_{m_p}$, and this sum is direct by (7.6.3). This proves that $\text{l.dh}_\Lambda(\Lambda/L) \leqslant 0$ because Λ/L is isomorphic to a direct summand of Λ and is therefore projective. That $\text{l.gl.dim}\,\Lambda = 0$ now follows from Theorem 15.

It is known that if Λ is a direct sum of simple left ideals then it is also a direct sum of simple right ideals. Such rings are called *semi-simple*. Thus it would follow from our last theorem that $\text{l.gl.dim}\,\Lambda = 0$ *if and only if* Λ *is semi-simple and when that is the case we also have* $\text{r.gl.dim}\,\Lambda = 0$. Since the structure of semi-simple rings is completely known, so also is that of rings of global dimension zero. However, to elaborate this would be to repeat what is well known and belongs to another field. Accordingly, we shall not pursue the matter beyond this point.

7.7 Noetherian rings

We turn now to the consideration of an important class of module and this, we shall find, will focus our attention on a special kind of ring.

Definition. A Λ-module M will be said to be Λ-*Noetherian* if every submodule of M can be generated by a finite number of elements.

Since M is to be regarded as one of its own submodules this implies, of course, that M is finitely generated. Our first result is an immediate consequence of the definition.

Proposition 5. *If M is Λ-Noetherian then so are all the submodules and factor (residue) modules of M.*

There is another form of the Noetherian property which should be noted. In order to have a concise statement we make another definition.

Definition. A Λ-module M is said to satisfy the *ascending chain condition for submodules* if, whenever

$$N_1 \subseteq N_2 \subseteq N_3 \subseteq \ldots$$

is an infinite increasing sequence of submodules of M, there exists an integer r, depending on the sequence, such that $N_s = N_r$ for all $s \geqslant r$.

Proposition 6. *A Λ-module M is Λ-Noetherian if and only if it satisfies the ascending chain condition for submodules.*

Proof. Suppose first that M is Noetherian and let $N_1 \subseteq N_2 \subseteq \ldots$ be an infinite ascending sequence of submodules. The *union* of the N_i is another submodule N (say), and, by hypothesis, N will be finitely generated. Let $u_1, u_2, \ldots, u_p$ be elements which generate N, then we can choose an integer r so that each of the elements belongs to N_r. If now $s \geqslant r$ then $N \subseteq N_r \subseteq N_s \subseteq N$, and therefore $N_r = N_s$. This shows that the ascending chain condition holds in M.

Next suppose that M satisfies the ascending chain condition and let N be a submodule of M. We shall assume that N is not finitely generated and obtain a contradiction. Choose u_1 in N and let N_1 be the submodule generated by u_1. Then $N \neq N_1$ and so we can find u_2 in N such that u_2 does not belong to N_1. Let N_2 be the submodule generated by u_1 and u_2, then $N_1 \subset N_2$ (strict inclusion) and $N_2 \neq N$. Next we choose u_3 in N so that $u_3 \notin N_2$ and let N_3 be the module generated by u_1, u_2 and u_3. Proceeding in this way we build up an infinite *strictly increasing* sequence of submodules and this violates the chain condition.

Proposition 7. *If* $0 \to A \to M \to B \to 0$ *is an exact sequence of* Λ-*modules and any two of* A, M, B *are* Λ-*Noetherian then so is the third.*

Proof. By Proposition 5, the only case which requires discussion is that in which A and B are known to be Noetherian and we seek to show that M is Noetherian. We shall therefore confine our attention to this case and, for the purpose of the proof, assume (as we may) that A is a submodule of M and that $B = M/A$.

Let $N_1 \subseteq N_2 \subseteq N_3 \subseteq \ldots$ be an increasing sequence of submodules of M. Since A and M/A are Noetherian, we can choose r so that, for all $s \geqslant r$,

$$A \cap N_s = A \cap N_r, \quad (A+N_s)/A = (A+N_r)/A.$$

Assume now that $s \geqslant r$ then $A \cap N_s = A \cap N_r$ and $A+N_s = A+N_r$. Let x belong to N_s. We have $x \in A+N_s = A+N_r$ and so $x = a+y$, where a belongs to A and y to N_r. Further,

$$a = x-y \in N_s \cap A = N_r \cap A$$

which shows, in particular, that a belongs to N_r. Accordingly

$$x = y+a \in N_r$$

and this implies that $N_s \subseteq N_r$. Since the opposite inclusion is obvious we have $N_r = N_s$, and this, by Proposition 6, completes the proof.

Corollary. *If A and B are Λ-Noetherian modules then so is their direct sum.*

This follows by applying the proposition to the canonical exact sequence $0 \to B \to (A+B) \to A \to 0$.

We come now to the concept of a Noetherian ring.

Definition. If Λ, when regarded as a left (right) Λ-module, is Λ-Noetherian, then we say that Λ is a left (right) *Noetherian ring.*

In other words, to say that Λ is a left Noetherian ring is to say that each left ideal of Λ is finitely generated or (equivalently) that the left ideals of Λ satisfy the ascending chain condition. It is known that a ring may be left Noetherian without being right Noetherian.

Theorem 17. *If Λ is left (right) Noetherian and M is a left (right) Λ-module which is generated by a finite number of elements, then all the submodules of M are finitely generated and therefore M is Λ-Noetherian.*

Proof. Let n be the smallest number of elements which will generate M. We shall use induction with respect to n. If $n=1$ then M is cyclic and therefore $M \approx \Lambda/I$, where I is a certain left ideal of Λ. The Noetherian character of M is now established by appealing to Proposition 5.

Suppose now that $n \geqslant 2$ and that the theorem has been established for all smaller values of the inductive variable. Let $u_1, u_2, \ldots, u_n$ generate M and let A be the submodule generated by u_1, then A is Noetherian, by the case $n=1$, and M/A is Noetherian by the inductive hypothesis. But $0 \to A \to M \to M/A \to 0$ is exact, hence M is Noetherian, by Proposition 7.

Proposition 8. *Let Λ be a left (right) Noetherian ring and let M be a finitely generated left (right) Λ-module. Then there exists a free resolution* $\mathbf{F}$ *of M in which each component module has a finite base.*

Proof. By Theorem 2 of section (1.8), there exists a free module F_0, with a finite base, for which there is an epimorphism $F_0 \to M$. Let M_1 be the kernel of this epimorphism then, by Theorem 17, M_1 is finitely generated. Accordingly, there exists a free module F_1, also with a finite base, and an epimorphism $F_1 \to M_1$. By combining $F_1 \to M_1$ with the inclusion map $M_1 \to F_0$, we obtain an exact sequence

$$F_1 \to F_0 \to M \to 0$$

and, proceeding in this way, we construct a free resolution of M with the required properties.

This result is one of several which frequently enable us, when dealing with Noetherian rings, to restrict our attention to finitely generated modules. This can be a great advantage. The next two results are of a similar kind.

Proposition 9. *Let Λ be a left Noetherian ring, A a finitely generated Λ-module and n $(n \geqslant 0)$ an integer. Then* $\mathrm{l.dh}_\Lambda(A) < n$ *if and only if* $\mathrm{Ext}^n_\Lambda(A, B) = 0$ *for all finitely generated left Λ-modules B.*

Proof. We assume that $\mathrm{Ext}^n_\Lambda(A, B) = 0$ for all finitely generated Λ-modules B and prove, by induction on n, that $\mathrm{l.dh}_\Lambda(A) < n$. The converse has already been established.

Consider first the case $n = 0$. Taking $B = A$ we obtain

$$\mathrm{Ext}^0_\Lambda(A, A) = 0$$

and therefore $\mathrm{Hom}_\Lambda(A, A) = 0$. This, of course, implies that $A = 0$ and establishes the result in this case.

For larger values of n we construct an exact sequence $F \to A \to 0$, where F is a free module with a finite base, and put $A_1 = \mathrm{Ker}(F \to A)$. Then the sequence $0 \to A_1 \to F \to A \to 0$ is exact and A_1, being isomorphic to a submodule of the Noetherian module F, is finitely generated. For the case $n = 1$ we observe that, since $\mathrm{Ext}^1_\Lambda(A, A_1) = 0$, the sequence

$$\mathrm{Hom}_\Lambda(A, F) \to \mathrm{Hom}_\Lambda(A, A) \to 0$$

is exact. In particular, the identity map $A \to A$ can be expressed as a combined homomorphism $A \to F \to A$ and this, by Proposition 1 of section (1.10), means that $F \to A$ is direct. Accordingly A is isomorphic to a direct summand of F and therefore, using Theorem 2 of section (5.1), it is projective. But this implies that $\mathrm{l.dh}_\Lambda(A) \leqslant 0$, which is just what we wish to prove.

Next assume that $n \geqslant 2$ and that the required result has been proved for all smaller values of the inductive variable. Let B be an arbitrary finitely generated Λ-module then, from the exact sequence

$$\mathrm{Ext}^{n-1}_\Lambda(F, B) \to \mathrm{Ext}^{n-1}_\Lambda(A_1, B) \to \mathrm{Ext}^n_\Lambda(A, B)$$

and the facts that $\mathrm{Ext}^{n-1}_\Lambda(F, B) = 0$ (see Proposition 2) and $\mathrm{Ext}^n_\Lambda(A, B) = 0$, it follows that $\mathrm{Ext}^{n-1}_\Lambda(A_1, B) = 0$. The inductive hypothesis now shows that $\mathrm{l.dh}_\Lambda(A_1) < n-1$, and from this it follows (Theorem 13) that $\mathrm{l.dh}_\Lambda(A) < n$. This completes the proof.

Proposition 10. *Let Λ be a left Noetherian ring; then in order that the left gobal dimension of Λ should not exceed n $(n \geqslant 0)$ it is necessary and sufficient that* $\mathrm{Ext}^{n+1}_\Lambda(A, B) = 0$ *whenever A and B are finitely generated left Λ-modules.*

Proof. Necessity has already been established in Theorem 14. Assume therefore that the condition is satisfied. Let A be a cyclic left Λ-module, then $\mathrm{Ext}_\Lambda^{n+1}(A, B) = 0$ whenever B is finitely generated hence, by Proposition 9, $\mathrm{l.dh}_\Lambda(A) \leqslant n$. The relation $\mathrm{l.gl.dim}\,\Lambda \leqslant n$ now follows from Theorem 15.

7.8 Commutative Noetherian rings

At this point we shall digress from the main investigations to establish a few facts which will prove useful when, at a later stage, we make a detailed study of commutative Noetherian rings of finite global dimension. We return to the main discussion in section (7.9).

In the commutative case the distinction between left Noetherian and right Noetherian disappears and we speak simply of a Noetherian ring. Let Λ be a commutative Noetherian ring and let A and B be finitely generated Λ-modules. To be definite, let A be generated by $a_1, a_2, \ldots, a_r$ and B by $b_1, b_2, \ldots, b_s$. Then the Λ-module $A \otimes_\Lambda B$ is generated by the elements $a_i \otimes b_j$ and so, in particular, it is finitely generated. We assert that $\mathrm{Hom}_\Lambda(A, B)$ is also a finitely generated Λ-module. To see this we observe first that, by Theorem 6 of section (2.5), $\mathrm{Hom}_\Lambda(\Lambda, B) \approx B$. Next, if F is Λ-free and has a finite base, then F is isomorphic to a direct sum in which each term is a copy of Λ. Accordingly, by Theorem 8 of section (2.5), $\mathrm{Hom}_\Lambda(F, B)$ is isomorphic to a finite direct sum of the form $B + B + \ldots + B$, and therefore it is finitely generated as a Λ-module. But, if F is suitably chosen, we have an epimorphism $F \to A$, and hence, since Hom_Λ is left exact, an exact sequence $0 \to \mathrm{Hom}_\Lambda(A, B) \to \mathrm{Hom}_\Lambda(F, B)$. The fact that $\mathrm{Hom}_\Lambda(A, B)$ is a finitely generated Λ-module now follows from Theorem 17.

Proposition 11. *If Λ is a commutative Noetherian ring and A and B are finitely generated Λ-modules then, for each $n \geqslant 0$, the Λ-modules $\mathrm{Tor}_n^\Lambda(A, B)$ and $\mathrm{Ext}_\Lambda^n(A, B)$ are both finitely generated.*

Proof. Using Proposition 8, we see that we can choose a free resolution $\mathbf{F}$ of A in which each component F_n has a finite base. By Theorem 4, the $\mathrm{Tor}_n^\Lambda(A, B)$ are effectively the homology modules of the complex

$$\cdots \to F_{n+1} \otimes_\Lambda B \to F_n \otimes_\Lambda B \to F_{n-1} \otimes_\Lambda B \to \cdots$$

and, by our observation, $F_n \otimes_\Lambda B$ is a finitely generated Λ-module and therefore (Theorem 17) a Λ-Noetherian module. Hence $\mathrm{Tor}_n^\Lambda(A, B)$ is isomorphic to a factor module of a submodule of a Noetherian module. This shows that it is itself Noetherian and, in particular, we see that it is finitely generated.

The proof that $\mathrm{Ext}^n_\Lambda(A, B)$ is finitely generated is similar.

7.9 Global dimension of Noetherian rings

After the digression of the last section we return to the discussion of non-commutative rings with the idea of relating their right and left global dimensions.

Lemma 3. *Let A be a right Λ-module and let n $(n \geq 0)$ be an integer. If now $\mathrm{Tor}^\Lambda_n(A, B) = 0$ for all left Λ-modules B, then also*

$$\mathrm{Tor}^\Lambda_{n+1}(A, B) = 0$$

for all B.

Proof. Let B be a left Λ-module and let us construct an exact sequence $0 \to B' \to P \to B \to 0$, where P is Λ-projective. This yields an exact sequence

$$\mathrm{Tor}^\Lambda_{n+1}(A, P) \to \mathrm{Tor}^\Lambda_{n+1}(A, B) \to \mathrm{Tor}^\Lambda_n(A, B')$$

in which the first term is null by Proposition 1, and the last term is null by hypothesis. Accordingly, $\mathrm{Tor}^\Lambda_{n+1}(A, B) = 0$ and this establishes the lemma.

Definition. If A is a right Λ-module and n $(n \geq -1)$ is an integer, then we say that *the weak homological dimension of A does not exceed n*, and we write $\mathrm{w.r.dh}_\Lambda(A) \leq n$, if $\mathrm{Tor}^\Lambda_m(A, B) = 0$ whenever B is a left Λ-module and m is an integer strictly greater than n.

In other words, to say that the weak homological dimension of A does not exceed n is to assert that all of $\mathrm{Tor}^\Lambda_{n+1}(A, B), \mathrm{Tor}^\Lambda_{n+2}(A, B), \ldots$ vanish identically as functors of B. However, in view of the last lemma, we may observe that

$$\mathrm{w.r.dh}_\Lambda(A) \leq n \rightleftarrows \mathrm{Tor}^\Lambda_{n+1}(A, B) = 0 \quad \text{for all } B, \tag{7.9.1}$$

where the double arrow is used to signify the equivalence of the two statements which it separates. The *weak homological dimension* of the right Λ-module A is now defined as the smallest integer k such that $\mathrm{w.r.dh}_\Lambda(A) \leq k$ and we then write $\mathrm{w.r.dh}_\Lambda(A) = k$. Should, however, it happen that for no value of m does the functor $\mathrm{Tor}^\Lambda_m(A, B)$ vanish identically in B, then we say that the weak homological dimension of A is 'plus infinity' and write $\mathrm{w.r.dh}_\Lambda(A) = \infty$.

There is, of course, a result corresponding to Lemma 3 in which the roles of right and left modules are interchanged. This leads to the concept of the weak homological dimension of a left Λ-module B, which we abbreviate to $\text{w.l.dh}_\Lambda(B)$, and in place of (7.9.1) we then obtain

$$\text{w.l.dh}_\Lambda(B) \leqslant n \rightleftarrows \text{Tor}_{n+1}^\Lambda(A,B) = 0 \quad \text{for all } A. \tag{7.9.2}$$

Now let A be a variable right Λ-module and B a variable left Λ-module. If $\sup_A \text{w.r.dh}_\Lambda(A) \leqslant n$, then Tor_{n+1}^Λ takes only null values and the converse of this statement is also true. Since a similar observation is valid for *left* Λ-modules, we have

$$\sup_A \text{w.r.dh}_\Lambda(A) = \sup_B \text{w.l.dh}_\Lambda(B). \tag{7.9.3}$$

Definition. The common value of $\sup_A \text{w.r.dh}_\Lambda(A)$ and $\sup_B \text{w.l.dh}_\Lambda(B)$ will be called the *weak global dimension* of Λ and it will be denoted by $\text{w.gl.dim}\,\Lambda$.

Note that, since $\text{Tor}_0^\Lambda(\Lambda,\Lambda) \approx \Lambda$ which is not null, we have $\text{w.gl.dim}\,\Lambda \geqslant 0$. Also if n $(n \geqslant 0)$ is an integer, then

$$\text{w.gl.dim}\,\Lambda \leqslant n \rightleftarrows \text{Tor}_{n+1}^\Lambda \quad \text{takes only null values.} \tag{7.9.4}$$

It is important to note that a ring has only a *single* weak global dimension, whereas it has both a left and a right global dimension.

Theorem 18. *If A is a right Λ-module and B is a left Λ-module, then*

$$\text{w.r.dh}_\Lambda(A) \leqslant \text{r.dh}_\Lambda(A) \quad \textit{and} \quad \text{w.l.dh}_\Lambda(B) \leqslant \text{l.dh}_\Lambda(B).$$

Further, $\text{w.gl.dim}\,\Lambda$ *does not exceed either* $\text{r.gl.dim}\,\Lambda$ *or* $\text{l.gl.dim}\,\Lambda$.

Proof. It will be enough to prove the first assertion since the second can be proved similarly and the remainder can be established by taking upper bounds. Moreover, we may suppose that $\text{r.dh}_\Lambda(A) = n$ (a finite integer), since otherwise the relation to be established is obvious. This being supposed, there exists a projective resolution $\mathbf{P}$ of A such that $P_m = 0$ for all $m > n$. Now, by Theorem 4, the $\text{Tor}_m^\Lambda(A,B)$ are isomorphic to the homology modules of the complex

$$\cdots \to P_2 \otimes_\Lambda B \to P_1 \otimes_\Lambda B \to P_0 \otimes_\Lambda B \to 0 \to 0 \to \cdots,$$

and this shows that $\text{Tor}_{n+1}^\Lambda(A,B) = 0$. But in this relation B is arbitrary, consequently, by (7.9.1), $\text{w.r.dh}_\Lambda(A) \leqslant n$. This completes the proof.

For the rest of the chapter we shall be concerned with conditions which allow us to state that the inequalities described in Theorem 18 are in fact equalities. For this we need two lemmas.

Let A and C be right Λ-modules and let M be a Z-module. Then, as we saw in section (5.3), $\mathrm{Hom}_Z(C, M)$ can be given the structure of a *left* Λ-module in such a way that, if ϕ belongs to $\mathrm{Hom}_Z(C, M)$, then $(\lambda\phi)c = \phi(c\lambda)$ whenever c belongs to C and λ to Λ. Indeed, we may say more, for, as is easily verified, this construction turns $\mathrm{Hom}_Z(C, M)$ into an additive functor of C and M taking values in the category of left Λ-modules. It is now possible to form $A \otimes_\Lambda \mathrm{Hom}_Z(C, M)$ and this will be a Z-module. It will be shown that there exists a natural Z-homomorphism from

$$A \otimes_\Lambda \mathrm{Hom}_Z(C, M) \quad \text{to} \quad \mathrm{Hom}_Z(\mathrm{Hom}_\Lambda(A, C), M).$$

To this end let $a \in A$ and $\phi \in \mathrm{Hom}_Z(C, M)$, then there exists a Z-homomorphism $\chi : \mathrm{Hom}_\Lambda(A, C) \to M$ defined by $\chi(\psi) = \phi\psi(a)$, whenever ψ belongs to $\mathrm{Hom}_\Lambda(A, C)$. Write $\mu^*(a, \phi) = \chi$, then we can verify, without difficulty, that, with an obvious notation,

$$\mu^*(a_1 + a_2, \phi) = \mu^*(a_1, \phi) + \mu^*(a_2, \phi),$$

$$\mu^*(a, \phi_1 + \phi_2) = \mu^*(a, \phi_1) + \mu^*(a, \phi_2)$$

and

$$\mu^*(a\lambda, \phi) = \mu^*(a, \lambda\phi).$$

This shows that μ^* induces a Z-homomorphism

$$\mu : A \otimes_\Lambda \mathrm{Hom}_Z(C, M) \to \mathrm{Hom}_Z(\mathrm{Hom}_\Lambda(A, C), M), \tag{7.9.5}$$

where

$$[\mu(a \otimes \phi)]\psi = \phi\psi(a). \tag{7.9.6}$$

Again, each of $A \otimes_\Lambda \mathrm{Hom}_Z(C, M)$ and $\mathrm{Hom}_Z(\mathrm{Hom}_\Lambda(A, C), M)$ is an additive functor which is covariant in A, M and contravariant in C. Hence, if we have Λ-homomorphisms $A \to A'$, $C' \to C$ and a Z-homomorphism $M \to M'$, then there results a diagram

$$\begin{array}{ccc} A \otimes_\Lambda \mathrm{Hom}_Z(C, M) & \xrightarrow{\mu} & \mathrm{Hom}_Z(\mathrm{Hom}_\Lambda(A, C), M) \\ \downarrow & & \downarrow \\ A' \otimes_\Lambda \mathrm{Hom}_Z(C', M') & \xrightarrow{\mu'} & \mathrm{Hom}_Z(\mathrm{Hom}_\Lambda(A', C'), M') \end{array}$$

which one can verify, in a straightforward manner, to be commutative. Accordingly, (7.9.5) defines a natural transformation of functors and so we shall refer to μ as the canonical transformation

$$A \otimes_\Lambda \mathrm{Hom}_Z(C, M) \to \mathrm{Hom}_Z(\mathrm{Hom}_\Lambda(A, C), M).$$

Lemma 4. *If A is a free module with a finite base, then the canonical mapping*

$$\mu : A \otimes_\Lambda \mathrm{Hom}_Z(C, M) \to \mathrm{Hom}_Z(\mathrm{Hom}_\Lambda(A, C), M)$$

is an isomorphism.

Proof. First consider the case in which $A = \Lambda$. By Theorem 1 of section (2.3) and Theorem 6 of section (2.5), we have isomorphisms

$$\Lambda \otimes_\Lambda \mathrm{Hom}_Z(C, M) \approx \mathrm{Hom}_Z(C, M) \approx \mathrm{Hom}_Z(\mathrm{Hom}_\Lambda(\Lambda, C), M).$$

However, the combined mapping coincides with the canonical mapping, and so the assertion is established in this special case. For the more general situation described in the lemma, A has a complete representation (as a direct sum) $\Lambda_i \to A \to \Lambda_i$ $(1 \leqslant i \leqslant n)$, where, for each i, $\Lambda_i = \Lambda$ though the mappings corresponding to different values of i need not be the same. We now construct, in an obvious manner, a commutative diagram.

$$\begin{array}{ccc} \sum_i \{\Lambda_i \otimes_\Lambda \mathrm{Hom}_Z(C, M)\} & \longrightarrow & A \otimes_\Lambda \mathrm{Hom}_Z(C, M) \\ \downarrow & & \downarrow \\ \sum_i \{\mathrm{Hom}_Z(\mathrm{Hom}_\Lambda(\Lambda_i, C), M)\} & \longrightarrow & \mathrm{Hom}_Z(\mathrm{Hom}_\Lambda(A, C), M) \end{array}$$

in which the sums are direct. By Theorem 1 of section (3.9), both horizontal maps are isomorphisms, while the left vertical map is an isomorphism by the special case already considered. It follows that the right vertical map is also an isomorphism and this completes the proof.

Lemma 5. *Let Λ be a right Noetherian ring and suppose that A is a finitely generated right Λ-module while M is a Z-injective module. Then, for any right Λ-module C and any $n \geqslant 0$, we have a module isomorphism*

$$\mathrm{Tor}_n^\Lambda(A, \mathrm{Hom}_Z(C, M)) \approx \mathrm{Hom}_Z(\mathrm{Ext}_\Lambda^n(A, C), M).$$

Proof. Put

$$U(A) = A \otimes_\Lambda \mathrm{Hom}_Z(C, M) \quad \text{and} \quad V(A) = \mathrm{Hom}_Z(\mathrm{Hom}_\Lambda(A, C), M),$$

then $U(A)$ and $V(A)$ are covariant functors of A and the mapping $\mu : U(A) \to V(A)$ of (7.9.5) is a natural transformation. We can now construct, using Proposition 8, a free resolution $\mathbf{F}$ of A in which each module F_n has a finite base. Next the transformation μ produces a translation

$$\begin{array}{ccccccc} \cdots \to U(F_{n+1}) & \longrightarrow & U(F_n) & \longrightarrow & U(F_{n-1}) \to \cdots \\ \downarrow & & \downarrow & & \downarrow \\ \cdots \to V(F_{n+1}) & \longrightarrow & V(F_n) & \longrightarrow & V(F_{n-1}) \to \cdots \end{array}$$

and Lemma 4 shows that all the vertical maps are isomorphisms. It therefore follows that $H_n(U(\mathbf{F})) \approx H_n(V(\mathbf{F}))$. Put

$$W(D) = \mathrm{Hom}_Z(D, M),$$

where D denotes a variable Z-module, then $V(\mathbf{F}) = W(\mathbf{X})$, where $\mathbf{X} = \mathrm{Hom}_\Lambda(\mathbf{F}, C)$. But M is Z-injective and therefore, by Theorem 7 of section (5.2), the functor W is exact. It is accordingly possible to make use of Theorem 1 of section (6.1) and this shows that $H_n(W(\mathbf{X}))$ is isomorphic to $W(H^n(\mathbf{X}))$. Collecting results we have

$$\begin{aligned} H_n(\mathbf{F} \otimes_\Lambda \mathrm{Hom}_Z(C, M)) &= H_n(U(\mathbf{F})) \approx H_n(V(\mathbf{F})) \\ &= H_n(W(\mathbf{X})) \approx W(H^n(\mathbf{X})) \\ &= W(H^n(\mathrm{Hom}_\Lambda(\mathbf{F}, C))). \end{aligned}$$

Thus, changing the notation and appealing to Theorems 4 and 9, we find that

$$\mathrm{Tor}_n^\Lambda(A, \mathrm{Hom}_Z(C, M)) \approx \mathrm{Hom}_Z(\mathrm{Ext}_\Lambda^n(A, C), M).$$

It is now possible to prove the theorems for which the lemmas have been a preparation.

Theorem 19. *Let Λ be a right Noetherian ring and A a finitely generated right Λ-module. Then* $\mathrm{w.r.dh}_\Lambda(A) = \mathrm{r.dh}_\Lambda(A)$.

Remark. It should be noted that, in the case of left Noetherian rings, a corresponding result holds for finitely generated left Λ-modules.

Proof. By Theorem 18 we need only show that $\mathrm{r.dh}_\Lambda(A) \leqslant \mathrm{w.r.dh}_\Lambda(A)$ and for this purpose we may suppose that $\mathrm{w.r.dh}_\Lambda(A)$ is a finite integer n ($n \geqslant -1$). Let C be an arbitrary right Λ-module and M an arbitrary injective Z-module then, by Lemma 5 and (7.9.1),

$$\mathrm{Hom}_Z(\mathrm{Ext}_\Lambda^{n+1}(A, C), M) \approx \mathrm{Tor}_{n+1}^\Lambda(A, \mathrm{Hom}_Z(C, M)) = 0.$$

But, by Theorem 8 of section (5.2), it is possible to select M so that $\mathrm{Ext}_\Lambda^{n+1}(A, C)$ is one of its submodules and, by doing this, we see that $\mathrm{Ext}_\Lambda^{n+1}(A, C) = 0$. Since, however, this holds for all C it follows (Theorem 12) that $\mathrm{r.dh}_\Lambda(A) \leqslant n$. This completes the proof.

Theorem 20. *If Λ is a right Noetherian ring then*

$$\mathrm{w.gl.dim}\,\Lambda = \mathrm{r.gl.dim}\,\Lambda,$$

while if Λ *is left Noetherian then* w.gl.dim Λ = l.gl.dim Λ. *In particular, if* Λ *is both right and left Noetherian then*

$$\text{r.gl.dim}\,\Lambda = \text{w.gl.dim}\,\Lambda = \text{l.gl.dim}\,\Lambda.$$

Proof. It is clearly enough to prove the first assertion and, by Theorem 18, this will follow if we show that r.gl.dim $\Lambda \leqslant$ w.gl.dim Λ. Let A be an arbitrary cyclic right Λ-module, then, by Theorem 19,

$$\text{r.dh}_\Lambda(A) = \text{w.r.dh}_\Lambda(A) \leqslant \text{w.gl.dim}\,\Lambda,$$

and now the required relation follows from Theorem 15 after it has been adapted to make it applicable to right modules.

8

SOME USEFUL IDENTITIES

Notation. Λ, Γ and Δ denote rings, which need not be commutative, but which possess identity elements. Z denotes the ring of integers.

8.1 Bimodules

In Chapter 2 we established enough of the elementary properties of $A \otimes_\Lambda C$ and $\mathrm{Hom}_\Lambda(B, C)$ to enable us to proceed, for a considerable distance, with the general theory. However, when dealing with special problems, we shall require certain additional results, and it is these which will now be considered. The kind of situation envisaged is that in which we wish to assert that two apparently different constructions with modules lead to the same result. These situations normally arise only when we have to deal with comparatively elaborate operations with modules; and usually, before such an operation can be carried out, one at least of the modules concerned has to have a double structure. For example, in order to be able to form $A \otimes_\Lambda (B \otimes_\Gamma C)$ we require that $B \otimes_\Gamma C$ should be a left Λ-module. If either B or C is a Λ-module, *in addition to being a* Γ*-module*, then we can hope to transfer this property to $B \otimes_\Gamma C$, but clearly something of this kind is needed.

In this way we come to the consideration of objects which are modules with respect to each of a pair of rings. Roughly speaking, this is what we understand by a bimodule the precise definition being as follows:

Definition. If an additive group M is both a Λ-module and a Γ-module, and if, whenever λ belongs to Λ and γ to Γ, multiplication (of elements of M) by λ commutes with multiplication by γ, then M is said to be a *bimodule* or, more explicitly, a (Λ, Γ)*-module.*

Suppose, for example, that the additive group M is a left Λ-module and a right Γ-module, then M is a (Λ, Γ)-module provided that $\lambda(x\gamma) = (\lambda x)\gamma$ whenever $\lambda \in \Lambda$, $\gamma \in \Gamma$ and $x \in M$. On the other hand, if M is a left Λ-module and also a left Γ-module, then we require that $\lambda(\gamma x) = \gamma(\lambda x)$, for all such λ, γ and x, before M can be classed as a bimodule.

In all we have *four* types of (Λ, Γ)-module because Λ can act on the right or the left and so also can Γ. If either Λ or Γ is commutative, then the four types reduce to *two*, while if Λ and Γ are both commutative we have essentially only *one* type of bimodule.

Definition. If M and N are (Λ, Γ)-modules of the same type then a mapping $M \to N$, which is both a Λ-homomorphism and a Γ-homomorphism, is called a *bihomomorphism* or a (Λ, Γ)-*homomorphism.*

It should be noted that the (Λ, Γ)-modules of a given type and their bihomomorphisms form a category.

The concept of a bimodule will now be illustrated by means of two examples. Let M be a Λ-module and suppose (for definiteness) that Λ operates on the left. If now k is an integer, $\lambda \in \Lambda$ and $x \in M$, then $\lambda(kx) = k(\lambda x)$. This shows that every Λ-module can be regarded as a (Λ, Z)-module, and, indeed, we have made use of this fact in earlier chapters though without using the terminology of the theory of bimodules. Again, if Γ is a *commutative* ring and N is a Γ-module, then $\gamma_1(\gamma_2 y) = \gamma_2(\gamma_1 y)$ whenever γ_1, γ_2 are in Γ and y belongs to N. This shows that (for a commutative ring Γ) every Γ-module can be regarded as a *special* kind of (Γ, Γ)-module, and indeed, here we have the real reason why tensor products and groups of homomorphisms can be given additional structure when the ground-ring is commutative.

8.2 General principles

Let $T(A)$ be an *additive* functor where, with the notation of section (3.9), A varies in $\mathscr{G}_\Lambda$ and the values of $T(A)$ are in $\mathscr{G}_\Delta$. If now, in a special case, A is not only a Λ-module but rather a (Λ, Γ)-module, then it is natural to ask whether anything more can be said about $T(A)$.

In answering this question, it is convenient to treat covariant and contravariant functors separately, at least to begin with. Consider first of all an additive *covariant* functor $U(A)$ from $\mathscr{G}_\Lambda$ to $\mathscr{G}_\Delta$. Let us agree that, when speaking of a (Λ, Γ)-module A, it is to be understood that we mean the type to be such that $U(A)$ is defined; also the elements of Γ will be supposed (for definiteness) to multiply those of the bimodule on the left. For a fixed γ in Γ, the mapping $a \to \gamma a$ defines a Λ-homomorphism of such a bimodule A into itself. Denote this Λ-homomorphism by (γ) then it is clear that, when γ and γ' are both in Γ,

$$(\gamma + \gamma') = (\gamma) + (\gamma'), \quad (\gamma\gamma') = (\gamma)(\gamma'),$$

and it is seen that

$$U((\gamma+\gamma')) = U((\gamma))+U((\gamma')), \quad U((\gamma\gamma')) = U((\gamma))\,U((\gamma')),$$
$$U((1)) = \text{identity}. \quad (8.2.1)$$

For each x belonging to $U(A)$ put

$$\gamma x = [U((\gamma))]x,$$

so that γx denotes the effect of operating on x with $U((\gamma))$, then it follows, from (8.2.1), that $U(A)$ is a Γ-module as well as a Δ-module. But $U((\gamma))$ is a Δ-homomorphism, consequently, for $U(A)$, multiplication by an element of Γ always commutes with multiplication by an element of Δ. Accordingly, we have shown that $U(A)$ is a (Δ, Γ)-module. (In the case where A has the elements of Γ operating on the *right*, a similar discussion shows that $U(A)$ is a (Δ, Γ)-module with the elements of Γ acting as *right* operators.) Thus, to sum up, *when the additive functor U is covariant and A is a (Λ, Γ)-module, then $U(A)$ can be given the structure of a (Δ, Γ)-module where, as Γ-modules, A and $U(A)$ are of the same type.*

In the case of an additive *contravariant* functor $V(A)$ say, again from $\mathscr{G}_\Lambda$ to $\mathscr{G}_\Delta$, a slight modification is necessary. If, as above, A is a bimodule having the elements of Γ as *left* multipliers, then we write

$$y\gamma = [V((\gamma))]y \quad (y \in V(A)).$$

With this definition it is found that $V(A)$ becomes a bimodule with the elements of Γ multiplying on the *right* and so we have, on this occasion, a change of type in the structure as Γ-module. (In the situation where the bimodule A is a *right* Γ-module, $V(A)$ will be a *left* Γ-module.) Thus, to sum up again, we may say that *when V is an additive contravariant functor and A is a (Λ, Γ)-module, then $V(A)$ can be regarded as a (Δ, Γ)-module where, considered as Γ-modules, A and $V(A)$ are of contrasting types.*

Now assume that A and A' are (Λ, Γ)-modules of the same type and that $f: A \to A'$ is a (Λ, Γ)-homomorphism. Then, for any γ in Γ, the diagram†

$$\begin{array}{ccc} A & \xrightarrow{f} & A' \\ {\scriptstyle(\gamma)}\downarrow & & \downarrow{\scriptstyle(\gamma)} \\ A & \xrightarrow{f} & A' \end{array}$$

is commutative and therefore so also are the two further diagrams

$$\begin{array}{ccc} U(A) & \longrightarrow & U(A') \\ {\scriptstyle U((\gamma))}\downarrow & & \downarrow{\scriptstyle U((\gamma))} \\ U(A) & \longrightarrow & U(A') \end{array} \qquad \begin{array}{ccc} V(A') & \longrightarrow & V(A) \\ {\scriptstyle V((\gamma))}\downarrow & & \downarrow{\scriptstyle V((\gamma))} \\ V(A') & \longrightarrow & V(A) \end{array}$$

† As a temporary expedient, we use (γ) to denote two different mappings.

This means that $U(A) \to U(A')$ and $V(A') \to V(A)$ are (Δ, Γ)-homomorphisms. Combining the covariant and contravariant cases together, we can summarize in the following way: *when $T(A)$ is an additive functor and we restrict A to the category of (Λ, Γ)-modules and their (Λ, Γ)-homomorphisms, then we may regard T as a functor of these bimodules which takes values in the category of (Δ, Γ)-modules.*

Next consider an additive functor $T(A, A_1, A_2, \ldots, A_k)$ of several variables, where A varies in $\mathscr{G}_\Lambda$, A_r in $\mathscr{G}_{\Lambda_r}$ $(1 \leqslant r \leqslant k)$ and the values of T are in $\mathscr{G}_\Delta$. Among $A_1, A_2, \ldots, A_k$ let the covariant variables be A_i $(i \in I)$ and the contravariant ones A_j $(j \in J)$. If now A is a (Λ, Γ)-module, γ belongs to Γ and $A_i \to A_i'$ $(i \in I)$, $A_j' \to A_j$ $(j \in J)$ are Λ_i-homomorphisms, Λ_j-homomorphisms respectively, then the diagram

$$\begin{array}{ccc} T(A, A_1, \ldots, A_k) & \longrightarrow & T(A, A_1', \ldots, A_k') \\ \downarrow & & \downarrow \\ T(A, A_1, \ldots, A_k) & \longrightarrow & T(A, A_1', \ldots, A_k') \end{array}$$

is commutative, where the vertical maps are induced by the Λ-homomorphism $(\gamma) : A \to A$. This means simply that

$$T(A, A_1, \ldots, A_k) \to T(A, A_1', \ldots, A_k')$$

is a (Δ, Γ)-homomorphism of the (Δ, Γ)-modules $T(A, A_1, \ldots, A_k)$ and $T(A, A_1', \ldots, A_k')$. Combining this with our earlier result concerning a variation in the single variable A, we arrive at the following conclusion: *if f is a (Λ, Γ)-homomorphism and, for each r $(1 \leqslant r \leqslant k)$, f_r is a Λ_r-homomorphism, then $T(f, f_1, \ldots, f_k)$ is a (Δ, Γ)-homomorphism.*

In the above, we have examined what happens when the *first* variable is restricted to vary in a category of bimodules, but, of course, what has been said applies equally well to each of the other variables.

After these very general observations, let us look at some examples. If A is a right Λ-module and B is a left Λ-module, then we can form $A \otimes_\Lambda B$ and this is simply a Z-module, that is to say, an additive abelian group. Should, however, A happen to be a (Λ, Γ)-module with the elements of Γ as left multipliers, then $A \otimes_\Lambda B$ is a (Z, Γ)-module, or, to put it both more fully and more simply, a left Γ-module. Further, if $\gamma \in \Gamma$ and $x \in A \otimes_\Lambda B$, then γx is the result of operating on x with $(\gamma) \otimes i_B$, where i_B denotes the identity map of B. In other words, $A \otimes_\Lambda B$ is a left Γ-module, and, whenever $a \in A$, $b \in B$ and $\gamma \in \Gamma$, we have

$$\gamma(a \otimes b) = (\gamma a) \otimes b. \tag{8.2.2}$$

In the same way, when A is a bimodule with the elements of Γ operating on the right, $A \otimes_\Lambda B$ is a right Γ-module and

$$(a \otimes b)\gamma = (a\gamma) \otimes b. \tag{8.2.3}$$

Of course, similar remarks apply to the case in which B, instead of A, is a (Λ, Γ)-module.

Again, if Λ is commutative, then, by regarding A as a (Λ, Λ)-module, we can turn $A \otimes_\Lambda B$ into a Λ-module and we can also do this by regarding B as a (Λ, Λ)-module. However, (8.2.2) and similar relations show that these two structures are in fact the same and coincide with that defined in section (2.2).

Next consider $\mathrm{Hom}_\Lambda(A, B)$ where now (for definiteness) we suppose that A and B are both left Λ-modules. If A is a (Λ, Γ)-module and the elements of Γ multiply those of A on the left, then, since $\mathrm{Hom}_\Lambda(A, B)$ is *contravariant* in A, it acquires the structure of a right Γ-module. Indeed, if $f \in \mathrm{Hom}_\Lambda(A, B)$, then $f\gamma = [\mathrm{Hom}((\gamma), i_B)]f$, which means that

$$(f\gamma)a = f(\gamma a) \tag{8.2.4}$$

for all elements a of A. On the other hand, when the elements of A are multiplied on the right by those of Γ, then $\mathrm{Hom}_\Lambda(A, B)$ is a left Γ-module and instead of (8.2.4) we have

$$(\gamma f)a = f(a\gamma). \tag{8.2.5}$$

Now suppose that it is B which is a (Λ, Γ)-module. For this case $\mathrm{Hom}_\Lambda(A, B)$ is a Γ-module of the same type as B and

$$(\gamma f)a = \gamma(f(a)) \tag{8.2.6}$$

when both are left Γ-modules, while

$$(f\gamma)a = (f(a))\gamma \tag{8.2.7}$$

when both are right Γ-modules.

When Λ is commutative, the procedures just described can be used to turn $\mathrm{Hom}_\Lambda(A, B)$ into a Λ-module in two different ways. However, they both give the same structure and this turns out to be identical with the one already introduced in section (1.4).

Finally, still assuming that Λ is commutative, we note that it is possible to turn $\mathrm{Tor}_n^\Lambda(A, B)$ into a Λ-module either by regarding A or by regarding B as a (Λ, Λ)-module; in addition, we have already, in section (7.1), given yet another method for endowing $\mathrm{Tor}_n^\Lambda(A, B)$ with such a structure. But if $\lambda \in \Lambda$, then, by (7.1.3),

$$\mathrm{Tor}_n^\Lambda[(\lambda), i_B] = \mathrm{Tor}_n^\Lambda[\lambda i_A, i_B] = \lambda\,\mathrm{Tor}_n^\Lambda[i_A, i_B],$$

which, properly interpreted, asserts that the first and third structures are the same. Of course, in a similar way, we can also show that the second and third structures coincide.

The same kind of observations apply to the extension functors $\mathrm{Ext}_n^{\Lambda}(A, B)$ in situations where Λ is commutative.

8.3 The associative law for tensor products

In our first theorem we have to deal with a right Λ-module A, a (Λ, Γ)-module B (with Λ operating on the left and Γ on the right) and a left Γ-module C. These modules are to be regarded as capable of variation within the categories to which they belong. Since, by the results of the last section, $B \otimes_\Gamma C$ is a left Λ-module and $A \otimes_\Lambda B$ is a right Γ-module, we can form both

$$A \otimes_\Lambda (B \otimes_\Gamma C) \quad \text{and} \quad (A \otimes_\Lambda B) \otimes_\Gamma C.$$

Further, each of these expressions is a covariant functor of A, B and C with values in the category of additive abelian groups. It will now be shown that these functors are equivalent.

Theorem 1. *There exists a Z-isomorphism*

$$\mu : A \otimes_\Lambda (B \otimes_\Gamma C) \approx (A \otimes_\Lambda B) \otimes_\Gamma C$$

which is uniquely defined by the property that $\mu[a \otimes (b \otimes c)] = (a \otimes b) \otimes c$ *whenever* $a \in A$, $b \in B$ *and* $c \in C$. *Moreover, when the modules are allowed to vary,* μ *is a functor equivalence.*

Proof. To begin with, let a be a fixed element of A and, for all b in B and c in C, put $\phi(b, c) = (a \otimes b) \otimes c$. Then, with an obvious notation,

$$\phi(b_1 + b_2, c) = \phi(b_1, c) + \phi(b_2, c), \quad \phi(b, c_1 + c_2) = \phi(b, c_1) + \phi(b, c_2),$$

and, when γ belongs to Γ,

$$\phi(b\gamma, c) = (a \otimes b\gamma) \otimes c = (a \otimes b)\gamma \otimes c = (a \otimes b) \otimes \gamma c = \phi(b, \gamma c).$$

This shows that ϕ determines a Z-homomorphism

$$\psi_a : B \otimes_\Gamma C \to (A \otimes_\Lambda B) \otimes_\Gamma C$$

such that
$$\psi_a(b \otimes c) = (a \otimes b) \otimes c. \tag{8.3.1}$$

Further, by the definitions,

$$\psi_a[\lambda(b \otimes c)] = \psi_a(\lambda b \otimes c) = (a \otimes \lambda b) \otimes c \quad (\lambda \in \Lambda),$$

so that
$$\psi_a[\lambda(b \otimes c)] = (a\lambda \otimes b) \otimes c. \tag{8.3.2}$$

For each element x of $B \otimes_\Gamma C$, put $\theta(a, x) = \psi_a(x)$. We contend that

$$\theta(a, x_1 + x_2) = \theta(a, x_1) + \theta(a, x_2),$$

$$\theta(a_1 + a_2, x) = \theta(a_1, x) + \theta(a_2, x)$$

and
$$\theta(a\lambda, x) = \theta(a, \lambda x).$$

The first of these relations follows immediately from the fact that ψ_a is a homomorphism. Next, by (8.3.1) and (8.3.2), the other two equations hold when x has the form $b \otimes c$ and now the general cases follow by summation.

The properties of θ ensure the existence of a homomorphism $\mu : A \otimes_\Lambda (B \otimes_\Gamma C) \to (A \otimes_\Lambda B) \otimes_\Gamma C$ for which $\mu(a \otimes x) = \psi_a(x)$ and therefore $\mu[a \otimes (b \otimes c)] = (a \otimes b) \otimes c$ whenever $a \in A$, $b \in B$ and $c \in C$. In precisely the same way we can establish the existence of a homomorphism

$$\nu : (A \otimes_\Lambda B) \otimes_\Gamma C \to A \otimes_\Lambda (B \otimes_\Gamma C)$$

for which $\nu[(a \otimes b) \otimes c] = a \otimes (b \otimes c)$. Clearly $\mu\nu$ and $\nu\mu$ are identity maps, consequently μ is an isomorphism. The assertion that μ sets up a functor equivalence is now obvious.

8.4 Tensor products over commutative rings

If Λ is a commutative ring and A, B, C are all Λ-modules, then both $A \otimes_\Lambda (B \otimes_\Lambda C)$ and $(A \otimes_\Lambda B) \otimes_\Lambda C$ are defined and, by Theorem 1, they are isomorphic. We could, if we wished, use this fact to define a tensor product $A \otimes_\Lambda B \otimes_\Lambda C$ in which no grouping of the modules is specified. However, this is rather a clumsy procedure, and it becomes even more awkward when a larger number of terms is involved. On the whole it is simpler to go back and make the necessary adjustments to the original definition.

To be more explicit, assume that Λ is a commutative ring and that $A_1, A_2, \ldots, A_k$ are all Λ-modules. Denote by $Z(A_1, A_2, \ldots, A_k)$ the free Z-module generated by all ordered sets $(a_1, a_2, \ldots, a_k)$ where now (and in the remainder of this section) a_r denotes an element of A_r. Next, among the submodules of $Z(A_1, A_2, \ldots, A_k)$, we let

$$Y(A_1, A_2, \ldots, A_k)$$

designate that one which is generated by all elements having one or other of the two forms

$$(a_1, \ldots, a_i + a_i', \ldots, a_k) - (a_1, \ldots, a_i, \ldots, a_k) - (a_1, \ldots, a_i', \ldots, a_k)$$

and
$$(a_1, \ldots, \lambda a_i, \ldots, a_j, \ldots, a_k) - (a_1, \ldots, a_i, \ldots, \lambda a_j, \ldots, a_k).$$

Here i and j vary freely in the range 1 to k and λ can be any element of Λ. Now put

$$A_1 \otimes_\Lambda A_2 \otimes_\Lambda \dots \otimes_\Lambda A_k = Z(A_1, A_2, \dots, A_k)/Y(A_1, A_2, \dots, A_k),$$

and denote by $a_1 \otimes a_2 \otimes \dots \otimes a_k$ the natural image of $(a_1, a_2, \dots, a_k)$ in $A_1 \otimes_\Lambda \dots \otimes_\Lambda A_k$. *At this stage*, we may say that $A_1 \otimes_\Lambda \dots \otimes_\Lambda A_k$ is a Z-module and, by the construction, we have identities of the types

$$a_1 \otimes \dots \otimes (a_i + a_i') \otimes \dots \otimes a_k$$
$$= (a_1 \otimes \dots \otimes a_i \otimes \dots \otimes a_k) + (a_1 \otimes \dots \otimes a_i' \otimes \dots \otimes a_k) \quad (8.4.1)$$

and

$$a_1 \otimes \dots \otimes \lambda a_i \otimes \dots \otimes a_j \otimes \dots \otimes a_k$$
$$= a_1 \otimes \dots \otimes a_i \otimes \dots \otimes \lambda a_j \otimes \dots \otimes a_k. \quad (8.4.2)$$

Further, every element of $A_1 \otimes_\Lambda \dots \otimes_\Lambda A_k$ can be expressed as a finite sum each of whose terms has the form $a_1 \otimes a_2 \otimes \dots \otimes a_k$.

Next consider a set $(f_1, f_2, \dots, f_k)$ of Λ-homomorphisms, where $f_r : A_r \to A_r'$. These determine a Z-homomorphism

$$f_1 \otimes \dots \otimes f_k : A_1 \otimes_\Lambda \dots \otimes_\Lambda A_k \to A_1' \otimes_\Lambda \dots \otimes_\Lambda A_k' \quad (8.4.3)$$

for which

$$(f_1 \otimes \dots \otimes f_k)(a_1 \otimes \dots \otimes a_k) = f_1(a_1) \otimes \dots \otimes f_k(a_k), \quad (8.4.4)$$

and now it is clear that $A_1 \otimes_\Lambda \dots \otimes_\Lambda A_k$ is an additive functor which is covariant in each of its variables. But Λ is commutative, consequently our general principles provide us with k methods of turning $A_1 \otimes_\Lambda \dots \otimes_\Lambda A_k$ into a Λ-module. However, by (8.4.2), we obtain only a single structure (and not k different structures) as Λ-module for our extended tensor product, and for this structure

$$\lambda(a_1 \otimes \dots \otimes a_i \otimes \dots \otimes a_k) = a_1 \otimes \dots \otimes \lambda a_i \otimes \dots \otimes a_k. \quad (8.4.5)$$

It follows at once, from (8.4.4), that $f_1 \otimes \dots \otimes f_k$ is not only a Z-homomorphism but also a Λ-homomorphism, and we may note, too, that if $\lambda \in \Lambda$ then

$$\lambda(f_1 \otimes \dots \otimes f_i \otimes \dots \otimes f_k) = f_1 \otimes \dots \otimes \lambda f_i \otimes \dots \otimes f_k \quad (8.4.6)$$

for $1 \leqslant i \leqslant k$. Hence $A_1 \otimes_\Lambda \dots \otimes_\Lambda A_k$ can (and normally should) be regarded as a covariant Λ-linear functor whose values lie in the category of Λ-modules.

The proof of the next theorem is omitted because it is entirely simple and straightforward.

Theorem 2. *Let Λ be a commutative ring, $(i_1, i_2, \ldots, i_k)$ a permutation of the numbers $(1, 2, \ldots, k)$ and $A_1, A_2, \ldots, A_k$ a sequence of Λ-modules. Then there exists a Λ-isomorphism*

$$\mu : A_1 \otimes_\Lambda A_2 \otimes_\Lambda \cdots \otimes_\Lambda A_k \approx A_{i_1} \otimes_\Lambda A_{i_2} \otimes_\Lambda \cdots \otimes_\Lambda A_{i_k}$$

such that $\quad \mu(a_1 \otimes a_2 \otimes \ldots \otimes a_k) = a_{i_1} \otimes a_{i_2} \otimes \ldots \otimes a_{i_k};$

and when the modules $A_1, A_2, \ldots, A_k$ vary, the isomorphism μ defines a functor equivalence.

Theorem 3. *Let Λ be a commutative ring and let $A_1, \ldots, A_k, B_1, \ldots, B_p$ be Λ-modules. Then there exists a Λ-isomorphism*

$$\mu : A_1 \otimes_\Lambda \cdots \otimes_\Lambda A_k \otimes_\Lambda B_1 \otimes_\Lambda \cdots \otimes_\Lambda B_p$$
$$\approx (A_1 \otimes_\Lambda \cdots \otimes_\Lambda A_k) \otimes_\Lambda (B_1 \otimes_\Lambda \cdots \otimes_\Lambda B_p) \quad (8.4.7)$$

such that

$$\mu(a_1 \otimes \ldots \otimes a_k \otimes b_1 \otimes \ldots \otimes b_p) = (a_1 \otimes \ldots \otimes a_k) \otimes (b_1 \otimes \ldots \otimes b_p).$$

Further, μ is a natural equivalence between the two sides of (8.4.7) when these are regarded as functors.

It is convenient to establish a preliminary lemma.

Lemma 1. *Let the assumptions be as in Theorem 3. If in*

$$A_1 \otimes_\Lambda A_2 \otimes_\Lambda \cdots \otimes_\Lambda A_k$$

we have

$$a'_1 \otimes \ldots \otimes a'_k + a''_1 \otimes \ldots \otimes a''_k + \ldots + a^*_1 \otimes \ldots \otimes a^*_k = 0,$$

then, for arbitrary elements $b_1, b_2, \ldots, b_p$ of $B_1, B_2, \ldots, B_p$ respectively,

$$a'_1 \otimes \ldots \otimes a'_k \otimes b_1 \otimes \ldots \otimes b_p + a''_1 \otimes \ldots \otimes a''_k \otimes b_1 \otimes \ldots \otimes b_p + \ldots$$
$$+ a^*_1 \otimes \ldots \otimes a^*_k \otimes b_1 \otimes \ldots \otimes b_p = 0.$$

Proof. It is only necessary to observe that the mapping

$$(a_1, \ldots, a_k) \to a_1 \otimes \ldots \otimes a_k \otimes b_1 \otimes \ldots \otimes b_p$$

induces a Λ-homomorphism

$$A_1 \otimes_\Lambda \cdots \otimes_\Lambda A_k \to A_1 \otimes_\Lambda \cdots \otimes_\Lambda A_k \otimes_\Lambda B_1 \otimes_\Lambda \cdots \otimes_\Lambda B_p$$

in which $a_1 \otimes \ldots \otimes a_k$ is mapped into $a_1 \otimes \ldots \otimes a_k \otimes b_1 \otimes \ldots \otimes b_p$.

***Proof of Theorem* 3.** It is immediately clear that there exists a Λ-homomorphism

$$\mu : A_1 \otimes_\Lambda \cdots \otimes_\Lambda A_k \otimes_\Lambda B_1 \otimes_\Lambda \cdots \otimes_\Lambda B_p$$
$$\to (A_1 \otimes_\Lambda \cdots \otimes_\Lambda A_k) \otimes_\Lambda (B_1 \otimes_\Lambda \cdots \otimes_\Lambda B_p)$$

with the property that

$$\mu(a_1 \otimes \dots \otimes a_k \otimes b_1 \otimes \dots \otimes b_p) = (a_1 \otimes \dots \otimes a_k) \otimes (b_1 \otimes \dots \otimes b_p).$$

Now let x belong to $A_1 \otimes_\Lambda \dots \otimes_\Lambda A_k$ and y to $B_1 \otimes_\Lambda \dots \otimes_\Lambda B_p$. We can express these elements as finite sums of the forms

$$x = \Sigma(a_1 \otimes \dots \otimes a_k), \quad y = \Sigma(b_1 \otimes \dots \otimes b_p)$$

and then (using Lemma 1 and its counterpart with the roles of the A_i and B_j interchanged)

$$\theta(x, y) = \Sigma\Sigma(a_1 \otimes \dots \otimes a_k \otimes b_1 \otimes \dots \otimes b_p) \tag{8.4.8}$$

depends only on x and y and not on their representations as sums. (It is to be understood, of course, that the typical term of the double sum in (8.4.8) is obtained by combining a term of the sum representing x with a term from the sum representing y.) Further, with an obvious notation,

$$\theta(x_1 + x_2, y) = \theta(x_1, y) + \theta(x_2, y),$$

$$\theta(x, y_1 + y_2) = \theta(x, y_1) + \theta(x, y_2),$$

$$\theta(\lambda x, y) = \lambda\theta(x, y) = \theta(x, \lambda y)$$

which shows that θ induces a Λ-homomorphism

$$\nu : (A_1 \otimes_\Lambda \dots \otimes_\Lambda A_k) \otimes_\Lambda (B_1 \otimes_\Lambda \dots \otimes_\Lambda B_p) \to A_1 \otimes_\Lambda \dots \otimes_\Lambda A_k \otimes_\Lambda B_1 \otimes_\Lambda \dots \otimes_\Lambda B_p$$

such that

$$\nu[(a_1 \otimes \dots \otimes a_k) \otimes (b_1 \otimes \dots \otimes b_p)] = a_1 \otimes \dots \otimes a_k \otimes b_1 \otimes \dots \otimes b_p.$$

Accordingly $\mu\nu$ and $\nu\mu$ are identity maps and the theorem follows for obvious reasons.

8.5 Mixed identities

The equivalences between functors or, as we may say instead, the *functor identities* considered in the last two sections, have been of a fairly obvious kind. We turn now to one which is less transparent and which concerns $\mathrm{Hom}_\Gamma(A \otimes_\Lambda B, C)$. *In this discussion, neither Λ nor Γ is assumed to be commutative.*

First consider conditions under which $\mathrm{Hom}_\Gamma(A \otimes_\Lambda B, C)$ has a meaning. Let A belong to the category of right Λ-modules and C to the category $\mathscr{G}_\Gamma$. Here it does not matter whether $\mathscr{G}_\Gamma$ consists of right Γ-modules or of left Γ-modules provided that we make a definite choice and keep it fixed throughout what follows. Having decided this, B is chosen to be a (Λ, Γ)-module with the elements of Λ operating on the left and with the elements of Γ operating on the same side of B as they do on the modules of $\mathscr{G}_\Gamma$. (In different terms, when regarded as a Γ-module, B has to belong to $\mathscr{G}_\Gamma$.)

With the above stipulations, $A \otimes_\Lambda B$ now takes on the structure of a Γ-module belonging to $\mathscr{G}_\Gamma$ and therefore $\mathrm{Hom}_\Gamma(A \otimes_\Lambda B, C)$ has a definite meaning. Further, this expression is a functor of the modules A, C and the bimodule B and, as such, it is contravariant in the variables A, B and covariant in C. But, by the results of section (8.2), $\mathrm{Hom}_\Gamma(B, C)$ has the structure of a right Λ-module, hence we can form $\mathrm{Hom}_\Lambda(A, \mathrm{Hom}_\Gamma(B, C))$ and this, too, is contravariant in A, B and covariant in C. The next theorem shows that, in spite of their different forms, $\mathrm{Hom}_\Gamma(A \otimes_\Lambda B, C)$ and $\mathrm{Hom}_\Lambda(A, \mathrm{Hom}_\Gamma(B, C))$ are, in fact, identical functors.

Theorem 4. *Under the conditions described above, there exists a canonical isomorphism*

$$\mathrm{Hom}_\Gamma(A \otimes_\Lambda B, C) \approx \mathrm{Hom}_\Lambda(A, \mathrm{Hom}_\Gamma(B, C)) \tag{8.5.1}$$

in which, if ϕ of $\mathrm{Hom}_\Gamma(A \otimes_\Lambda B, C)$ corresponds to ψ of

$$\mathrm{Hom}_\Lambda(A, \mathrm{Hom}_\Gamma(B, C)),$$

then

$$\phi(a \otimes b) = [\psi(a)]\,b \tag{8.5.2}$$

for all a in A and b in B. Furthermore, this isomorphism effects an equivalence between $\mathrm{Hom}_\Gamma(A \otimes_\Lambda B, C)$ and $\mathrm{Hom}_\Lambda(A, \mathrm{Hom}_\Gamma(B, C))$ when these are regarded as functors.

It should be noted that (8.5.2) serves to characterize the isomorphism completely.

Proof. Let ϕ belong to $\mathrm{Hom}_\Gamma(A \otimes_\Lambda B, C)$ and let a be an element of A. The mapping $b \to \phi(a \otimes b)$ is a Γ-homomorphism $B \to C$ which depends on a; accordingly, we may denote it by $\psi(a)$. An easy verification now shows that

$$\psi(a_1 + a_2) = \psi(a_1) + \psi(a_2) \quad \text{and} \quad \psi(a_1 \lambda) = \psi(a_1)\,\lambda,$$

where a_1, a_2 are in A and λ belongs to Λ; hence the mapping $a \to \psi(a)$ is an element ψ of $\mathrm{Hom}_\Lambda(A, \mathrm{Hom}_\Gamma(B, C))$ and ψ satisfies (8.5.2). A further simple verification shows that the mapping $\phi \to \psi$ is a homomorphism

$$\mathrm{Hom}_\Gamma(A \otimes_\Lambda B, C) \to \mathrm{Hom}_\Lambda(A, \mathrm{Hom}_\Gamma(B, C)), \tag{8.5.3}$$

and, since it is clear that $\psi = 0$ implies $\phi = 0$, it follows that (8.5.3) is a monomorphism.

In order to see that (8.5.3) is also an epimorphism, let ψ be an arbitrarily assigned element of $\mathrm{Hom}_\Lambda(A, \mathrm{Hom}_\Gamma(B,C))$, then the mapping $(a,b) \to [\psi(a)]\, b$ induces (in the usual way) a Z-homomorphism $\phi : A \otimes_\Lambda B \to C$ with the property that $\phi(a \otimes b) = [\psi(a)]\, b$. From this property we can check at once that ϕ is, in fact, a Γ-homomorphism and that its image under (8.5.3) is ψ. This completes the proof that (8.5.3) is an isomorphism.

Finally, let $u : A' \to A$ be a Λ-homomorphism, $w : C \to C'$ a Γ-homomorphism and $v : B' \to B$ a bihomomorphism. If now ϕ and ψ correspond in the isomorphism (8.5.3), then their respective images on applying the mappings

$$\mathrm{Hom}_\Gamma(A \otimes_\Lambda B, C) \to \mathrm{Hom}_\Gamma(A' \otimes_\Lambda B', C')$$

and $$\mathrm{Hom}_\Lambda(A, \mathrm{Hom}_\Gamma(B,C)) \to \mathrm{Hom}_\Lambda(A', \mathrm{Hom}_\Gamma(B',C'))$$

are $$\phi' = w\phi(u \otimes v) \quad \text{and} \quad \psi' = \mathrm{Hom}(v,w)\,\psi u.$$

Thus if $a' \in A'$ and $b' \in B'$, then, using (8.5.2),

$$\begin{aligned}\phi'(a' \otimes b') &= w\phi(u(a') \otimes v(b')) = [w(\psi u(a'))]\, v(b') \\ &= [w(\psi u(a'))\, v]\, b' = [\psi'(a')]\, b'.\end{aligned}$$

This shows that ϕ' and ψ' correspond under the canonical mapping and thereby establishes that (8.5.3) is a functor equivalence.

In the formulation of Theorem 4 we allowed B to be a bimodule in order that $\mathrm{Hom}_\Gamma(A \otimes_\Lambda B, C)$ should have a meaning, but equally well we could have imposed a double structure on A. Let us take this opportunity to observe how the theorem looks in these slightly different circumstances.

Under the conditions proposed, B would be a left Λ-module, C a module of $\mathscr{G}_\Gamma$ and A a (Λ, Γ)-module, the type of A being such that it is a right Λ-module and, as Γ-module, belongs to $\mathscr{G}_\Gamma$. Then, when A, B and C vary in their respective categories, $\mathrm{Hom}_\Gamma(A \otimes_\Lambda B, C)$ and $\mathrm{Hom}_\Lambda(B, \mathrm{Hom}_\Gamma(A,C))$ are both functors that are contravariant in A, B and covariant in C. By making trivial modifications to the arguments used to prove Theorem 4, these functors can be shown to be identical in the following sense: *there exists a natural equivalence*

$$\mathrm{Hom}_\Gamma(A \otimes_\Lambda B, C) \approx \mathrm{Hom}_\Lambda(B, \mathrm{Hom}_\Gamma(A,C)) \tag{8.5.4}$$

which is such that an element ϕ *of* $\text{Hom}_\Gamma(A \otimes_\Lambda B, C)$ *corresponds to the element* ψ *of* $\text{Hom}_\Lambda(B, \text{Hom}_\Gamma(A, C))$ *provided that*

$$\phi(a \otimes b) = [\psi(b)]\, a \tag{8.5.5}$$

for all $a \in A$ *and* $b \in B$.

8.6 Rings and modules of fractions

One of the most useful devices, in the theory of *commutative rings*, is that of forming rings and modules of fractions for, in this way, one can frequently reduce the study of a general problem to a similar one, in which the ring under consideration has only one maximal ideal. This device is of such common occurrence that it is given a special name. It is called the *method of reduction to the local case.*

In this section, we shall describe the process of forming fractions and study it in relation to torsion functors. Accordingly we suppose, from now on, that Λ is a *commutative* ring and that S *is a multiplicatively closed set of elements of* Λ *containing* 1 *but not containing* 0. By saying that S is *multiplicatively closed* we mean, of course, that if c and c' belong to S, then their product cc' again belongs to S. The assumption that 1 belongs to S is actually not essential (for a multiplicatively closed set remains multiplicatively closed when 1 is added to it) but it is an advantage and results in no real loss of generality.

Let A be a Λ-module and consider *formal quotients* of the type $\frac{a}{c}$, where $a \in A$ and $c \in S$. If $\frac{a_1}{c_1}$ and $\frac{a_2}{c_2}$ are two such quotients, we shall write $\frac{a_1}{c_1} \sim \frac{a_2}{c_2}$ if $cc_2a_1 = cc_1a_2$ for some element c in S, and then it is an easy matter to verify that this is, in fact, an equivalence relation. In future, the equivalence class to which a quotient $\frac{a}{c}$ belongs will be denoted by $\left[\frac{a}{c}\right]$ and the totality of these equivalence classes will be designated by A_S. The next step is to make A_S into a Λ-module and this is done by putting

$$\left[\frac{a_1}{c_1}\right] + \left[\frac{a_2}{c_2}\right] = \left[\frac{c_2a_1 + c_1a_2}{c_1c_2}\right], \tag{8.6.1}$$

$$\lambda\left[\frac{a}{c}\right] = \left[\frac{\lambda a}{c}\right]. \tag{8.6.2}$$

It must, of course, be verified that these definitions are effective in the sense that they depend only on the classes and not on the selected representatives, and also it is necessary to check that the axioms for a Λ-module are satisfied.† However, these verifications present no difficulty and we leave them to the reader.

Assume that $\phi: A \to A'$ is a homomorphism of Λ-modules then, if $\frac{a_1}{c_1} \sim \frac{a_2}{c_2}$, it is clear that $\frac{\phi(a_1)}{c_1} \sim \frac{\phi(a_2)}{c_2}$. Accordingly there is a well-defined mapping

$$A_S \to A'_S, \tag{8.6.3}$$

in which $\left[\frac{a}{c}\right] \to \left[\frac{\phi(a)}{c}\right]$. This mapping is obviously a Λ-homomorphism. It is now a simple matter to check that A_S, considered as belonging to the category of Λ-modules, is an additive covariant functor of A. Further, we have a canonical Λ-homomorphism

$$A \to A_S \tag{8.6.4}$$

in which $a \to \left[\frac{a}{1}\right]$, and this homomorphism defines a natural transformation of functors.

The construction which was applied to A can, of course, be applied to the Λ-module Λ and, if this is done, it yields Λ_S whose typical element has the form $\left[\frac{\lambda}{c}\right]$, where $\lambda \in \Lambda$ and $c \in S$. It is now easy to confirm that Λ_S can be given the structure of a commutative ring by setting

$$\left[\frac{\lambda_1}{c_1}\right]\left[\frac{\lambda_2}{c_2}\right] = \left[\frac{\lambda_1\lambda_2}{c_1c_2}\right]. \tag{8.6.5}$$

This ring, which has $\left[\frac{1}{1}\right]$ as its identity element, is called the *ring of fractions* (*or quotients*) *of* Λ *with respect to* S. The canonical mapping

$$\Lambda \to \Lambda_S, \tag{8.6.6}$$

where $\lambda \to \left[\frac{\lambda}{1}\right]$, is now a *ring homomorphism*; that is to say, if $f(\lambda)$ denotes the image of λ, then

$$f(\lambda_1+\lambda_2) = f(\lambda_1)+f(\lambda_2), \quad f(\lambda_1\lambda_2) = f(\lambda_1)f(\lambda_2)$$

and $f(1)$ is the identity element of Λ_S.

Once again, let A be a Λ-module. It is possible to turn A_S into a Λ_S-module in such a way that

$$\left[\frac{\lambda}{c}\right]\left[\frac{a_1}{c_1}\right] = \left[\frac{\lambda a_1}{cc_1}\right]. \tag{8.6.7}$$

† A_S is called a *module of fractions* (*or quotients*).

Further, as has already been observed, every Λ-homomorphism $\phi : A \to A'$ determines a homomorphism $\phi_S : A_S \to A'_S$, where

$$\phi_S\left(\left[\frac{a}{c}\right]\right) = \left[\frac{\phi(a)}{c}\right],$$

and it is now a matter for trivial verification that ϕ_S is a Λ_S-homomorphism. More precisely, it is seen that A_S *can be regarded as an additive covariant functor of A whose values belong to the category of Λ_S-modules.* We shall need to establish a number of properties of this functor.

Let us start by observing that Λ_S is a (Λ, Λ_S)-module, consequently it is possible, for every Λ-module A, to make $A \otimes_\Lambda \Lambda_S$ into a Λ_S-module in such a way that (with a self-explanatory notation)

$$\left(a \otimes \left[\frac{\lambda}{c}\right]\right)\left[\frac{\lambda'}{c'}\right] = a \otimes \left[\frac{\lambda\lambda'}{cc'}\right].$$

Indeed, $A \otimes_\Lambda \Lambda_S$ can be regarded as an additive covariant functor of A with values in the category of Λ_S-modules.

Theorem 5. *If A is a Λ-module then there exists a Λ_S-isomorphism*

$$u : A \otimes_\Lambda \Lambda_S \approx A_S$$

for which

$$u\left(a \otimes \left[\frac{\lambda}{c}\right]\right) = \left[\frac{\lambda a}{c}\right], \tag{8.6.8}$$

where $a \in A$, $\lambda \in \Lambda$ and $c \in S$. Moreover, this isomorphism defines an equivalence between $A \otimes_\Lambda \Lambda_S$ and A_S, regarded as functors of A.

Proof. If, with our previous notation, $\left[\frac{\lambda}{c}\right] = \left[\frac{\lambda'}{c'}\right]$, then, for any a in A, we have $\left[\frac{\lambda a}{c}\right] = \left[\frac{\lambda' a}{c'}\right]$. It is now immediately clear that we can construct a Z-homomorphism $u : A \otimes_\Lambda \Lambda_S \to A_S$ for which (8.6.8) holds and then a trivial verification shows that u is a Λ_S-homomorphism. Next we observe that if $x \in A_S$ and $x = \left[\frac{a}{c}\right] = \left[\frac{a'}{c'}\right]$, then $c_0 c' a = c_0 c a'$ for some $c_0 \in S$ and therefore

$$a \otimes \left[\frac{1}{c}\right] = c_0 c' a \otimes \left[\frac{1}{c_0 c' c}\right] = c_0 c a' \otimes \left[\frac{1}{c_0 c' c}\right] = a' \otimes \left[\frac{1}{c'}\right].$$

Accordingly we can define a mapping $v : A_S \to A \otimes_\Lambda \Lambda_S$ for which

$$v\left[\frac{a}{c}\right] = a \otimes \left[\frac{1}{c}\right].$$

This mapping v is a Λ_S-homomorphism and uv and vu are both identity maps. It follows that u is an isomorphism and, since the remaining assertion of the theorem is now obvious, this completes the proof.

Lemma 2. *If $\phi : A \to B$ is a Λ-monomorphism, then the corresponding Λ_S-homomorphism $\phi_S : A_S \to B_S$ is also a monomorphism.*

Proof. With the usual notation let $\left[\frac{a}{c}\right]$ be an element of A_S whose image under ϕ_S is zero, then $\left[\frac{\phi(a)}{c}\right]$ is the zero element of B_S and therefore $c_0\phi(a) = \phi(c_0 a) = 0$ for a suitable element c_0 of S. But ϕ is a monomorphism, consequently $c_0 a = 0$ and therefore $\left[\frac{a}{c}\right] = 0$. This proves the lemma.

Remark. Suppose that A is a submodule of B and let us take $A \to B$ to be the inclusion map. The lemma then asserts that *the submodule of B_S consisting of elements of the form $\left[\frac{a}{c}\right]$, where $a \in A$ and $c \in S$, can be identified with the Λ_S-module A_S.*

We can now establish a most important property of the functor A_S.

Theorem 6. *A_S is an exact functor of the Λ-module A.*

Proof. Let $0 \to A' \to A \to A'' \to 0$ be an exact sequence of Λ-modules. Then $0 \to A'_S \to A_S$ is exact by Lemma 2 while $A'_S \to A_S \to A''_S \to 0$ is exact by Theorem 5 and the right exactness of tensor products.

Proposition 1. *Let A be a Λ-module and let the elements u_i $(i \in I)$ form a system of generators for A. Then the elements $\left[\frac{u_i}{1}\right]$ $(i \in I)$ form a system of generators for the Λ_S-module A_S. Furthermore, if the u_i are a Λ-base for A, then the $\left[\frac{u_i}{1}\right]$ are a Λ_S-base for A_S.*

Proof. The first assertion is obvious so we shall suppose that the u_i are a Λ-base for A and that $\sum_i \left[\frac{\lambda_i}{c_i}\right]\left[\frac{u_i}{1}\right] = 0$, where $\lambda_i \in \Lambda$, $c_i \in S$ and $\left[\frac{\lambda_i}{c_i}\right] = 0$ for almost all i. Choose distinct elements $i_1, i_2, \ldots, i_n$ of I so that $\left[\frac{\lambda_i}{c_i}\right] = 0$ whenever i does not belong to this finite set, and put $c = c_{i_1} c_{i_2} \ldots c_{i_n}$, $c_r = c_{i_1} \ldots c_{i_{r-1}} c_{i_{r+1}} \ldots c_{i_n}$ where $1 \leqslant r \leqslant n$. Then

$$0 = \sum_{r=1}^{n} \left[\frac{\lambda_{i_r}}{c_{i_r}}\right]\left[\frac{u_{i_r}}{1}\right] = \sum_{r=1}^{n} \left[\frac{\lambda_{i_r} c_r u_{i_r}}{c}\right] = \left[\frac{\sum_{r=1}^{n} \lambda_{i_r} c_r u_{i_r}}{c}\right],$$

consequently we can find $c' \in S$ such that $\sum_{r=1}^{n} c'\lambda_{i_r} c_r u_{i_r} = 0$. But the u_i are a base for A, accordingly $c'\lambda_{i_r} c_r = 0$ which shows that

$$\left[\frac{\lambda_{i_r}}{c_{i_r}}\right] = 0 \quad \text{for } 1 \leqslant r \leqslant n.$$

Thus $\left[\frac{\lambda_i}{c_i}\right] = 0$ for *all* values of i, and this completes the proof.

Corollary. *If P is Λ-projective then P_S is Λ_S-projective.*

Proof. Since P is a direct summand of some Λ-free module, we can construct a split exact sequence $0 \to P \to F \to L \to 0$ where F is Λ-free. By Theorem 2 of section (3.9), this gives rise to a split exact sequence $0 \to P_S \to F_S \to L_S \to 0$, hence, since the proposition shows that F_S is Λ_S-free, P_S is isomorphic to a direct summand of a Λ_S-free module. The corollary now follows.

The next theorem shows that the application of torsion functors commutes with the formation of fractions.

Theorem 7. *Let A and B be Λ-modules, then, for each $n \geqslant 0$, we have a Λ_S-isomorphism*

$$[\mathrm{Tor}_n^{\Lambda}(A, B)]_S \approx \mathrm{Tor}_n^{\Lambda_S}(A_S, B_S).$$

Proof. Using Theorem 1 we construct isomorphisms

$$\begin{aligned}(A \otimes_\Lambda \Lambda_S) \otimes_{\Lambda_S} (B \otimes_\Lambda \Lambda_S) &\approx A \otimes_\Lambda [\Lambda_S \otimes_{\Lambda_S} (B \otimes_\Lambda \Lambda_S)] \\ &\approx A \otimes_\Lambda (B \otimes_\Lambda \Lambda_S) \approx (A \otimes_\Lambda B) \otimes_\Lambda \Lambda_S,\end{aligned} \tag{8.6.9}$$

where the combined isomorphism

$$\mu : (A \otimes_\Lambda \Lambda_S) \otimes_{\Lambda_S} (B \otimes_\Lambda \Lambda_S) \approx (A \otimes_\Lambda B) \otimes_\Lambda \Lambda_S \tag{8.6.10}$$

is such that

$$\mu\left\{\left(a \otimes \left[\frac{\lambda}{c}\right]\right) \otimes \left(b \otimes \left[\frac{\lambda'}{c'}\right]\right)\right\} = (a \otimes b) \otimes \left[\frac{\lambda\lambda'}{cc'}\right].$$

It follows first that μ is a Λ_S-isomorphism and secondly that μ defines a functor equivalence between the two sides of (8.6.10) regarded as functors of A and B with values in the category of Λ_S-modules. But A_S and $A \otimes_\Lambda \Lambda_S$ are equivalent functors and so are B_S and $B \otimes_\Lambda \Lambda_S$, hence, from (8.6.10), we obtain a further equivalence namely

$$\nu : A_S \otimes_{\Lambda_S} B_S \approx (A \otimes_\Lambda B) \otimes_\Lambda \Lambda_S. \tag{8.6.11}$$

Now let **F** be a free resolution of the Λ-module A so that

$$\cdots \to F_1 \to F_0 \to A \to 0$$

is an exact sequence of Λ-modules. By Theorem 6,

$$\cdots \to (F_1)_S \to (F_0)_S \to A_S \to 0$$

is an exact sequence of Λ_S-modules which, by Proposition 1, determines a free resolution of A_S. Moreover, the equivalence ν of (8.6.11) shows that the two complexes (over Λ_S)

$$\cdots \to (F_n)_S \otimes_{\Lambda_S} B_S \to (F_{n-1})_S \otimes_{\Lambda_S} B_S \to \cdots$$

and $$\cdots \to (F_n \otimes_\Lambda B) \otimes_\Lambda \Lambda_S \to (F_{n-1} \otimes_\Lambda B) \otimes_\Lambda \Lambda_S \to \cdots$$

have isomorphic homology modules. The nth homology module of the first complex is just $\mathrm{Tor}_n^{\Lambda_S}(A_S, B_S)$. To discuss the second, let **X** be the complex

$$\cdots \to F_n \otimes_\Lambda B \to F_{n-1} \otimes_\Lambda B \to \cdots$$

then we have shown that $\mathrm{Tor}_n^{\Lambda_S}(A_S, B_S)$ and $H_n(\mathbf{X} \otimes_\Lambda \Lambda_S)$ are Λ_S-isomorphic. But, since $A \otimes_\Lambda \Lambda_S$ is an exact functor of A (Theorems 5 and 6), it follows, from Theorem 1 of section (6.1), that

$$H_n(\mathbf{X} \otimes_\Lambda \Lambda_S) \approx H_n(\mathbf{X}) \otimes_\Lambda \Lambda_S,$$

this being an isomorphism with respect to Λ_S. But $H_n(\mathbf{X})$ and $\mathrm{Tor}_n^\Lambda(A, B)$ are isomorphic as Λ-modules and now the proof is complete.

It is convenient to establish one further result concerning modules of fractions in order to have it available for use in the next chapter. Before stating this result it is necessary to observe that any Λ_S-module M can be regarded as a Λ-module if we put

$$\lambda x = \left[\frac{\lambda}{1}\right] x \quad (\lambda \in \Lambda,\ x \in M).$$

Proposition 2. *If M is a Λ_S-module and we consider it as a Λ-module, then the canonical mapping $\omega : M \to M_S$ (see (8.6.4)) is a Λ_S-isomorphism.*

Proof. Let x belong to M and $\left[\frac{\lambda}{c}\right]$ to Λ_S, where $\lambda \in \Lambda$ and $c \in S$. Then

$$c\left(\left[\frac{\lambda}{c}\right] x\right) = \left[\frac{c}{1}\right]\left(\left[\frac{\lambda}{c}\right] x\right) = \left[\frac{\lambda}{1}\right] x = \lambda x,$$

hence $$\left[\frac{\lambda}{1}\right] \omega(x) = \lambda\omega(x) = \omega(\lambda x) = c\omega\left(\left[\frac{\lambda}{c}\right] x\right) = \left[\frac{c}{1}\right] \omega\left(\left[\frac{\lambda}{c}\right] x\right),$$

which yields

$$\left[\frac{\lambda}{c}\right]\omega(x) = \omega\left(\left[\frac{\lambda}{c}\right]x\right)$$

on multiplication of both sides by $\left[\frac{1}{c}\right]$. This shows that ω is Λ_S-linear. Further,

$$\omega\left(\left[\frac{\lambda}{c}\right]x\right) = \left[\frac{\lambda}{c}\right]\omega(x) = \left[\frac{\lambda}{c}\right]\left[\frac{x}{1}\right] = \left[\frac{\lambda x}{c}\right],$$

which makes it clear that ω is an epimorphism.

To complete the proof, assume that $\omega(x) = 0$, i.e. that $\left[\frac{x}{1}\right] = 0$, where $x \in M$. For a suitable element $c' \in S$ we have $c'x = 0$ and therefore

$$0 = \left[\frac{1}{c'}\right](c'x) = \left[\frac{1}{c'}\right]\left(\left[\frac{c'}{1}\right]x\right) = x,$$

which shows that ω is a monomorphism and thereby establishes the proposition.

9

COMMUTATIVE NOETHERIAN RINGS OF FINITE GLOBAL DIMENSION

Notation. In this chapter, Λ denotes a *commutative* ring with an identity element and Z denotes the ring of integers. In section (9.3), and in the later sections, the letter Q is used to denote a *local ring*, that is to say, a commutative Noetherian ring (with an identity element) having only one maximal ideal. The maximal ideal of Q is designated by $\mathfrak{m}$ and the residue class field $Q/\mathfrak{m}$ by K.

9.1 Some special cases

The purpose of the present chapter is to determine the structure of all commutative Noetherian rings which have finite global dimension, and subsequently a complete solution of this problem will be given in terms of certain well-known concepts of ideal theory. At a later stage we shall review, without giving proofs, the parts of ideal theory which are relevant for this purpose, but, to begin with, a number of special rings will be discussed by *ad hoc* methods. Indeed, until we come to section (9.4), only a familiarity with the simpler notions of ideal theory will be assumed.

Since Λ is understood throughout to be *commutative*, the distinction between right and left Λ-modules disappears. The homological dimension of a Λ-module A can therefore be written as $\mathrm{dh}_\Lambda(A)$, and the single global dimension of Λ as gl.dim Λ so that we have

$$\sup_A .\mathrm{dh}_\Lambda(A) = \text{gl.dim}\,\Lambda.$$

There is now also only one family of extension functors. Both $\mathrm{Ext}^n_\Lambda(A,B)$ and $\mathrm{Tor}^\Lambda_n(A,B)$ take values in the category of Λ-modules and, besides being additive functors, are Λ-linear in the sense of section (3.7).

Furthermore, it should also be noted that we have a Λ-isomorphism $\mathrm{Tor}^\Lambda_n(A,B) \approx \mathrm{Tor}^\Lambda_n(B,A)$ for arbitrary Λ-modules A and B. To see this, let $\mathbf{P}$ be a projective resolution of A then

$$\begin{array}{ccccccccc}
\cdots \longrightarrow & P_{n+1}\otimes_\Lambda B & \longrightarrow & P_n\otimes_\Lambda B & \longrightarrow & P_{n-1}\otimes_\Lambda B & \longrightarrow \cdots \\
& \downarrow & & \downarrow & & \downarrow & \\
\cdots \longrightarrow & B\otimes_\Lambda P_{n+1} & \longrightarrow & B\otimes_\Lambda P_n & \longrightarrow & B\otimes_\Lambda P_{n-1} & \longrightarrow \cdots
\end{array}$$

is a commutative diagram, the vertical maps being the canonical Λ-isomorphisms described in Theorem 2 of section (8.4). The required result now follows by passing to the homology modules. One consequence of this observation is the result that the weak homological dimension of A is the same whether we regard it as a left Λ-module or as a right Λ-module, and therefore we may use $\text{w.dh}_\Lambda(A)$ to denote this dimension, without any danger of ambiguity.

A natural starting point is, perhaps, to examine the rings for which $\text{gl.dim}\,\Lambda = 0$. We have already discussed these rings in the non-commutative case (see Theorem 16 of section (7.6)) and, with the aid of our earlier results, it is very easy to give a full description appropriate to the present more restricted situation. We have, in fact,

Theorem 1. *The commutative ring Λ has global dimension zero if, and only if, it is ring-isomorphic to the direct sum of a finite number of commutative fields.*

Proof. Assume first that $\text{gl.dim}\,\Lambda = 0$, then, by Theorem 16 of section (7.6),

$$\Lambda = S_1 + S_2 + \ldots + S_n, \tag{9.1.1}$$

where the sum is direct and the S_r are simple ideals. Let

$$1 = e_1 + e_2 + \ldots + e_n \quad (e_r \in S_r)$$

then, since

$$\begin{aligned} e_r &= e_r e_1 + \ldots + e_r e_{r-1} + e_r e_r + e_r e_{r+1} + \ldots + e_r e_n \\ &= 0 + \ldots + 0 + e_r + \ldots + 0, \end{aligned}$$

we see that $e_r^2 = e_r$ and $e_r e_t = 0$ $(r \neq t)$. Also, if $x \in S_r$ then, for $t \neq r$, $xe_t \in S_r \cap S_t = 0$ consequently $x = xe_1 + xe_2 + \ldots + xe_n = xe_r$. It follows that $S_r = \Lambda e_r$ and, in particular, we see that $e_r \neq 0$.

The facts that $S_r = \Lambda e_r$ and $e_r^2 = e_r$ show that S_r is a commutative ring with e_r as its identity element, while the relations $e_r e_t = 0$ $(r \neq t)$ make it clear that, in (9.1.1), Λ is exhibited as the direct sum of these n rings. Further, if $x \in S_r$ and $x \neq 0$, then

$$S_r = \Lambda x = \Lambda(e_r x) = S_r x,$$

and this shows that S_r is a field. Accordingly, Λ is a finite direct sum of fields.

Assume now that $K_1, K_2, \ldots, K_n$ are commutative fields and let Λ be their direct sum, the ring operations in Λ being defined in terms of the corresponding operations with components. If $1 \leqslant i \leqslant n$ and α_r denotes a variable element of K_r, then the elements $(\alpha_1, \alpha_2, \ldots, \alpha_n)$, for which $\alpha_j = 0$ for all $j \neq i$, form a simple ideal S_i of Λ, and Λ is the direct sum of $S_1, S_2, \ldots, S_n$. That gl.dim $\Lambda = 0$ now follows by appealing, for a second time, to Theorem 16 of section (7.6).

It is an obvious consequence of Theorem 1 that, if Λ is a field, then every Λ-module is projective. However, this falls short of the full truth which is recorded in

Theorem 2. *If K is a commutative field then every K-module is free.*

Remark. It may be noted that the theorem remains true even if the requirement that K be commutative is dropped. In fact, the proof, which follows, works equally well in the non-commutative case.

Proof. Let $A \neq 0$ be a K-module, then the existence of a maximal linearly independent subset $[a_i]_{i \in I}$ (say) of A follows by a simple application of Zorn's lemma. Let $\alpha \in A$ then the family $[a_i]_{i \in I}$, when augmented by α, can no longer retain the property of linear independence. Accordingly there exists a relation

$$k\alpha + k_1 a_{i_1} + k_2 a_{i_2} + \ldots + k_n a_{i_n} = 0,$$

where (i) $i_1, i_2, \ldots, i_n$ are distinct elements of I and (ii) $k, k_1, \ldots, k_n$ belong to K and are not all zero. However, $a_{i_1}, a_{i_2}, \ldots, a_{i_n}$ are linearly independent, consequently $k \neq 0$ and therefore, multiplying through by k^{-1}, we see that α belongs to the submodule generated by $[a_i]_{i \in I}$. But α was an arbitrary element of A, hence the submodule must coincide with A itself. Finally, the linear independence of the a_i now shows that they form a K-base for A.

The next special ring we propose to discuss is the ring Z of integers. This ring has the property that if a, b are two of its elements and $ab = 0$, then either $a = 0$ or $b = 0$. In other words, Z is an *integral domain*. A second property possessed by Z is that every Z-ideal can be generated by a single element or, to use the customary terminology, every ideal is a *principal ideal*. Indeed, if $I \neq 0$ is a Z-ideal, then I is easily seen to be generated by the smallest positive integer which belongs to it. Let us make the following

Definition. A commutative ring P (with an identity element) which is an integral domain and has the further property that every ideal is principal (i.e. can be generated by a single element), will be called a *principal ideal domain*.

Observe that if $\mathfrak{a} \neq 0$ is an ideal of a principal ideal domain P then, for some $\alpha \in P$, $\mathfrak{a} = P\alpha$. The mapping $p \to p\alpha$ is then a P-isomorphism of P on to $\mathfrak{a}$, because P is an integral domain. Accordingly, *every ideal is a P-free module.*

Theorem 3. *Let P be a principal ideal domain and let A be a free P-module. Then any submodule M of A is also P-free.*

Proof. Let $[a_i]_{i\in I}$ be a base of A and let us well order the elements of I, the symbol $\leqslant$ being used to denote the ordering relation. If $j \in I$ we shall denote by A_j the submodule generated by the elements $[a_i]_{i\leqslant j}$. These elements form a base for A_j and the union of all the A_j is A itself.

Let $x \in A_j \cap M$, then x has a unique representation of the form $x = \sum_{i\leqslant j} p_i a_i$, where $p_i \in P$ and almost all the p_i $(i \leqslant j)$ are zero. Further, the mapping $x \rightarrow p_j$ is a P-homomorphism of $A_j \cap M$ on to an ideal $\mathfrak{a}_j$ and its kernel B_j (say) consists of all the elements of M whose representations, in terms of the base $[a_i]_{i\in I}$ of A, only involve non-zero coefficients for values of i which *strictly* precede j.

Now, since P is a principal ideal domain, either $\mathfrak{a}_j = 0$ or $\mathfrak{a}_j$ is isomorphic to P, hence, in any event, $\mathfrak{a}_j$ is a free P-module and therefore the exact sequence

$$0 \rightarrow B_j \rightarrow (A_j \cap M) \rightarrow \mathfrak{a}_j \rightarrow 0$$

splits. Accordingly we may write

$$A_j \cap M = B_j + C_j \quad \text{(direct sum)},$$

where C_j is a suitable submodule of $A_j \cap M$ which is isomorphic to the free module $\mathfrak{a}_j$. If therefore we show that M is the (internal) direct sum of the C_j $(j \in I)$, this will complete the proof.

Suppose first that $c_{j_1} + c_{j_2} + \ldots + c_{j_n} = 0$, where $j_1 < j_2 < \ldots < j_n$ and $c_{j_r} \in C_{j_r}$. It is clear that $c_{j_1} + \ldots + c_{j_{n-1}} \in A_{j_{n-1}} \cap M \subseteq B_{j_n}$, whence, since $B_{j_n} \cap C_{j_n} = 0$, it follows that $c_{j_n} = 0$. By repeated applications of this argument it follows that all the c_{j_r} are zero, and now our observations show that the smallest submodule of M containing all the C_j, that is, $\sum_{j\in I} C_j$, is the direct sum of the C_j. Clearly $\sum_{j\in I} C_j \subseteq M$, and it only remains for us to show that the opposite inclusion holds.

Assume the contrary, then, for some $i \in I$, $A_i \cap M \nsubseteq \sum_{j\in I} C_j$. From now on we suppose that *i is the first such element of I*. Choose $x \in A_i \cap M$ so that $x \notin \sum_{j\in I} C_j$ and let $x = \sum_{\omega\leqslant i} p_\omega a_\omega$. We can find $y \in C_i$ so that $x - y \in B_i$, hence $x - y$ is of the form $\sum_{\omega<i} p'_\omega a_\omega$. But only a finite number of the p'_ω are different from zero consequently, for some $i_0 < i$, $x - y \in A_{i_0} \cap M$.

By the choice of i, this implies that $x - y \in \sum_{j \in I} C_j$ and, since $y \in C_i \subseteq \sum_{j \in I} C_j$, we now find that $x \in \sum_{j \in I} C_j$, contrary to our earlier hypothesis. Thus the required contradiction has been obtained and the proof is complete.

Theorem 4. *If P is a principal ideal domain then* gl.dim $P \leqslant 1$.

Proof. Let A be an arbitrary P-module and let us construct an exact sequence $0 \to A' \to F \to A \to 0$, where F is P-free. By Theorem 3, A' is also free and therefore, using Theorem 13 of section (7.5),

$$\mathrm{dh}_P(A) \leqslant \mathrm{dh}_P(A') + 1 = 1.$$

Theorem 5. *The global dimension of the ring Z of integers is unity.*

To see this we have only to observe that, on the one hand, Z is a principal ideal domain while, on the other, it is not a direct sum of fields.

The last special ring we shall treat in this section is the polynomial ring $K[X_1, X_2, \ldots, X_n]$, where K is a commutative field and the X_i are indeterminates. This will be shown to have global dimension equal to n, and the proof will be by induction on the number of variables. To make this induction possible, we require a result which can be regarded as a kind of *reduction principle*, but before this can be stated and proved it will be necessary to make a number of observations of a general character.

Suppose that we have a ring-epimorphism $\phi : \Lambda \to \Omega$ of a commutative ring Λ on to a commutative ring Ω. This implies, of course, that $\phi(1)$ is the identity element of Ω. If now A is an Ω-module, then it can be given the structure of a Λ-module by writing

$$\lambda a = \phi(\lambda) a \quad (\lambda \in \Lambda,\ a \in A). \tag{9.1.2}$$

Bearing this in mind, let A and B be Ω-modules, then a mapping $f : A \to B$ is a homomorphism of Ω-modules if and only if it is a homomorphism of Λ-modules, hence $\mathrm{Hom}_\Lambda(A, B)$ and $\mathrm{Hom}_\Omega(A, B)$ consist of the same elements. Furthermore, $\mathrm{Hom}_\Lambda(A, B)$ and $\mathrm{Hom}_\Omega(A, B)$ have natural structures as Λ-module and Ω-module respectively, the two structures being connected by the relation

$$\lambda f = \phi(\lambda) f, \tag{9.1.3}$$

where $\lambda \in \Lambda$ and f is an arbitrary element of the coincident sets $\mathrm{Hom}_\Lambda(A, B)$ and $\mathrm{Hom}_\Omega(A, B)$.

Next let A be an Ω-module and M a Λ-module. If $\lambda \in \mathrm{Ker}(\phi)$ and i_A and i_M are the identity maps of A and M respectively, then

$$\lambda\, \mathrm{Ext}^n_\Lambda(i_A, i_M) = \mathrm{Ext}^n_\Lambda(\lambda i_A, i_M) = \mathrm{Ext}^n_\Lambda(0, i_M) = 0.$$

But $\mathrm{Ext}_\Lambda^n(i_A, i_M)$ is the identity map of $\mathrm{Ext}_\Lambda^n(A, M)$, consequently $\lambda u = 0$ for all $u \in \mathrm{Ext}_\Lambda^n(A, M)$. It follows that if $\phi(\lambda_1) = \phi(\lambda_2)$, then $\lambda_1 u = \lambda_2 u$, and now we see that we can regard $\mathrm{Ext}_\Lambda^n(A, M)$ as an Ω-module by writing

$$\phi(\eta)\, u = \eta u \quad (\eta \in \Lambda,\ u \in \mathrm{Ext}_\Lambda^n(A, M)). \tag{9.1.4}$$

Accordingly, *for fixed M, the family $\{\mathrm{Ext}_\Lambda^n(A, M)\}_{n \geq 0}$ forms a connected right sequence of contravariant functors, these functors being defined on the category of Ω-modules A and taking values which are also Ω-modules.*

If $\mathfrak{a}$ is a Λ-ideal, we shall understand by $\mathfrak{a}M$ the submodule of M generated by all products αy, where $\alpha \in \mathfrak{a}$ and $y \in M$. Should $\phi(\lambda_1) = \phi(\lambda_2)$, then, for each $x \in M/[\mathrm{Ker}(\phi)]\, M$, $\lambda_1 x = \lambda_2 x$, and therefore we can turn $M/[\mathrm{Ker}(\phi)]\, M$ into an Ω-module by writing

$$\phi(\lambda)\, x = \lambda x.$$

If we do this then *the* $\mathrm{Ext}_\Omega^n(A, M/[\mathrm{Ker}(\phi)]\, M)$ *also form a connected right sequence of contravariant functors of the Ω-module A, whose values are again Ω-modules.*

We propose now to compare the two sequences for a case in which $\Omega = \Lambda/(\alpha)$ and $\phi : \Lambda \to \Omega = \Lambda/(\alpha)$ is the canonical ring-homomorphism. Here α denotes an element of Λ.

Theorem 6. *Let α ($\alpha \in \Lambda$) be neither a unit nor a zero-divisor, and let M be a fixed Λ-module with the property that $\alpha y = 0$ for $y \in M$ only when $y = 0$. Put $\Omega = \Lambda/(\alpha)$ and for each Ω-module A write*

$$T^n(A) = \mathrm{Ext}_\Lambda^{n+1}(A, M), \quad U^n(A) = \mathrm{Ext}_\Omega^n(A, M/\alpha M),$$

these being regarded as contravariant functors of A taking values which are also Ω-modules. Then $\{T^n\}_{n \geq 0}$ and $\{U^n\}_{n \geq 0}$ are isomorphic as connected sequences of functors.

Proof. Since multiplication by α annihilates only the zero element of M,

$$0 \to M \xrightarrow{\psi} M \to M/\alpha M \to 0 \quad (\psi = \alpha i_M)$$

is an exact sequence and from it we obtain a further exact sequence, namely,

$$\mathrm{Hom}_\Lambda(A, M) \to \mathrm{Hom}_\Lambda(A, M/\alpha M) \to \mathrm{Ext}_\Lambda^1(A, M) \to \mathrm{Ext}_\Lambda^1(A, M).$$

Now if $f \in \mathrm{Hom}_\Lambda(A, M)$ and $a \in A$, then

$$\alpha f(a) = f(\alpha a) = f(0) = 0,$$

consequently, by the hypothesis concerning α and M, $f(a) = 0$. This shows that $\mathrm{Hom}_\Lambda(A, M) = 0$. On the other hand, the mapping $\mathrm{Ext}^1_\Lambda(A, M) \to \mathrm{Ext}^1_\Lambda(A, M)$ is, with an obvious notation,

$$\mathrm{Ext}^1_\Lambda(i_A, \alpha i_M) = \mathrm{Ext}^1_\Lambda(\alpha i_A, i_M) = \mathrm{Ext}^1_\Lambda(0, i_M) = 0,$$

hence we arrive at an exact sequence

$$0 \to \mathrm{Hom}_\Lambda(A, M/\alpha M) \to \mathrm{Ext}^1_\Lambda(A, M) \to 0,$$

which means that $\mathrm{Hom}_\Lambda(A, M/\alpha M) \to \mathrm{Ext}^1_\Lambda(A, M)$ is an isomorphism. But, as explained above, $\mathrm{Hom}_\Lambda(A, M/\alpha M) = \mathrm{Hom}_\Omega(A, M/\alpha M)$, and so we have an isomorphism $\mathrm{Hom}_\Omega(A, M/\alpha M) \to \mathrm{Ext}^1_\Lambda(A, M)$ which is a natural equivalence if we regard each module as a functor of A taking values which are Ω-modules. In other words, we have a natural equivalence $U^0 \approx T^0$.

Let $0 \to B \to P \to A \to 0$ be an exact sequence of Ω-modules where P is Ω-projective. Then, for $n \geqslant 0$,

$$U^n(P) \to U^n(B) \to U^{n+1}(A) \to 0$$

is exact, and $\quad T^n(P) \to T^n(B) \to T^{n+1}(A) \to T^{n+1}(P)$

is also exact. The required result will therefore follow (from the corollary to Theorem 11 of section (6.5)) if we can show that $T^{n+1}(P) = 0$. Since T^{n+1} is an additive functor, it will suffice to prove that $T^{n+1}(F) = 0$, where F is an arbitrary Ω-free module.

Let F have $[v_i]_{i \in I}$ for a base over Ω and let us construct a Λ-free module Φ with a base $[u_i]_{i \in I}$ (say), the set I of parameters being the same in the two cases. The sequence

$$0 \to \Phi \xrightarrow{\chi} \Phi \to F \to 0$$

of Λ-modules, where χ is α times the identity map $\Phi \to \Phi$ and $\Phi \to F$ is obtained by mapping u_i into v_i, is exact, because α is not a zero-divisor. Accordingly

$$0 \to \mathrm{Ext}^2_\Lambda(F, M) \to 0 \to 0 \to \mathrm{Ext}^3_\Lambda(F, M) \to 0 \to 0 \to \cdots$$

is exact and therefore $T^{n+1}(F) = \mathrm{Ext}^{n+2}_\Lambda(F, M) = 0$ for all $n \geqslant 0$ as required.

Corollary. *If α is neither a unit nor a zero-divisor, L is a submodule of a Λ-free module, and A is an arbitrary Ω-module, then we have Ω-isomorphisms*

$$\mathrm{Ext}^{n+1}_\Lambda(A, L) \approx \mathrm{Ext}^n_\Omega(A, L/\alpha L) \quad (\Omega = \Lambda/(\alpha)) \qquad (9.1.5)$$

for all $n \geqslant 0$.

Proof. The corollary follows at once from the theorem if we note that, because α is not a zero-divisor, it cannot annihilate, by multiplication, a non-zero element of a Λ-free module.

We shall use the term *proper ideal* to mean an ideal of a ring which is different from the ring itself. By a *maximal ideal* $\mathfrak{m}$ of Λ, is to be understood a proper ideal having the property that there is no other proper ideal which strictly contains it. If $\mathfrak{n}$ is an arbitrary proper ideal then a simple application of Zorn's lemma shows *that* $\mathfrak{n}$ *is contained in at least one maximal ideal.* For Noetherian rings this is still easier to see for, if the contrary were true, there would exist an unending strictly increasing sequence $\mathfrak{n} \subset \mathfrak{n}_1 \subset \mathfrak{n}_2 \subset \ldots$ of ideals and this would violate the ascending chain condition (see Proposition 6 of section (7.7)).

As is customary in ideal theory, if $\alpha_1, \alpha_2, \ldots, \alpha_s$ are elements of Λ then we shall use $(\alpha_1, \alpha_2, \ldots, \alpha_s)$ to denote the ideal which they generate.

Suppose that $\mathfrak{m}$ is a maximal ideal of Λ and that λ belongs to Λ but not to $\mathfrak{m}$. Then λ and $\mathfrak{m}$ together generate the whole ring and therefore $\lambda\lambda' + \mu = 1$ for suitable elements $\lambda' \in \Lambda$ and $\mu \in \mathfrak{m}$. This shows that *the residue class ring* $\Lambda/\mathfrak{m}$ *is a field* and, indeed, this property characterizes maximal ideals.

Consider, for the moment, a polynomial ring $K[X_1, X_2, \ldots, X_n]$, where K is a commutative field. It is well known† that such a ring is Noetherian, for this is the famous 'Basis Theorem' of Hilbert. In the special case where K is *algebraically closed*, it is also known that *the maximal ideals of* $K[X_1, X_2, \ldots, X_n]$ *are precisely those which can be generated by a set of* n *polynomials of the form* $X_1 - \alpha_1$, $X_2 - \alpha_2, \ldots,$ $X_n - \alpha_n$, *where* $\alpha_1, \alpha_2, \ldots, \alpha_n$ *belong to* K. This is essentially a restatement of another famous theorem, also due to Hilbert, known as the 'Zero's Theorem'.

Lemma 1. *If* N *is a finitely generated* Λ*-module and* $\mathfrak{m}N = N$ *for every maximal ideal* $\mathfrak{m}$ *of* Λ, *then* $N = 0$.

Proof. Let $y_1, y_2, \ldots, y_s$ be elements of N which generate it and let $\mathfrak{n}$ be the Λ-ideal consisting of the elements $\alpha \in \Lambda$ such that $\alpha y = 0$ for all $y \in N$. Clearly it will suffice to prove that $\mathfrak{n} = \Lambda$. We shall assume the contrary and obtain a contradiction.

Since $\mathfrak{n} \neq \Lambda$, there exists a maximal ideal $\mathfrak{m}$ such that $\mathfrak{n} \subseteq \mathfrak{m}$. Now $y_i \in N = \mathfrak{m}N$ consequently we have relations

$$y_i = \mu_{i1}y_1 + \mu_{i2}y_2 + \ldots + \mu_{is}y_s \quad (1 \leqslant i \leqslant s,\ \mu_{ij} \in \mathfrak{m}),$$

† For further particulars, see the notes on Chapter 9.

or, if δ_{ij} is equal to 1 when $i = j$ and is zero otherwise, $\Sigma_j(\mu_{ij} - \delta_{ij}) y_j = 0$. Denote by Δ the determinant

$$\begin{vmatrix} \mu_{11}-1 & \mu_{12} & \mu_{13} & \cdots & \mu_{1s} \\ \mu_{21} & \mu_{22}-1 & \mu_{23} & \cdots & \mu_{2s} \\ \cdots & \cdots & \cdots & \cdots & \cdots \\ \mu_{s1} & \mu_{s2} & \mu_{s3} & \cdots & \mu_{ss}-1 \end{vmatrix}$$

then $\Delta y_j = 0$ for $1 \leqslant j \leqslant s$, which shows that $\Delta \in \mathfrak{n}$. But $\Delta = (-1)^s + \mu$, where $\mu \in \mathfrak{m}$, consequently $1 \in \mathfrak{m}$, which contradicts the fact that $\mathfrak{m}$ is a proper ideal.

Theorem 7. *Let $\Lambda = K[X_1, X_2, \ldots, X_n]$ be a polynomial ring in n variables $X_1, X_2, \ldots, X_n$, with coefficients in a commutative field K. Then* gl.dim $\Lambda = n$.

Proof. If $n = 0$ then $\Lambda = K$ and so gl.dim $\Lambda = 0$ by Theorem 1. Let us suppose therefore that $n \geqslant 1$ and let us write

$$\Lambda^* = K[X_1, X_2, \ldots, X_{n-1}];$$

then Λ^* can be identified, in an obvious manner, with the residue class ring $\Lambda/(X_n)$. Note that X_n is neither a unit nor a zero-divisor in Λ.

We can regard K as a Λ^*-module by defining the product of the polynomial $\phi(X_1, X_2, \ldots, X_{n-1})$ and $k \in K$ to be $\phi(0, 0, \ldots, 0)\, k$. If we do this, then, by the corollary to Theorem 6,

$$\mathrm{Ext}^n_\Lambda(K, \Lambda) \approx \mathrm{Ext}^{n-1}_{\Lambda^*}(K, \Lambda/(X_n)) = \mathrm{Ext}^{n-1}_{\Lambda^*}(K, \Lambda^*).$$

If we repeat this argument with Λ^* replacing $K[X_1, X_2, \ldots, X_n]$ and then with $K[X_1, X_2, \ldots, X_{n-2}]$ replacing Λ^* and so on, we arrive at Z-isomorphisms

$$\mathrm{Ext}^n_\Lambda(K, \Lambda) \approx \mathrm{Ext}^0_K(K, K) \approx \mathrm{Hom}_K(K, K) \approx K.$$

In particular, we see that $\mathrm{Ext}^n_\Lambda(K, \Lambda) \neq 0$ and this proves that gl.dim $\Lambda \geqslant n$.

It remains to be proved that gl.dim $\Lambda \leqslant n$. *To begin with we shall assume that K is algebraically closed.* The proof will be by induction on n, so it will be supposed that $n \geqslant 1$ and that the required upper bound for the global dimension has been established for polynomial rings in $n-1$ variables. Let A and B be arbitrary *finitely generated* Λ-modules then, by Proposition 10 of section (7.7), it is enough to show that $\mathrm{Ext}^{n+1}_\Lambda(A, B) = 0$.

Construct exact sequences

$$0 \to L \to F \to A \to 0 \quad \text{and} \quad 0 \to M \to \Phi \to B \to 0$$

of finitely generated Λ-modules, where F and Φ are Λ-free. Then, by the corollary to Theorem 6,

$$\mathrm{Ext}_{\Lambda}^{n+1}(L/X_nL, \Phi) \approx \mathrm{Ext}_{\Lambda^*}^{n}(L/X_nL, \Phi/X_n\Phi) = 0 \quad (9.1.6)$$

and $$\mathrm{Ext}_{\Lambda}^{n+2}(L/X_nL, M) \approx \mathrm{Ext}_{\Lambda^*}^{n+1}(L/X_nL, M/X_nM) = 0, \quad (9.1.7)$$

since, by the inductive hypothesis, gl.dim $\Lambda^* \leqslant n-1$. But we have an exact sequence

$$\mathrm{Ext}_{\Lambda}^{n+1}(L/X_nL, \Phi) \to \mathrm{Ext}_{\Lambda}^{n+1}(L/X_nL, B) \to \mathrm{Ext}_{\Lambda}^{n+2}(L/X_nL, M),$$

consequently, in view of (9.1.6) and (9.1.7), $\mathrm{Ext}_{\Lambda}^{n+1}(L/X_nL, B) = 0$. Further, if ψ denotes X_n times the identity map of L, the sequence $0 \to L \xrightarrow{\psi} L \to L/X_nL \to 0$ is exact and so itself yields an exact sequence

$$\mathrm{Ext}_{\Lambda}^{n}(L, B) \to \mathrm{Ext}_{\Lambda}^{n}(L, B) \to \mathrm{Ext}_{\Lambda}^{n+1}(L/X_nL, B) = 0.$$

This means that $\mathrm{Ext}_{\Lambda}^{n}(L, B) \to \mathrm{Ext}_{\Lambda}^{n}(L, B)$, which is X_n times the identity map of $\mathrm{Ext}_{\Lambda}^{n}(L, B)$, is an epimorphism, consequently

$$\mathrm{Ext}_{\Lambda}^{n}(L, B) = X_n \mathrm{Ext}_{\Lambda}^{n}(L, B),$$

and so, *a fortiori*, $$\mathrm{Ext}_{\Lambda}^{n}(L, B) = \mathfrak{m}\,\mathrm{Ext}_{\Lambda}^{n}(L, B), \quad (9.1.8)$$

where $\mathfrak{m}$ is the maximal ideal $(X_1, X_2, \ldots, X_n)$. But, because K is algebraically closed, given *any* maximal ideal $\mathfrak{m}'$ we have

$$\Lambda = K[X_1', X_2', \ldots, X_n'],$$

where the X_i' are algebraically independent over K and

$$\mathfrak{m}' = (X_1', X_2', \ldots, X_n').$$

Accordingly (9.1.8) holds for every maximal ideal of Λ, hence, by Lemma 1 and Proposition 11 of section (7.8), $\mathrm{Ext}_{\Lambda}^{n}(L, B) = 0$. Now from $0 \to L \to F \to A \to 0$ we obtain an exact sequence

$$\mathrm{Ext}_{\Lambda}^{n}(L, B) \to \mathrm{Ext}_{\Lambda}^{n+1}(A, B) \to 0,$$

from which it may be concluded that $\mathrm{Ext}_{\Lambda}^{n+1}(A, B) = 0$. However, this is just what we were seeking to prove, and so the theorem has been established for the case in which the coefficient field is algebraically closed.

Finally, let us examine the general case. Suppose that A is an arbitrarily assigned Λ-module and let us construct an exact sequence

$$0 \to D \to F_{n-1} \to F_{n-2} \to \cdots \to F_0 \to A \to 0, \qquad (9.1.9)$$

where each F_i is a Λ-free module. Clearly it is sufficient to prove that D is Λ-projective. To this end, let $\tilde{K}$ be the algebraic closure of K and put $\tilde{\Lambda} = \tilde{K}[X_1, X_2, \ldots, X_n]$. Further, for each Λ-module M write $\tilde{M} = M \otimes_\Lambda \tilde{\Lambda}$ and regard $\tilde{M}$ as a $\tilde{\Lambda}$-module, where multiplication by elements of $\tilde{\Lambda}$ is defined by

$$(y \otimes \tilde{\lambda})\tilde{\lambda}_1 = y \otimes \tilde{\lambda}\tilde{\lambda}_1 \quad (y \in M;\ \tilde{\lambda} \in \tilde{\Lambda},\ \tilde{\lambda}_1 \in \tilde{\Lambda}).$$

Since K is a subfield of $\tilde{K}$ and Λ is a subring of $\tilde{\Lambda}$, we can regard $\tilde{K}$ as a K-module and $\tilde{\Lambda}$ as a Λ-module. By Theorem 2, $\tilde{K}$ admits of a K-base, and it is clear that any such base will form a Λ-base for $\tilde{\Lambda}$. In particular, $\tilde{\Lambda}$ is Λ-free hence, by Theorem 4 of section (5.1), $\tilde{M} = M \otimes_\Lambda \tilde{\Lambda}$ is an exact functor of M. Accordingly, (9.1.9) yields an exact sequence

$$0 \to \tilde{D} \to \tilde{F}_{n-1} \to \tilde{F}_{n-2} \to \cdots \to \tilde{F}_0 \to \tilde{A} \to 0 \qquad (9.1.10)$$

of $\tilde{\Lambda}$-modules. But if F is Λ-free, say with $[u_i]_{i \in I}$ as base, then $\tilde{F}$ is $\tilde{\Lambda}$-free with the elements $[u_i \otimes 1]_{i \in I}$ as a base,† hence in (9.1.10), $\tilde{F}_0, \tilde{F}_1, \ldots, \tilde{F}_{n-1}$ are all $\tilde{\Lambda}$-free. Further, by the case already considered, $\mathrm{dh}_{\tilde{\Lambda}}(\tilde{A}) \leqslant n$ and so, by Theorem 11 of section (7.5), $\tilde{D}$ is $\tilde{\Lambda}$-projective. It is therefore possible to find a free $\tilde{\Lambda}$-module Θ such that $\Theta = \tilde{D} + \tilde{D}_1$ (direct sum) for a suitable $\tilde{\Lambda}$-module $\tilde{D}_1$. But, since Θ is $\tilde{\Lambda}$-free and $\tilde{\Lambda}$ is Λ-free, it follows that Θ is Λ-free and therefore that $\tilde{D}$ is Λ-projective. Further, if $[\omega_j]_{j \in J}$ is a Λ-base for $\tilde{\Lambda}$, the corollary to Theorem 3 of section (2.4) shows that the Λ-homomorphisms $\phi_j : D \to \tilde{D}$, defined by

$$\phi_j(x) = x \otimes \omega_j \quad (x \in D),$$

give an injective representation of $\tilde{D}$ as a direct sum of copies of D. That D is Λ-projective now follows from Proposition 3 of section (5.1).

9.2 Reduction of the general problem

The main purpose of the present chapter is to determine the structure of commutative Noetherian rings of finite global dimension and, at this point, we shall simplify the problem by reducing it to the case in which the ring has only one maximal ideal. Rings of this kind have a special name.

† See the corollary to Theorem 3 of section (2.4).

Definition. A (commutative) Noetherian ring Λ with only one maximal ideal will be called a *local ring.*

The method of reduction rests heavily on the theory of rings and modules of fractions as developed in section (8.6) and, in discussing these topics, we shall employ the notation that was introduced there.

Proposition 1. *Let S be a multiplicatively closed set of elements (of Λ) which contains 1 but does not contain the zero element. If now Λ is Noetherian then so is Λ_S.*

Proof. Let $\mathfrak{a}'$ be a Λ_S-ideal and denote by $\mathfrak{a}$ the set of all elements $\alpha \in \Lambda$ such that $\left[\frac{\alpha}{c}\right]$ belongs to $\mathfrak{a}'$ for some element c of S. An easy verification shows that $\mathfrak{a}$ is an ideal, consequently, since Λ is Noetherian, it is finitely generated, say by the elements $\alpha_1, \alpha_2, \ldots, \alpha_n$. Choose $c_1, c_2, \ldots, c_n$ in S so that, for each i $(1 \leqslant i \leqslant n)$, $\left[\frac{\alpha_i}{c_i}\right] \in \mathfrak{a}'$ and let $\left[\frac{\alpha}{c}\right]$ be an arbitrary element of $\mathfrak{a}'$. Then $\alpha = \lambda_1\alpha_1 + \lambda_2\alpha_2 + \ldots + \lambda_n\alpha_n$ for suitable elements $\lambda_1, \ldots, \lambda_n$ in Λ and

$$\left[\frac{\alpha}{c}\right] = \left[\frac{\lambda_1 c_1}{c}\right]\left[\frac{\alpha_1}{c_1}\right] + \ldots + \left[\frac{\lambda_n c_n}{c}\right]\left[\frac{\alpha_n}{c_n}\right].$$

This shows that $\left[\frac{\alpha_1}{c_1}\right], \ldots, \left[\frac{\alpha_n}{c_n}\right]$ generate $\mathfrak{a}'$ and now the proposition follows.

Theorem 8. *Let Λ be a Noetherian ring and S a multiplicatively closed set of elements of Λ containing 1 but not containing 0. Then, with the usual notation for rings of fractions,* gl.dim $\Lambda_S \leqslant$ gl.dim Λ. *In particular, if Λ is of finite global dimension then so is Λ_S.*

Proof. For the purpose of the proof we may suppose that gl.dim $\Lambda = n$, where n is a non-negative integer. Now Λ is Noetherian hence, by Proposition 1, Λ_S is Noetherian as well consequently, for each of the two rings, global dimension and weak global dimension coincide (Theorem 20, section (7.9)). Accordingly, if A and B are arbitrary Λ_S-modules, then the proof will be complete if we show that $\mathrm{Tor}_{n+1}^{\Lambda_S}(A, B) = 0$.

We can regard A as a Λ-module by putting

$$\lambda a = \phi(\lambda)\, a \quad (\lambda \in \Lambda,\ a \in A),$$

where $\phi : \Lambda \to \Lambda_S$ is the canonical ring-homomorphism given by (8.6.6) and B can be considered as a Λ-module in the same way. But, by Theorem 7 of section (8.6), we have a Λ_S-isomorphism

$$\mathrm{Tor}_{n+1}^{\Lambda_S}(A_S, B_S) \approx \{\mathrm{Tor}_{n+1}^{\Lambda}(A, B)\}_S,$$

which shows that $\mathrm{Tor}_{n+1}^{\Lambda_S}(A_S, B_S) = 0$, because $\mathrm{Tor}_{n+1}^{\Lambda}$ vanishes identically on account of the fact that the weak global dimension of Λ is n. Finally, by Proposition 2 of section (8.6), $A_S \approx A$ and $B_S \approx B$, these being Λ_S-isomorphisms, consequently $\mathrm{Tor}_{n+1}^{\Lambda_S}(A, B) = 0$ as required.

A closely related result with a similar proof reads as follows:

Theorem 9. *Let Λ be a Noetherian ring and S a multiplicatively closed set of elements of Λ containing* 1 *but not containing the zero element. If now A is a finitely generated Λ-module, then* $\mathrm{dh}_{\Lambda_S}(A_S) \leqslant \mathrm{dh}_{\Lambda}(A)$.

Proof. It may be supposed that $\mathrm{dh}_{\Lambda}(A) = n$, where n is a non-negative integer, since otherwise the assertion is obvious. Further, since our previous results show† that A_S is a finitely generated Λ_S-module while Λ_S itself is Noetherian, it follows‡ that

$$\mathrm{w.dh}_{\Lambda_S}(A_S) = \mathrm{dh}_{\Lambda_S}(A_S);$$

consequently, if B is an arbitrary Λ_S-module, it will be enough to show that $\mathrm{Tor}_{n+1}^{\Lambda_S}(A_S, B) = 0$.

Let us use the canonical homomorphism $\Lambda \to \Lambda_S$ to make B into a Λ-module. If this is done then, by Theorem 7 of section (8.6),

$$\mathrm{Tor}_{n+1}^{\Lambda_S}(A_S, B_S) \approx \{\mathrm{Tor}_{n+1}^{\Lambda}(A, B)\}_S = 0,$$

because $\mathrm{w.dh}_{\Lambda}(A) = \mathrm{dh}_{\Lambda}(A) = n$. To complete the proof, it is only necessary to observe that B and B_S are isomorphic as Λ_S-modules.§

It will be recalled that an ideal $\mathfrak{p}$ of Λ is called a *prime ideal* if, whenever $\alpha\beta \in \mathfrak{p}$, then at least one of α and β belongs to $\mathfrak{p}$. From this it follows that a proper ideal is a prime ideal when and only when the corresponding residue class ring is an integral domain. Now, for a maximal ideal $\mathfrak{m}$, $\Lambda/\mathfrak{m}$ is not only an integral domain but a field, consequently *every maximal ideal is a prime ideal.*

Suppose next that $\mathfrak{p}$ is a proper prime ideal and let c_1 and c_2 be elements of Λ which do not belong to $\mathfrak{p}$, then $c_1 c_2$ also does not belong to $\mathfrak{p}$. Accordingly, if S denotes the complement of $\mathfrak{p}$ in Λ, then S is a multiplicatively closed set with the property that $1 \in S$, while $0 \notin S$. It is therefore possible to form the ring Λ_S of fractions and, should

† See Proposition 1 of section (8.6).
‡ See Theorem 19 of section (7.9).
§ This is proved in Proposition 2 of section (8.6).

A be a Λ-module, the module A_S of fractions. However, when S arises as the complement of a prime ideal $\mathfrak{p}$, it is customary to write $\Lambda_\mathfrak{p}$ and $A_\mathfrak{p}$ rather than Λ_S and A_S, and to speak of these as the ring of fractions and module of fractions formed *with respect to* $\mathfrak{p}$. The two notations conflict to some extent, but the fact that we shall normally use $\mathfrak{p}$ to denote a general prime ideal (and $\mathfrak{m}$ to denote a maximal ideal) will be found sufficient to make the intention clear in each instance.

Proposition 2. *If $\mathfrak{p}$ is a proper prime ideal of a Noetherian ring Λ, then the ring $\Lambda_\mathfrak{p}$ of fractions is a local ring.*

Proof. Consider the elements of $\Lambda_\mathfrak{p}$ which are of the form $\left[\frac{\lambda}{c}\right]$, where $\lambda \in \mathfrak{p}$ and $c \notin \mathfrak{p}$, the notation being that used in section (8.6). These elements form a $\Lambda_\mathfrak{p}$-ideal $\mathfrak{m}'$ (say) which does not contain $\left[\frac{1}{1}\right]$ and is therefore proper. Let $\mathfrak{a}'$ be an ideal of $\Lambda_\mathfrak{p}$ not contained in $\mathfrak{m}'$ then, for some $\lambda^* \notin \mathfrak{p}$ and $c^* \notin \mathfrak{p}$, we have $\left[\frac{\lambda^*}{c^*}\right] \in \mathfrak{a}'$. But $\left[\frac{c^*}{\lambda^*}\right]$ also belongs to $\Lambda_\mathfrak{p}$ consequently

$$\left[\frac{1}{1}\right] = \left[\frac{c^*}{\lambda^*}\right]\left[\frac{\lambda^*}{c^*}\right] \in \mathfrak{a}',$$

which shows that $\mathfrak{a}' = \Lambda_\mathfrak{p}$. Accordingly, every proper ideal of $\Lambda_\mathfrak{p}$ is contained in $\mathfrak{m}'$ and therefore $\mathfrak{m}'$ is the only maximal ideal. It remains to be shown that $\Lambda_\mathfrak{p}$ is Noetherian but this follows from Proposition 1.

Lemma 2. *If A is a Λ-module and if (with the notation explained above) $A_\mathfrak{m} = 0$ for every maximal ideal $\mathfrak{m}$ of A, then $A = 0$.*

Proof. Let $[\mathfrak{m}_i]_{i \in I}$ be the family composed of all the maximal ideals of Λ and let $a \in A$. It will suffice to show that $a = 0$. Now, in $A_{\mathfrak{m}_i}$, $\left[\frac{a}{1}\right] = 0$ consequently, for a suitable $\lambda_i \notin \mathfrak{m}_i$, we have $\lambda_i a = 0$. Now the λ_i together generate the whole ring, for otherwise they would all be contained in some maximal ideal, say in $\mathfrak{m}_{i_0}$, which is impossible because $\lambda_{i_0} \notin \mathfrak{m}_{i_0}$. In particular, we see that the identity element 1 of Λ belongs to the ideal generated by the λ_i, and so it follows that $1a = 0$. This proves the lemma.

We are now ready to reduce the general problem to the local case.

Theorem 10. *If Λ is a Noetherian ring then*

$$\text{gl.dim}\,\Lambda = \sup_{\mathfrak{m}}.\text{gl.dim}\,\Lambda_\mathfrak{m},$$

where $\mathfrak{m}$ ranges over all the maximal ideals of Λ.

Proof. By Theorem 8, gl.dim $\Lambda_{\mathfrak{m}} \leqslant$ gl.dim Λ for every maximal ideal $\mathfrak{m}$, consequently it will be enough to show that

$$\text{gl.dim}\,\Lambda \leqslant \sup_{\mathfrak{m}}.\text{gl.dim}\,\Lambda_{\mathfrak{m}},$$

and for this we may restrict our attention to the situation in which sup.gl.dim $\Lambda_{\mathfrak{m}} = n$, where n is a non-negative integer. Before proceeding, let us note that Λ and all the $\Lambda_{\mathfrak{m}}$ are Noetherian, so that, by Theorem 20 of section (7.9), we do not have to distinguish between global dimension and weak global dimension. Accordingly, if A and B are arbitrary Λ-modules then it will suffice to prove that

$$\text{Tor}_{n+1}^{\Lambda}(A, B) = 0.$$

Now, by Theorem 7 of section (8.6),

$$[\text{Tor}_{n+1}^{\Lambda}(A, B)]_{\mathfrak{m}} \approx \text{Tor}_{n+1}^{\Lambda_{\mathfrak{m}}}(A_{\mathfrak{m}}, B_{\mathfrak{m}}),$$

and this latter is a null module because w.gl.dim $\Lambda_{\mathfrak{m}} \leqslant n$. Thus

$$[\text{Tor}_{n+1}^{\Lambda}(A,B)]_{\mathfrak{m}} = 0$$

for every maximal ideal $\mathfrak{m}$, and so the required relation follows from Lemma 2.

The next theorem reduces, to the local case, the problem of finding the homological dimension of a finitely generated Λ-module.

Theorem 11. *Let Λ be a Noetherian ring and A a finitely generated Λ-module. Then*

$$\text{dh}_{\Lambda}(A) = \sup_{\mathfrak{m}}.\text{dh}_{\Lambda_{\mathfrak{m}}}(A_{\mathfrak{m}}),$$

where $\mathfrak{m}$ ranges over all the maximal ideals of Λ.

Proof. By Theorem 9, it will suffice if, assuming that

$$\sup.\text{dh}_{\Lambda_{\mathfrak{m}}}(A_{\mathfrak{m}}) = n,$$

where n $(n \geqslant -1)$ is an integer, we deduce that $\text{dh}_{\Lambda}(A) \leqslant n$. Accordingly, let B be an arbitrary Λ-module, then

$$[\text{Tor}_{n+1}^{\Lambda}(A, B)]_{\mathfrak{m}} \approx \text{Tor}_{n+1}^{\Lambda_{\mathfrak{m}}}(A_{\mathfrak{m}}, B_{\mathfrak{m}}) = 0,$$

and now it follows, from Lemma 2, that $\text{Tor}_{n+1}^{\Lambda}(A, B) = 0$. Accordingly, $\text{w.dh}_{\Lambda}(A) \leqslant n$ and since, in the present instance,

$$\text{w.dh}_{\Lambda}(A) = \text{dh}_{\Lambda}(A),$$

this completes the proof.

In view of the fact that, whenever $\mathfrak{m}$ is a maximal ideal, $\Lambda_{\mathfrak{m}}$ is a local ring, we shall now turn our attention to modules over local rings.

9.3 Modules over local rings

In this section and in the later sections of Chapter 9, we shall employ the letter Q to denote a local ring.† The maximal ideal of Q will be written as $\mathfrak{m}$ and for the residue field $Q/\mathfrak{m}$ we shall normally use the letter K.

Let A be a Q-module then $A/\mathfrak{m}A$ can be regarded as a K-module, or, as we customarily say, a *vector space* over K by writing

$$\phi(q)x = qx \quad (q \in Q,\ x \in A/\mathfrak{m}A),$$

where $\phi: Q \to Q/\mathfrak{m}$ is the canonical homomorphism. If A is a finitely generated Q-module, then $A/\mathfrak{m}A$ will be of finite dimension as a vector space over K.

Proposition 3. *Let A be a finitely generated Q-module and let $a_1, a_2, \ldots, a_s$ be elements of A. Further, let $\bar{a}_i$ $(1 \leqslant i \leqslant s)$ be the natural image of a_i under the mapping $A \to A/\mathfrak{m}A$. Then the elements a_i generate A as a Q-module if and only if $\bar{a}_1, \bar{a}_2, \ldots, \bar{a}_s$ generate $A/\mathfrak{m}A$ as a K-space.*

Proof. We shall assume that $A/\mathfrak{m}A = K\bar{a}_1 + K\bar{a}_2 + \ldots + K\bar{a}_s$ and deduce that $A = Qa_1 + Qa_2 + \ldots + Qa_s$. This will be sufficient, since the converse is obvious.

Let $a \in A$ and let $\bar{a}$ be its image in $A/\mathfrak{m}A$, then we can find $q_1, q_2, \ldots, q_s$ in Q such that $\bar{a} = q_1\bar{a}_1 + q_2\bar{a}_2 + \ldots + q_s\bar{a}_s$, hence

$$a \in (Qa_1 + Qa_2 + \ldots + Qa_s) + \mathfrak{m}A.$$

Thus if $B = Qa_1 + Qa_2 + \ldots + Qa_s$, then $A = B + \mathfrak{m}A$, and therefore $\mathfrak{m}(A/B) = (B + \mathfrak{m}A)/B = A/B$. But A/B is finitely generated and $\mathfrak{m}$ is the only maximal ideal of Q, consequently (Lemma 1) $A/B = 0$. Thus $A = B = Qa_1 + Qa_2 + \ldots + Qa_s$ and the proof is complete.

Corollary. *Let A be a finitely generated Q-module. If $a_1, a_2, \ldots, a_s$ generate A and have the further property that no proper subset of them also generates the whole module, then s is equal to the dimension of $A/\mathfrak{m}A$ as a vector space over K.*

Proof. With the previous notation $A/\mathfrak{m}A = K\bar{a}_1 + K\bar{a}_2 + \ldots + K\bar{a}_s$ and the proposition shows that, in this last relation, no $\bar{a}_i$ is superfluous. The corollary therefore follows.

Lemma 3. *Let A be a finitely generated Q-module such that*

$$\operatorname{Ext}^1_Q(A, K) = 0.$$

† In earlier chapters, Q was used to designate an injective module but it is not needed for this purpose in the present one.

Then A is Q-free. Further, if $a_1, a_2, \ldots, a_s$ are elements of A whose images $\bar{a}_1, \bar{a}_2, \ldots, \bar{a}_s$ in $A/\mathfrak{m}A$ form a K-base for $A/\mathfrak{m}A$, then the a_i form a Q-base for A.

Proof. Choose elements $a_1, a_2, \ldots, a_s$ so that the $\bar{a}_i$ form a K-base for $A/\mathfrak{m}A$ then, by Proposition 3, they generate A. Let F be a free Q-module with a base $v_1, v_2, \ldots, v_s$ of s elements, then we can construct an exact sequence $0 \to B \to F \to A \to 0$ where, in the mapping $F \to A$, v_i is mapped into a_i. It will be shown that $B = 0$ and this will establish the lemma. Note that the epimorphism $F \to A$ induces an epimorphism $F/\mathfrak{m}F \to A/\mathfrak{m}A$.

Since Hom_Q is left exact, our mappings give rise to a commutative diagram

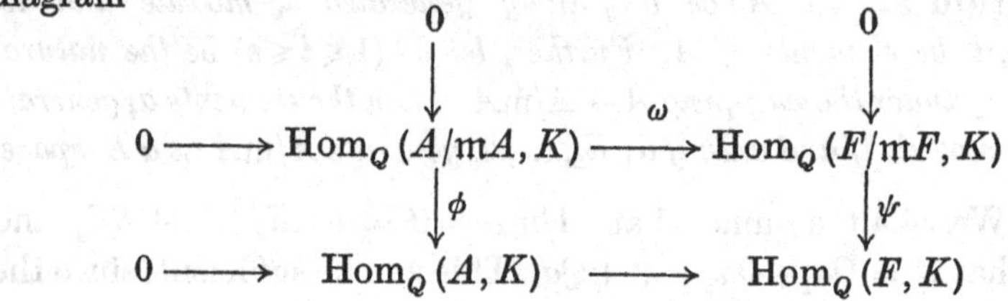

$$\begin{array}{ccccc} & & 0 & & 0 \\ & & \downarrow & & \downarrow \\ 0 & \longrightarrow & \mathrm{Hom}_Q(A/\mathfrak{m}A, K) & \xrightarrow{\omega} & \mathrm{Hom}_Q(F/\mathfrak{m}F, K) \\ & & \downarrow \phi & & \downarrow \psi \\ 0 & \longrightarrow & \mathrm{Hom}_Q(A, K) & \longrightarrow & \mathrm{Hom}_Q(F, K) \end{array}$$

in which the rows and columns are exact. Further, if $f \in \mathrm{Hom}_Q(A, K)$, $\mu \in \mathfrak{m}$ and $a \in A$, then $f(\mu a) = \mu f(a) = 0$, which shows that f vanishes on $\mathfrak{m}A$. It follows that there is a Q-homomorphism $g : A/\mathfrak{m}A \to K$ such that f is the combined mapping $A \to A/\mathfrak{m}A \to K$. Thus we have established that ϕ is an epimorphism as well as a monomorphism, consequently it is an isomorphism. Similarly, ψ is an isomorphism.

Again, $A/\mathfrak{m}A$ and $F/\mathfrak{m}F$ are K-modules, in fact each is an s-dimensional vector space over K. We can therefore regard $F/\mathfrak{m}F \to A/\mathfrak{m}A$ as a mapping of one vector space on to another vector space having the same dimension. But such a mapping will have the zero element as its kernel, and therefore $F/\mathfrak{m}F \to A/\mathfrak{m}A$ must be an isomorphism of Q-modules. From this it follows that

$$\omega : \mathrm{Hom}_Q(A/\mathfrak{m}A, K) \to \mathrm{Hom}_Q(F/\mathfrak{m}F, K)$$

is also an isomorphism

We have now shown that each of the mappings ϕ, ψ and ω is an isomorphism, consequently

$$\mathrm{Hom}_Q(A, K) \to \mathrm{Hom}_Q(F, K)$$

is an isomorphism as well.

By hypothesis $\mathrm{Ext}_Q^1(A, K) = 0$ and so we have an exact sequence

$$\mathrm{Hom}_Q(A, K) \to \mathrm{Hom}_Q(F, K) \to \mathrm{Hom}_Q(B, K) \to 0,$$

in which the first mapping is an isomorphism. Accordingly,

$$\mathrm{Hom}_Q(B, K) = 0.$$

But $0 \to \mathrm{Hom}_Q(B/\mathfrak{m}B, K) \to \mathrm{Hom}_Q(B, K)$ is exact and therefore

$$\mathrm{Hom}_K(B/\mathfrak{m}B, K) = \mathrm{Hom}_Q(B/\mathfrak{m}B, K) = 0.$$

However, $B/\mathfrak{m}B$ is K-free so this can only mean that $B/\mathfrak{m}B = 0$, or, equivalently, that $B = \mathfrak{m}B$. But B, being a submodule of F, is finitely generated, consequently $B = 0$ by Lemma 1.

We can now amplify Lemma 3 a little so as to obtain

Theorem 12. *Let Q be a local ring and A a finitely generated Q-module. Then A is Q-projective if and only if A is Q-free; and in order that A be Q-free (Q-projective) it is necessary and sufficient that* $\mathrm{Ext}_Q^1(A, K) = 0$. *Furthermore, when* $\mathrm{Ext}_Q^1(A, K) = 0$, *a set $a_1, a_2, \ldots, a_s$ of elements of A will form a Q-base for A if and only if their natural images in $A/\mathfrak{m}A$ form a K-base for that space.*

Proof. If A is Q-projective then $\mathrm{Ext}_Q^1(A, K) = 0$, and so it follows, by Lemma 3, that A is Q-free. To complete the proof, we have only to show that if the a_i form a Q-base for A, then their images form a K-base for $A/\mathfrak{m}A$ and this is obvious.

There is an analogue of Theorem 12 in which Tor_1^Q takes over the role of Ext_Q^1. We formulate this as follows:

Theorem 13. *Let Q be a local ring and let A be a finitely generated Q-module. Then A is Q-free if and only if* $\mathrm{Tor}_1^Q(A, K) = 0$.

Proof. It will be assumed that $\mathrm{Tor}_1^Q(A, K) = 0$, and it will be deduced that A is Q-free. This will be sufficient since the converse is clear.

Choose $a_1, a_2, \ldots, a_s$ so that their natural images in $A/\mathfrak{m}A$ form a K-base for that space, and then construct an exact sequence

$$0 \to B \to F \to A \to 0,$$

with F a Q-free module, *exactly as in the proof of* Lemma 3. The theorem will follow if we show that $B = 0$, and, since B is finitely generated, it will be enough to prove that $B = \mathfrak{m}B$ or, better still, that $B/\mathfrak{m}B = 0$.

Since $\otimes_Q$ is a right exact functor and $F \to A$ induces an epimorphism $F/\mathfrak{m}F \to A/\mathfrak{m}A$, the diagram

$$\begin{array}{ccccc} F \otimes_Q K & \longrightarrow & A \otimes_Q K & \longrightarrow & 0 \\ \downarrow \phi & & \downarrow \psi & & \\ (F/\mathfrak{m}F) \otimes_Q K & \xrightarrow{\omega} & (A/\mathfrak{m}A) \otimes_Q K & \longrightarrow & 0 \\ \downarrow & & \downarrow & & \\ 0 & & 0 & & \end{array}$$

is commutative and its rows and columns are exact. Further, the definition of a tensor product shows that, because $F/\mathfrak{m}F$ is a K-module, $(F/\mathfrak{m}F) \otimes_Q K$ is simply $(F/\mathfrak{m}F) \otimes_K K$ with the structure as Q-module induced by the natural homomorphism $Q \to K$; and, of course, a similar remark applies to $(A/\mathfrak{m}A) \otimes_Q K$. Thus

$$\omega : (F/\mathfrak{m}F) \otimes_Q K \to (A/\mathfrak{m}A) \otimes_Q K$$

is none other than the K-epimorphism

$$(F/\mathfrak{m}F) \otimes_K K \to (A/\mathfrak{m}A) \otimes_K K.$$

But $(F/\mathfrak{m}F) \otimes_K K \approx F/\mathfrak{m}F$ and $(A/\mathfrak{m}A) \otimes_K K \approx A/\mathfrak{m}A$, hence the domain and range of ω are both s-dimensional vector spaces and therefore we may conclude that *ω is an isomorphism.*

We assert next that *the mappings ϕ and ψ are also isomorphisms* and to establish this we need only show that they are monomorphisms. As the case of ψ is typical, it is the only one we need consider. It should be noted that, in proving $A \otimes_Q K \to (A/\mathfrak{m}A) \otimes_Q K$ is an isomorphism, we do not require any special assumptions concerning A. It need not even be finitely generated.

Assume that $\psi(x) = 0$, and let us write $x = a \otimes 1_K$. Then

$$0 = \psi(x) = \bar{a} \otimes 1_K,$$

where $\bar{a}$ is the image of a in $A/\mathfrak{m}A$. But

$$(A/\mathfrak{m}A) \otimes_Q K = (A/\mathfrak{m}A) \otimes_K K \approx A/\mathfrak{m}A,$$

and by means of this we deduce (from the relation $0 = \bar{a} \otimes 1_K$) that $\bar{a} = 0$. Thus we can write $a = \Sigma \mu_i a_i$, where $\mu_i \in \mathfrak{m}$ and $a_i \in A$; consequently

$$x = a \otimes 1_K = \Sigma(\mu_i a_i \otimes 1_K) = \Sigma(a_i \otimes \mu_i 1_K) = 0,$$

since $\mu_i 1_K = 0$. This establishes that ψ is an isomorphism.

Our combined observations now show that $F \otimes_Q K \to A \otimes_Q K$ is an isomorphism. Further, in view of the fact that $\mathrm{Tor}_1^Q(A, K) = 0$, we have an exact sequence

$$0 \to B \otimes_Q K \to F \otimes_Q K \to A \otimes_Q K,$$

which allows us to conclude that $B \otimes_Q K = 0$. Finally

$$B \otimes_Q K \to (B/\mathfrak{m}B) \otimes_Q K \to 0$$

is exact, consequently

$$0 = (B/\mathfrak{m}B) \otimes_Q K = (B/\mathfrak{m}B) \otimes_K K \approx B/\mathfrak{m}B,$$

and this completes the proof.

Since the demonstration that ψ is an isomorphism did not make use of any special properties of the Q-module A, we obtain the result described below as Lemma 4. This will be needed on a future occasion.

Lemma 4. *If A is a Q-module then the mapping*

$$A \otimes_Q K \to (A/\mathfrak{m}A) \otimes_Q K$$

is an isomorphism.

Theorem 14. *Let Q be a local ring, A a finitely generated Q-module, and r $(r \geqslant -1)$ an integer. Then*

$$\mathrm{dh}_Q(A) \leqslant r \text{ if and only if } \mathrm{Tor}_{r+1}^Q(A, K) = 0. \qquad (9.3.1)$$

Proof. It will be assumed that $\mathrm{Tor}_{r+1}^Q(A, K) = 0$, and from this it will be deduced that $\mathrm{dh}_Q(A) \leqslant r$. The converse assertion does not need to be discussed because it is an immediate consequence of Theorem 18 of section (7.9).

Suppose first that $r = -1$, then $A \otimes_Q K = 0$, hence also

$$(A/\mathfrak{m}A) \otimes_Q K = 0,$$

because $(A/\mathfrak{m}A) \otimes_Q K$ is a homomorphic image of $A \otimes_Q K$. But

$$(A/\mathfrak{m}A) \otimes_Q K = (A/\mathfrak{m}A) \otimes_K K \approx A/\mathfrak{m}A,$$

hence $A/\mathfrak{m}A = 0$ and therefore $A = 0$. This disposes of the case $r = -1$. For $r = 0$ there is nothing new to prove because, by Theorem 13, from $\mathrm{Tor}_1^Q(A, K) = 0$ it follows that A is Q-free.

We turn now to the general case and assume that $r \geqslant 1$. Since A is finitely generated and Q is Noetherian, we can construct an exact sequence

$$0 \to A_r \to F_{r-1} \to F_{r-2} \to \cdots \to F_0 \to A \to 0,$$

where each of $F_0, F_1, \ldots, F_{r-1}$ is a free module with a finite base and A_r is a finitely generated Q-module. Put $A_0 = A$ and, for $1 \leqslant i \leqslant r-1$, let $\mathrm{Im}(F_i \to F_{i-1}) = A_i$, then we have exact sequences

$$0 \to A_{i+1} \to F_i \to A_i \to 0 \quad (0 \leqslant i \leqslant r-1)$$

from which we derive further exact sequences, namely,

$$0 \to \mathrm{Tor}^Q_{r-i+1}(A_i, K) \to \mathrm{Tor}^Q_{r-i}(A_{i+1}, K) \to 0 \quad (0 \leqslant i \leqslant r-1).$$

In other words, we have isomorphisms

$$\mathrm{Tor}^Q_1(A_r, K) \approx \mathrm{Tor}^Q_2(A_{r-1}, K) \approx \ldots \approx \mathrm{Tor}^Q_{r+1}(A_0, K) = 0,$$

and now it follows, from Theorem 13, that A_r is Q-free. Accordingly, $\mathrm{dh}_Q(A) \leqslant r$ and the proof is complete.

Theorem 15 (below) can be obtained from Theorem 12 by a method so similar to that used to derive Theorem 14 from Theorem 13 that we omit the details.

Theorem 15. *Let Q be a local ring, A a finitely generated Q-module and r $(r \geqslant -1)$ an integer. Then*

$$\mathrm{dh}_Q(A) \leqslant r \textit{ if and only if } \mathrm{Ext}^{r+1}_Q(A, K) = 0. \tag{9.3.2}$$

The next theorem shows that no Q-module can have a larger homological dimension than the residue field K.

Theorem 16. *If Q is a local ring then* $\mathrm{gl.dim}\, Q = \mathrm{dh}_Q(K)$.

Proof. It follows from the definition of global dimension that $\mathrm{dh}_Q(K) \leqslant \mathrm{gl.dim}\, Q$. To establish the opposite inequality we shall suppose that $\mathrm{dh}_Q(K) = r$, r being a finite integer, and deduce that $\mathrm{gl.dim}\, Q \leqslant r$.

Let A be a cyclic Q-module, then, since $\mathrm{dh}_Q(K) = r$,

$$\mathrm{Tor}^Q_{r+1}(A, K) = 0$$

by Theorem 18 of section (7.9). It therefore follows, by Theorem 14, that $\mathrm{dh}_Q(A) \leqslant r$ and thence, by Theorem 15 of section (7.6), that $\mathrm{gl.dim}\, Q \leqslant r$.

Theorem 14 shows that $\mathrm{dh}_Q(K) \leqslant r$ if and only if $\mathrm{Tor}^Q_{r+1}(K, K) = 0$. Combining this remark with the result just proved we obtain

Theorem 17. *If Q is a local ring and p $(p \geqslant 0)$ is an integer, then*

$$\mathrm{gl.dim}\, Q \leqslant p \textit{ if and only if } \mathrm{Tor}^Q_{p+1}(K, K) = 0. \tag{9.3.3}$$

In a similar manner, but using Theorem 15 instead of Theorem 14, one can also establish

Theorem 18. *If Q is a local ring and p $(p \geqslant 0)$ is an integer, then*

$$\mathrm{gl.dim}\, Q \leqslant p \textit{ if and only if } \mathrm{Ext}^{p+1}_Q(K, K) = 0. \tag{9.3.4}$$

Theorem 16 shows that, to determine the global dimension of Q, it suffices to find the homological dimension of the residue field of the maximal ideal $\mathfrak{m}$. This fact will now be used to advance our knowledge of commutative Noetherian rings which may have more than one maximal ideal.

Theorem 19. *Let Λ be a commutative Noetherian ring and let $\mathfrak{m}$ be one of its maximal ideals. Then $\mathrm{dh}_\Lambda(\Lambda/\mathfrak{m}) = \mathrm{gl.dim}\,\Lambda_\mathfrak{m}$. Furthermore,*

$$\mathrm{gl.dim}\,\Lambda = \sup_{\mathfrak{m}}.\mathrm{dh}_\Lambda(\Lambda/\mathfrak{m}), \qquad (9.3.5)$$

where, when taking the upper bound, $\mathfrak{m}$ is to range over all the maximal ideals of Λ.

Proof. Let $\mathfrak{m}^*$ also be a maximal ideal and let S^* be the complement of $\mathfrak{m}^*$ in Λ then, by Theorem 6 of section (8.6), $0 \to \mathfrak{m} \to \Lambda \to \Lambda/\mathfrak{m} \to 0$ gives rise to the exact sequence

$$0 \to \mathfrak{m}_{S^*} \to \Lambda_{S^*} \to (\Lambda/\mathfrak{m})_{S^*} \to 0.$$

First consider the case $\mathfrak{m} \neq \mathfrak{m}^*$. Then $\mathfrak{m}$ is not contained in $\mathfrak{m}^*$ and so $\mathfrak{m}_{S^*}$ contains an element $\left[\frac{\mu}{1}\right]$ where $\mu \notin \mathfrak{m}^*$. But this means that $\mu \in S^*$, consequently $\mathrm{Im}\,(\mathfrak{m}_{S^*} \to \Lambda_{S^*})$ contains an element which has an inverse in the ring Λ_{S^*}. Accordingly $\mathrm{Im}\,(\mathfrak{m}_{S^*} \to \Lambda_{S^*}) = \Lambda_{S^*}$ and therefore $(\Lambda/\mathfrak{m})_{S^*} = 0$. Changing the notation, we can write this result as

$$[\Lambda/\mathfrak{m}]_{\mathfrak{m}^*} = 0 \quad (\mathfrak{m}^* \neq \mathfrak{m}).$$

Next consider the case $\mathfrak{m}^* = \mathfrak{m}$. This time $\mathrm{Im}\,(\mathfrak{m}_{S^*} \to \Lambda_{S^*})$ consists of all elements $\left[\frac{\mu}{c}\right]$, where $\mu \in \mathfrak{m}$ and $c \notin \mathfrak{m}$, which means that it is just the maximal ideal of Λ_{S^*}. Accordingly $(\Lambda/\mathfrak{m})_{S^*}$ can be identified with the residue field of the local ring $\Lambda_{S^*} = \Lambda_\mathfrak{m}$, hence, if we denote this residue field by $K^{(\mathfrak{m})}$, we obtain

$$[\Lambda/\mathfrak{m}]_\mathfrak{m} = K^{(\mathfrak{m})}.$$

Theorem 11 now shows that

$$\mathrm{dh}_\Lambda(\Lambda/\mathfrak{m}) = \mathrm{dh}_{\Lambda_\mathfrak{m}}(K^{(\mathfrak{m})}),$$

and this is just $\mathrm{gl.dim}\,\Lambda_\mathfrak{m}$ by Theorem 16. Thus the first assertion of the theorem has been established and the second assertion follows from the first by virtue of Theorem 10.

We return now to the consideration of modules over a local ring Q.

Definition. Let M be a Q-module and q an element of Q. Then q is said to be a *zero-divisor in* M if there exists a non-zero element $x \in M$ such that $qx = 0$.

The above definition leads on to the important notion of a normal M-sequence.

Definition. Let M be a Q-module and $q_1, q_2, \ldots, q_p$ a sequence of elements of Q. The sequence is described as a *normal M-sequence* if

(i) $q_1, q_2, \ldots, q_p$ all belong to $\mathfrak{m}$,

and

(ii) for each value of i $(1 \leqslant i \leqslant p)$, q_i is not a zero-divisor in

$$M/(q_1, q_2, \ldots, q_{i-1})\,M.$$

Of course, when $i = 1$, condition (ii) is to be interpreted as stating that q_1 is not a zero-divisor in M.

Theorem 20. *Let $M \neq 0$ be a finitely generated Q-module, let $\alpha \in \mathfrak{m}$ and suppose that α is not a zero-divisor in M. Then*

$$\mathrm{dh}_Q\,(M/\alpha M) = \mathrm{dh}_Q\,(M) + 1,$$

where, of course, both sides may be infinite. More generally, if $\alpha_1, \alpha_2, \ldots, \alpha_p$ is a normal M-sequence, then

$$\mathrm{dh}_Q\,[M/(\alpha_1, \alpha_2, \ldots, \alpha_p)\,M] = \mathrm{dh}_Q\,(M) + p.$$

Proof. Let us observe that if $M_i = M/(\alpha_1, \ldots, \alpha_i)\,M$, then $M_i \neq 0$, because otherwise we should have $M = \mathfrak{m}M$ and therefore (Lemma 1) $M = 0$. Further, M_i is finitely generated, α_{i+1} belongs to $\mathfrak{m}$ and is not a zero-divisor in M_i, and finally $M_{i+1} = M_i/\alpha_{i+1}M_i$. This shows that the second assertion is a simple consequence of the first and so, from now on, the discussion will be limited to the initial contention of the theorem.

Let i_M and i_K denote the identity maps of M and K respectively then, because α is not a zero-divisor in M,

$$0 \to M \xrightarrow{\alpha i_M} M \to M/\alpha M \to 0$$

is an exact sequence. From it we obtain, for each integer $r \geqslant 0$, a further exact sequence namely

$$\mathrm{Tor}_{r+1}^Q\,(M, K) \to \mathrm{Tor}_{r+1}^Q\,(M, K) \to \mathrm{Tor}_{r+1}^Q\,(M/\alpha M, K)$$
$$\to \mathrm{Tor}_r^Q\,(M, K) \to \mathrm{Tor}_r^Q\,(M, K).$$

Now $\mathrm{Tor}_r^Q\,(M, K) \to \mathrm{Tor}_r^Q\,(M, K)$ is the mapping

$$\mathrm{Tor}_r^Q\,(\alpha i_M, i_K) = \mathrm{Tor}_r^Q\,(i_M, \alpha i_K) = \mathrm{Tor}_r^Q\,(i_M, 0) = 0,$$

and similarly $\mathrm{Tor}_{r+1}^{Q}(M,K)\to\mathrm{Tor}_{r+1}^{Q}(M,K)$ is also a null map. It follows that the sequence

$$0\to\mathrm{Tor}_{r+1}^{Q}(M,K)\to\mathrm{Tor}_{r+1}^{Q}(M/\alpha M,K)\to\mathrm{Tor}_{r}^{Q}(M,K)\to 0 \quad (9.3.6)$$

is exact.

Suppose first that $\mathrm{dh}_Q(M)=\infty$, then, by Theorem 14,

$$\mathrm{Tor}_{r}^{Q}(M,K)\neq 0 \quad \text{and so} \quad \mathrm{Tor}_{r+1}^{Q}(M/\alpha M,K)\neq 0,$$

this being true for all r. It follows that $\mathrm{dh}_Q(M/\alpha M)=\infty$. Next suppose that $\mathrm{dh}_Q(M)=n$. Then, again using Theorem 14, $\mathrm{Tor}_{n}^{Q}(M,K)\neq 0$, hence $\mathrm{Tor}_{n+1}^{Q}(M/\alpha M,K)\neq 0$ and therefore $\mathrm{dh}_Q(M/\alpha M)\geqslant n+1$. On the other hand, putting $r=n+1$ in (9.3.6), we see that

$$\mathrm{Tor}_{n+2}^{Q}(M/\alpha M,K)=0,$$

which shows that $\mathrm{dh}_Q(M/\alpha M)\leqslant n+1$. Accordingly,

$$\mathrm{dh}_Q(M/\alpha M)=n+1=1+\mathrm{dh}_Q(M)$$

as required.

As the proof of the next theorem is comparatively complicated it will be preceded by two lemmas, in preparation for which we make some preliminary observations.

The Q-module $\mathfrak{m}/\mathfrak{m}^2$, where $\mathfrak{m}^2=\mathfrak{m}\mathfrak{m}$, can be regarded as a vector space over K of finite dimension n (say). Let $u_1,u_2,\dots,u_n$ be elements of $\mathfrak{m}$ whose images in $\mathfrak{m}/\mathfrak{m}^2$ form a K-base for that space, then, if $q_1u_1+q_2u_2+\dots+q_nu_n\in\mathfrak{m}^2$ for $q_1,q_2,\dots,q_n$ in Q, it follows that each q_i is in $\mathfrak{m}$. Furthermore, by Proposition 3, $u_1,u_2,\dots,u_n$ will generate the ideal $\mathfrak{m}$.

With the aid of the u_i we shall construct a 0-sequence

$$F_n\xrightarrow{d_n}F_{n-1}\xrightarrow{d_{n-1}}F_{n-2}\to\cdots\to F_1\xrightarrow{d_1}Q\to K\to 0 \quad (9.3.7)$$

with rather special properties. Before, however, we get involved in the details, it may help to note that if $n=0$, then $\mathfrak{m}=\mathfrak{m}^2$ and so (Lemma 1) $\mathfrak{m}=0$. Thus in the case $n=0$, Q is a field and for this situation it will be found that our assertions are obvious.

F_s $(1\leqslant s\leqslant n)$ is defined to be the free Q-module generated by the $\binom{n}{s}$ symbols† $[i_1,i_2,\dots,i_s]$, where $i_1,i_2,\dots,i_s$ are integers satisfying $1\leqslant i_1<i_2<\dots<i_s\leqslant n$; and it will be convenient to put $F_0=Q$. For

† By $\binom{n}{s}$ we mean, of course, the binomial coefficient $n!/s!(n-s)!$.

$s \geqslant 1$ the homomorphism $d_s : F_s \to F_{s-1}$ is the Q-homomorphism defined by the formula

$$d_s[i_1, i_2, \ldots, i_s] = \sum_{p=1}^{s} (-1)^{p+1} u_{i_p}[i_1, \ldots, \hat{i_p}, \ldots, i_s],$$

where the ^ over i_p means that i_p is to be omitted from the set of integers in the bracket. When $s = 1$ the formula is to be interpreted as asserting that $d_1[i] = u_i$.

A simple verification now shows that, for $2 \leqslant s \leqslant n$, $d_{s-1}d_s = 0$, while it is clear that, if $Q \to K$ denotes the canonical mapping, then $F_1 \xrightarrow{d_1} Q \to K \to 0$ is *exact*. In particular, (9.3.7) is a 0-sequence.

By construction, $\operatorname{Im}(d_s) \subseteq \mathfrak{m}F_{s-1}$, consequently $\mathfrak{m}F_s$ is carried into $\mathfrak{m}^2F_{s-1}$ and therefore d_s $(1 \leqslant s \leqslant n)$ induces a homomorphism

$$d_s^* : F_s/\mathfrak{m}F_s \to \mathfrak{m}F_{s-1}/\mathfrak{m}^2F_{s-1}.$$

Lemma 5. *For each s $(1 \leqslant s \leqslant n)$ the induced homomorphism*

$$d_s^* : F_s/\mathfrak{m}F_s \to \mathfrak{m}F_{s-1}/\mathfrak{m}^2F_{s-1}$$

is a monomorphism.

Proof. Let
$$x = \sum_{1 \leqslant i_1 < \ldots < i_s \leqslant n} q_{i_1 \ldots i_s} [i_1, \ldots, i_s],$$

where $q_{i_1 \ldots i_s} \in Q$, be an element of F_s for which $d_s(x) \in \mathfrak{m}^2F_{s-1}$, then the lemma will follow if we show that $x \in \mathfrak{m}F_s$. As a step in this direction, we shall examine the coefficient of $[i'_1, i'_2, \ldots, i'_{s-1}]$ in $d_s(x)$, where $i'_1, i'_2, \ldots, i'_{s-1}$ are arbitrary integers satisfying $1 \leqslant i'_1 < i'_2 < \ldots < i'_{s-1} \leqslant n$. Let i^* $(1 \leqslant i^* \leqslant n)$ be different from all of $i'_1, i'_2, \ldots, i'_{s-1}$ and choose $p^* = p^*(i^*)$ so that

$$i'_1 < \ldots < i'_{p^*} < i^* < i'_{p^*+1} < \ldots < i'_{s-1}.$$

Then the term involving $[i'_1, \ldots, i'_{p^*}, i^*, i'_{p^*+1}, \ldots, i'_{s-1}]$, in the representation of x, contributes to $d_s(x)$ a term

$$(-1)^{p^*} q_{i'_1 \ldots i^* \ldots i'_{s-1}} u_{i^*}[i'_1, i'_2, \ldots, i'_{s-1}].$$

Accordingly
$$\sum_{i^*} (-1)^{p^*} q_{i'_1 \ldots i^* \ldots i'_{s-1}} u_{i^*}$$

belongs to $\mathfrak{m}^2$ because it is the coefficient of $[i'_1, i'_2, \ldots, i'_{s-1}]$ in the representation of the element $d_s(x)$ of $\mathfrak{m}^2F_{s-1}$. It follows therefore, by the choice of $u_1, u_2, \ldots, u_n$, that $q_{i'_1 \ldots i^* \ldots i'_{s-1}} \in \mathfrak{m}$ and this implies in turn (on account of the freedom available in selecting $i'_1, i'_2, \ldots, i'_{s-1}$ and i^*) that $q_{i_1 i_2 \ldots i_s} \in \mathfrak{m}$ for all sets $i_1, i_2, \ldots, i_s$. The lemma now follows.

The sequence (9.3.7), as constructed, is only a 0-sequence but it will now be converted into an *exact* sequence by an adjustment to the modules F_s.

Lemma 6. *There exists an exact sequence*

$$X_n \xrightarrow{\delta_n} X_{n-1} \xrightarrow{\delta_{n-1}} X_{n-2} \to \cdots \to X_1 \xrightarrow{\delta_1} Q \to K \to 0$$

with the following properties:

(i) X_s $(1 \leqslant s \leqslant n)$ *is a free Q-module with a finite base of at least* $\binom{n}{s}$ *elements*;

(ii) F_s *is a direct summand of the Q-module* X_s;

(iii) *for all* s $(1 \leqslant s \leqslant n)$, δ_s *coincides with* d_s *on* F_s;

(iv) *for all* s $(1 \leqslant s \leqslant n)$ *the mapping*

$$X_s/\mathfrak{m}X_s \to \operatorname{Im}(\delta_s)/\mathfrak{m}\operatorname{Im}(\delta_s)$$

induced by δ_s *is an isomorphism.*

Proof. The modules $X_1, X_2, \ldots, X_n$ and the mappings $\delta_1, \delta_2, \ldots, \delta_n$ will be constructed by a procedure which first produces X_1 and δ_1, then X_2 and δ_2 and so on. For each module X_s we shall have a direct sum representation $X_s = F_s + \Phi_s$, where Φ_s is a free module with a finite base, and it will be convenient to obtain, in succession, $\Phi_1, \Phi_2, \ldots, \Phi_n$, together with their appropriate homomorphisms.

We can begin the construction by taking $\Phi_1 = 0$. Assume now that $1 \leqslant s < n$ and that $\Phi_1, \Phi_2, \ldots, \Phi_s$, together with $\delta_1, \delta_2, \ldots, \delta_s$, have been constructed to fulfil the requirements of the lemma so far as they relate only to $X_1, X_2, \ldots, X_s$, Q and K. For $1 \leqslant i \leqslant s$ put $L_i = \operatorname{Ker}(\delta_i)$, then, by virtue of (ii) and (iii), $\operatorname{Im}(d_{s+1}) \subseteq \operatorname{Ker}(d_s) \subseteq \operatorname{Ker}(\delta_s) = L_s$, and so d_{s+1} induces a homomorphism

$$F_{s+1}/\mathfrak{m}F_{s+1} \to L_s/\mathfrak{m}L_s.$$

On the other hand, since, by (iv), δ_s induces an isomorphism

$$X_s/\mathfrak{m}X_s \approx \operatorname{Im}(\delta_s)/\mathfrak{m}\operatorname{Im}(\delta_s)$$

it follows that $L_s = \operatorname{Ker}(\delta_s) \subseteq \mathfrak{m}X_s$. Thus we have an inclusion map $L_s \to \mathfrak{m}X_s$, and this produces a homomorphism $L_s/\mathfrak{m}L_s \to \mathfrak{m}X_s/\mathfrak{m}^2X_s$. Again, from the direct sum representation $X_s = F_s + \Phi_s$ one obtains a homomorphism $X_s \to F_s$ which in turn induces one more homomorphism, namely, $\mathfrak{m}X_s/\mathfrak{m}^2X_s \to \mathfrak{m}F_s/\mathfrak{m}^2F_s$. Now the combined mapping

$$F_{s+1}/\mathfrak{m}F_{s+1} \to L_s/\mathfrak{m}L_s \to \mathfrak{m}X_s/\mathfrak{m}^2X_s \to \mathfrak{m}F_s/\mathfrak{m}^2F_s$$

is none other than the monomorphism discussed in Lemma 5, consequently $F_{s+1}/\mathfrak{m}F_{s+1} \to L_s/\mathfrak{m}L_s$ *is itself a monomorphism of* Q*-modules and therefore also a monomorphism of* K*-spaces.*

By the inductive hypothesis, X_s has a finite base consequently L_s is finitely generated and so $L_s/\mathfrak{m}L_s$ is a K-space of finite dimension. This space has $\mathrm{Im}\,[(F_{s+1}/\mathfrak{m}F_{s+1}) \to (L_s/\mathfrak{m}L_s)]$ as a subspace and, by our previous remarks, the subspace has the *same dimension* as $F_{s+1}/\mathfrak{m}F_{s+1}$. We now choose a Q-module N_s such that $\mathfrak{m}L_s \subseteq N_s \subseteq L_s$ and

$$L_s/\mathfrak{m}L_s = \mathrm{Im}\,[(F_{s+1}/\mathfrak{m}F_{s+1}) \to (L_s/\mathfrak{m}L_s)] + (N_s/\mathfrak{m}L_s), \qquad (9.3.8)$$

this being a direct sum of K-spaces.

Let $\zeta_1, \zeta_2, \ldots, \zeta_p$ be elements of N_s whose images in $N_s/\mathfrak{m}L_s$ are a K-base for that space. Let Φ_{s+1} be a free module generated by p symbols $z_1, z_2, \ldots, z_p$, and let $\Phi_{s+1} \to X_s$ be the homomorphism in which z_i is mapped into ζ_i. Put $X_{s+1} = F_{s+1} + \Phi_{s+1}$ (direct sum) and finally take for $\delta_{s+1} : X_{s+1} \to X_s$ the mapping which agrees with d_{s+1} on F_{s+1} and, on Φ_{s+1}, is the mapping $\Phi_{s+1} \to X_s$ just constructed.

The lemma will now follow (by induction) if we show that

(*a*) $X_{s+1} \xrightarrow{\delta_{s+1}} X_s \xrightarrow{\delta_s} X_{s-1}$ is exact, and

(*b*) the mapping $X_{s+1}/\mathfrak{m}X_{s+1} \to \mathrm{Im}\,(\delta_{s+1})/\mathfrak{m}\,\mathrm{Im}\,(\delta_{s+1})$ induced, by δ_{s+1}, is an isomorphism.

It is clear that

$$\mathrm{Im}\,(\delta_{s+1}) = \mathrm{Im}\,(d_{s+1}) + Q\zeta_1 + \ldots + Q\zeta_p \subseteq L_s,$$

whence

$$\frac{\mathrm{Im}\,(\delta_{s+1}) + \mathfrak{m}L_s}{\mathfrak{m}L_s} = \mathrm{Im}\,[(F_{s+1}/\mathfrak{m}F_{s+1}) \to (L_s/\mathfrak{m}L_s)] + (N_s/\mathfrak{m}L_s) = L_s/\mathfrak{m}L_s.$$

Thus $\mathrm{Im}\,(\delta_{s+1}) + \mathfrak{m}L_s = L_s$, and so

$$\mathfrak{m}[L_s/\mathrm{Im}\,(\delta_{s+1})] = [\mathrm{Im}\,(\delta_{s+1}) + \mathfrak{m}L_s]/\mathrm{Im}\,(\delta_{s+1}) = L_s/\mathrm{Im}\,(\delta_{s+1}),$$

which shows (Lemma 1) that $\mathrm{Im}\,(\delta_{s+1}) = L_s = \mathrm{Ker}\,(\delta_s)$, thereby establishing (*a*). Furthermore, the mapping in (*b*), which is certainly an epimorphism, can now be written as $X_{s+1}/\mathfrak{m}X_{s+1} \to L_s/\mathfrak{m}L_s$, and to complete the proof we need only show that it is a monomorphism. This will be done by demonstrating that

$$X_{s+1}/\mathfrak{m}X_{s+1} = (F_{s+1}/\mathfrak{m}F_{s+1}) + (\Phi_{s+1}/\mathfrak{m}\Phi_{s+1}) \quad \text{(direct sum)}$$

and

$$L_s/\mathfrak{m}L_s = \mathrm{Im}\,[(F_{s+1}/\mathfrak{m}F_{s+1}) \to (L_s/\mathfrak{m}L_s)] + (N_s/\mathfrak{m}L_s) \quad \text{(direct sum)}$$

have the same dimension as K-spaces. But we already know that the first terms of the right-hand sides have equal dimensions while both the second terms are of dimension p by construction. Accordingly, the lemma is proved.

Before stating the next theorem, let us observe that, if A is an arbitrary Q-module, then $\mathfrak{m}\,\mathrm{Tor}_s^Q(A,K) = 0$ for all s. (This is because, if $\mu \in \mathfrak{m}$, then μ times the identity mapping of $\mathrm{Tor}_s^Q(A,K)$ can be written, with our usual notation, as $\mathrm{Tor}_s^Q(i_A, \mu i_K)$, and this, of course, is a null mapping since $\mu i_K = 0$.) Accordingly, for each value of s, $\mathrm{Tor}_s^Q(A,K)$ has a structure as a K-space in which

$$\phi(q)x = qx \quad (q \in Q;\ x \in \mathrm{Tor}_s^Q(A,K)),$$

where ϕ is the natural mapping $Q \to K$.

Theorem 21. *Let Q be a local ring and let $\mathfrak{m}/\mathfrak{m}^2$, regarded as a vector space over the residue field K, be of dimension n. Then, for $0 \leqslant s \leqslant n$, $\mathrm{Tor}_s^Q(K,K)$, also regarded as a vector space over K, has dimension at least equal to the binomial coefficient $\binom{n}{s}$.*

Proof. There exists an exact sequence

$$\cdots \to X_{n+2} \to X_{n+1} \xrightarrow{\delta_{n+1}} X_n \xrightarrow{\delta_n} X_{n-1} \to \cdots \to X_1 \xrightarrow{\delta_1} Q \to K \to 0,$$

where $X_n \to \cdots \to X_1 \to Q \to K \to 0$ is the sequence of Lemma 6 while, for $m > n$, X_m is a free Q-module with a finite base. By Lemma 6, if $0 \leqslant s \leqslant n$, then δ_s induces an isomorphism $X_s/\mathfrak{m}X_s \approx \mathrm{Im}(\delta_s)/\mathfrak{m}\,\mathrm{Im}(\delta_s)$, and this shows that $\mathrm{Im}(\delta_{s+1}) = \mathrm{Ker}(\delta_s) \subseteq \mathfrak{m}X_s$.

Consider the 0-sequence

$$X_{n+1} \otimes_Q K \to X_n \otimes_Q K \to \cdots \to X_0 \otimes_Q K \to 0,$$

where $X_0 = Q$. Since $\mathrm{Im}(X_{s+1} \to X_s) \subseteq \mathfrak{m}X_s$ for $0 \leqslant s \leqslant n$ and K is annihilated by elements of $\mathfrak{m}$, we see that

$$\mathrm{Im}(X_{s+1} \otimes_Q K \to X_s \otimes_Q K) = 0;$$

consequently, computing the homology modules,

$$\mathrm{Tor}_s^Q(K,K) \approx X_s \otimes_Q K \quad (0 \leqslant s \leqslant n).$$

Further, using Lemma 4,

$$X_s \otimes_Q K \approx (X_s/\mathfrak{m}X_s) \otimes_Q K = (X_s/\mathfrak{m}X_s) \otimes_K K \approx X_s/\mathfrak{m}X_s,$$

and therefore $\mathrm{Tor}_s^Q(K,K)$ and $X_s/\mathfrak{m}X_s$ are isomorphic as K-spaces. But, since X_s $(0 \leqslant s \leqslant n)$ is Q-free and has a base of at least $\binom{n}{s}$ elements, the dimension of $X_s/\mathfrak{m}X_s$ as a vector space is at least equal to this number.

Corollary. *If* $\mathfrak{m}/\mathfrak{m}^2$ *is a* K*-space of dimension* n, *then* $\text{gl.dim}\, Q \geqslant n$.

For, by the theorem, $\text{Tor}_n^Q(K, K) \neq 0$, since it is a K-space of dimension at least equal to unity.

9.4 Some auxiliary results

The discussion has now reached a point where, in order to make further progress, we need to bring in certain results which belong primarily to the classical theory of ideals. These results, which are three in number, will be given without proof,† and they will be distinguished from our other theorems and propositions by having asterisks attached to the numerals giving their serial order.

As before Q will denote a local ring whose maximal ideal is $\mathfrak{m}$ and whose residue field is K. By the sth power $\mathfrak{m}^s$ of $\mathfrak{m}$ we shall mean, of course, the ideal generated by all products $\mu_1\mu_2 \ldots \mu_s$, where each μ_i belongs to $\mathfrak{m}$.

Definition. The smallest integer p, for which there exists a proper ideal (i.e. different from Q) generated by p elements and containing a power of $\mathfrak{m}$, will be called the *ideal-theoretic dimension* of Q and will be denoted by $\text{Dim}\, Q$.

We shall regard the zero ideal of Q as being generated by the empty set, consequently $\text{Dim}\, Q = 0$ *if and only if* $\mathfrak{m}^s = 0$ *for some value of* s. It may be mentioned, too, that the ideal-theoretic dimension can be defined in an entirely different (but completely equivalent) manner using prime ideals. In this alternative form, $\text{Dim}\, Q$ *is the largest integer* p *for which there exists a strictly decreasing chain*

$$\mathfrak{p}_0 \supset \mathfrak{p}_1 \supset \mathfrak{p}_2 \supset \ldots \supset \mathfrak{p}_p \quad (\mathfrak{p}_0 = \mathfrak{m})$$

of prime ideals. However, we shall make no direct reference to this latter characterization.

Let the dimension of $\mathfrak{m}/\mathfrak{m}^2$, as a vector space over K, be n, then, by Proposition 3, $\mathfrak{m}$ can be generated by n elements, consequently $\text{Dim}\, Q \leqslant n$. Further, by virtue of the corollary to Theorem 21, $n \leqslant \text{gl.dim}\, Q$ and therefore

$$\text{Dim}\, Q \leqslant n \leqslant \text{gl.dim}\, Q. \tag{9.4.1}$$

It is clear from this that if $\text{Dim}\, Q < n$ then $\text{Dim}\, Q \neq \text{gl.dim}\, Q$, but the position is simpler than this remark might suggest for, as we shall see later, $\text{Dim}\, Q = \text{gl.dim}\, Q$ whenever $\text{gl.dim}\, Q$ is finite.

† References are given in the notes on this chapter (at the end of the book) for the reader who would like to know where he can find proofs.

If $\mathfrak{a} \neq Q$ is a Q-ideal, then $Q/\mathfrak{a}$ is again a local ring because $Q/\mathfrak{a}$ is certainly Noetherian and every proper ideal is contained in $\mathfrak{m}/\mathfrak{a}$. Put differently, we may say that *the image of a local ring under a ring-homomorphism is again a local ring.* Questions now arise as to how the ideal-theoretic dimension will behave when a ring-homomorphism is applied and our first auxiliary result gives the answer to one question of this kind.

Proposition 4*. *If $\alpha \in \mathfrak{m}$ and is not a zero-divisor, then the ideal-theoretic dimensions of Q and $Q/(\alpha)$ satisfy*

$$\operatorname{Dim} Q = \operatorname{Dim}[Q/(\alpha)] + 1.$$

Returning to (9.4.1), we see that $\operatorname{Dim} Q$ is never greater than the dimension of $\mathfrak{m}/\mathfrak{m}^2$ as a vector space. If the two numbers are equal the ring has a special name.

Definition. If Q is such that its ideal-theoretic dimension is equal to the dimension of $\mathfrak{m}/\mathfrak{m}^2$ as a vector space over K, then Q is said to be a *regular* local ring.

It will be proved later that $\operatorname{gl.dim} Q$ is finite if and only if Q is regular. For the moment, however, we make only a simple observation. Suppose that Q is regular and that $\operatorname{Dim} Q = 0$; then $\mathfrak{m}/\mathfrak{m}^2 = 0$, hence (Lemma 1) $\mathfrak{m} = 0$ and therefore Q is a field. Thus we see that *a regular local ring of ideal-theoretic dimension zero is a field and, of course, conversely.*

The second of the auxiliary results we need concerns a characteristic property of regular local rings and can be stated as follows:

Proposition 5*. *Let Q be a regular local ring and let $\mathfrak{m}$ be generated by elements $u_1, u_2, \ldots, u_n$, where the number n of these elements is as small as possible. Then*

$$(0) \subset (u_1) \subset (u_1, u_2) \subset \ldots \subset (u_1, u_2, \ldots, u_n)$$

is a strictly increasing sequence of prime ideals. In particular $u_1, u_2, \ldots, u_n$ is a normal Q-sequence.

The fact that $u_1, u_2, \ldots, u_n$ is a normal Q-sequence follows in this way. If $q \in Q$ and $qu_{i+1} \in (u_1, u_2, \ldots, u_i)$ then $q \in (u_1, u_2, \ldots, u_i)$, because $(u_1, u_2, \ldots, u_i)$ is a prime ideal and u_{i+1} does not belong to it. This shows that u_{i+1} is not a zero-divisor in $Q/(u_1, \ldots, u_i)\, Q$, and now the assertion is proved.

Before we make further deductions it will be convenient to give the final auxiliary result. This has to do with the existence of elements which are not zero-divisors and applies to an arbitrary local ring.

Proposition 6*. *Let A be a finitely generated Q-module and suppose that the only submodule B of A such that $\mathfrak{m}B = 0$ is the zero-submodule. Then $\mathfrak{m}$ contains an element which is not a zero-divisor in A.*

9.5 Homological codimension

As an immediate application of Proposition 5* we shall derive

Theorem 22. *If Q is a regular local ring then* gl.dim Q *is finite and is equal to the ideal-theoretic dimension* Dim Q *of Q.*

Proof. Let $u_1, u_2, \ldots, u_n$ be elements which generate $\mathfrak{m}$ and suppose that they are chosen so that their number n is as small as possible, then, by the corollary to Proposition 3, n is equal to the dimension of $\mathfrak{m}/\mathfrak{m}^2$ as a vector space over K. But Q is regular, consequently the dimension of this space is Dim Q and so we have Dim $Q = n$. Again, by Proposition 5*, $u_1, u_2, \ldots, u_n$ is a normal Q-sequence and therefore (Theorem 20)

$$\mathrm{dh}_Q(K) = \mathrm{dh}_Q[Q/(u_1, \ldots, u_n)Q] = n + \mathrm{dh}_Q(Q) = n.$$

However we know that $\mathrm{dh}_Q(K) = \mathrm{gl.dim}\, Q$, and so the proof is complete.

We return now to the consideration of local rings which are not necessarily regular.

Definition. Let A be a Q-module and let p be an integer. Then we say that A has *homological codimension equal to* p and we write $\mathrm{codh}_Q(A) = p$ if there exists a normal A-sequence with p terms but no normal sequence with more than p terms.

Later it will be shown that, if A is finitely generated and $\mathrm{gl.dim}\, Q < \infty$, then $\mathrm{dh}_Q(A) + \mathrm{codh}_Q(A) = \mathrm{gl.dim}\, Q$, and it is this result which justifies the name given to the new concept.

The integer $\mathrm{codh}_Q(Q)$ plays an important role in the general theory, and we conclude this section by establishing two elementary results concerning it.

Proposition 7. *If Q is a local ring then* $\mathrm{codh}_Q(Q) \leqslant \mathrm{Dim}\, Q$.

Proof. Let $\mu_1, \mu_2, \ldots, \mu_s$ be elements of $\mathfrak{m}$ which form a normal Q-sequence and let Q_i $(0 \leqslant i \leqslant s)$ be the local ring $Q/(\mu_1, \ldots, \mu_i)$, where, of course, by Q_0 we mean Q itself. If $i < s$, denote by $\hat{\mu}_{i+1}$ the natural image of μ_{i+1} in Q_i, then $\hat{\mu}_{i+1}$ is not a zero-divisor in Q_i and

$$Q_{i+1} = Q_i/\hat{\mu}_{i+1}Q_i.$$

Since $\hat{\mu}_{i+1}$ is contained in the maximal ideal of Q_i it follows (Proposition 4*) that $\operatorname{Dim} Q_i = \operatorname{Dim} Q_{i+1} + 1$. Accordingly

$$\operatorname{Dim} Q = s + \operatorname{Dim} Q_s \geqslant s.$$

But $\mu_1, \mu_2, \ldots, \mu_s$ is an *arbitrary* normal Q-sequence and so the proposition follows.

Note that (9.4.1) can now be extended. Indeed we have, for any local ring Q,

$$\operatorname{codh}_Q(Q) \leqslant \operatorname{Dim} Q \leqslant n \leqslant \operatorname{gl.dim} Q, \tag{9.5.1}$$

where, as before, n is the dimension of $\mathfrak{m}/\mathfrak{m}^2$ as a vector space over K.

Lemma 7. *Let μ be an element of $\mathfrak{m}$ which is not a zero-divisor in the ring Q and put $Q^* = Q/(\mu)$. Then* $\operatorname{codh}_Q(Q) \geqslant \operatorname{codh}_{Q^*}(Q^*) + 1$.

Proof. Set $s = \operatorname{codh}_{Q^*}(Q^*)$ and let $\mu_1^*, \mu_2^*, \ldots, \mu_s^*$ be s elements of the maximal ideal $\mathfrak{m}^*$ of Q^* forming a normal sequence for the Q^*-module Q^*. Choose elements $\mu_1, \mu_2, \ldots, \mu_s$ in $\mathfrak{m}$ so that μ_i^* is the natural image of μ_i in Q^*, then, if $0 \leqslant j < s$, $Q^*/(\mu_1^*, \ldots, \mu_j^*) = Q/(\mu, \mu_1, \ldots, \mu_j)$, and, for this module, multiplication by μ_{j+1} has the same effect as multiplication by μ_{j+1}^*. This shows that μ_{j+1} is not a zero-divisor in $Q/(\mu, \mu_1, \ldots, \mu_j)$, hence $\mu, \mu_1, \ldots, \mu_s$ is a normal Q-sequence. Accordingly

$$\operatorname{codh}_Q(Q) \geqslant s + 1 = \operatorname{codh}_{Q^*}(Q^*) + 1,$$

as the lemma asserts.

9.6 Modules of finite homological dimension

As before Q denotes a local ring with maximal ideal $\mathfrak{m}$ and residue field K. Since $\operatorname{gl.dim} Q$ is the upper bound of $\operatorname{dh}_Q(A)$, where A ranges over all cyclic Q-modules, it might be expected that if $\operatorname{gl.dim} Q = \infty$ there would exist cyclic modules whose homological dimensions were finite but very large. However, this is not the case, as is shown by the next theorem.

Theorem 23. *Let A be a finitely generated Q-module such that $\operatorname{dh}_Q(A) < \infty$. Then $\operatorname{dh}_Q(A) \leqslant \operatorname{codh}_Q(Q)$. Further, there exists a cyclic Q-module whose homological dimension is precisely equal to $\operatorname{codh}_Q(Q)$.*

Proof. Put $s = \operatorname{codh}_Q(Q)$, and consider the final assertion of the theorem. By the definition of codimension, there exists a normal Q-sequence $\mu_1, \mu_2, \ldots, \mu_s$. Accordingly (Theorem 20)

$$\operatorname{dh}_Q[Q/(\mu_1, \mu_2, \ldots, \mu_s) Q] = s + \operatorname{dh}_Q(Q) = s,$$

and so $Q/(\mu_1, \ldots, \mu_s)Q$ is a cyclic module of homological dimension equal to $\text{codh}_Q(Q)$.

Let $\text{dh}_Q(A) = p < \infty$, then we wish to show that $p \leqslant s$. The method of proof uses induction with respect to s. Consider first the case $s = 0$. We shall assume that $p > 0$ and obtain a contradiction.

If $p > 1$ we construct an exact sequence $0 \to A_1 \to F \to A \to 0$, where A_1 and F are both finitely generated and F is free. By Theorem 13 of section (7.5), $\text{dh}_Q(A_1) = p - 1$. If $p - 1 > 1$ we repeat the process, and, in this way, eventually obtain a finitely generated Q-module B (say) such that $\text{dh}_Q(B) = 1$. Suppose that the dimension of $B/\mathfrak{m}B$ as a vector space over K is r. Choose $b_1, b_2, \ldots, b_r$ in B so that their images in $B/\mathfrak{m}B$ form a base for that space, then (Proposition 3) $b_1, b_2, \ldots, b_r$ generate B. Let Φ be a free module generated by r symbols $y_1, y_2, \ldots, y_r$ say, and let $0 \to D \to \Phi \to B \to 0$ be an exact sequence, where in $\Phi \to B$, y_i is mapped into b_i. D will be finitely generated and, since $\text{dh}_Q(B) = 1$, it will also be projective; consequently, by Theorem 12, D must be Q-free. Again, $\Phi \to B$ induces an epimorphism $\Phi/\mathfrak{m}\Phi \to B/\mathfrak{m}B$ which, since $\Phi/\mathfrak{m}\Phi$ and $B/\mathfrak{m}B$ are both r-dimensional K-spaces, is in fact an isomorphism. It follows that $\text{Ker}(\Phi \to B) \subseteq \mathfrak{m}\Phi$. Now, by hypothesis, $\text{codh}_Q(Q) = s = 0$, which means that every element of $\mathfrak{m}$ is a zero-divisor in Q. It therefore follows (Proposition 6*) that there exists an ideal $\mathfrak{a} \neq 0$ such that $\mathfrak{m}\mathfrak{a} = 0$. Choose $\alpha \in \mathfrak{a}$ so that $\alpha \neq 0$, then $\alpha \,\text{Ker}(\Phi \to B) = 0$ and therefore $\alpha D = 0$. But $D \neq 0$ because B is not projective, consequently, since D is free, it cannot be annihilated by multiplication by a non-zero element. We therefore have the contradiction required to establish the theorem for the case $s = 0$.

Now suppose that $s > 0$ and that the theorem has been proved for all local rings Q' for which $\text{codh}_{Q'}(Q') < s$. As before we write $p = \text{dh}_Q(A)$. Since we wish to show that $p \leqslant s$ we may suppose, in what follows, that $p \geqslant 1$.

Construct an exact sequence $0 \to L \to X \to A \to 0$, where X is a free module with a finite base, then L is a finitely generated Q-module and $\text{dh}_Q(L) = p - 1$. Having obtained L we next construct an exact sequence

$$0 \to \Phi_{p-1} \to \Phi_{p-2} \to \cdots \to \Phi_0 \to L \to 0,$$

where $\Phi_0, \Phi_1, \ldots, \Phi_{p-2}$ are free modules each with a finite base. Then Φ_{p-1} is finitely generated and, by Theorem 11 of section (7.5), it is projective; consequently (Theorem 12) it is also Q-free. Now $\text{codh}_Q(Q) = s > 0$, consequently there exists an element

$\mu \in \mathfrak{m}$ which is not a zero-divisor in Q. Consider the commutative diagram

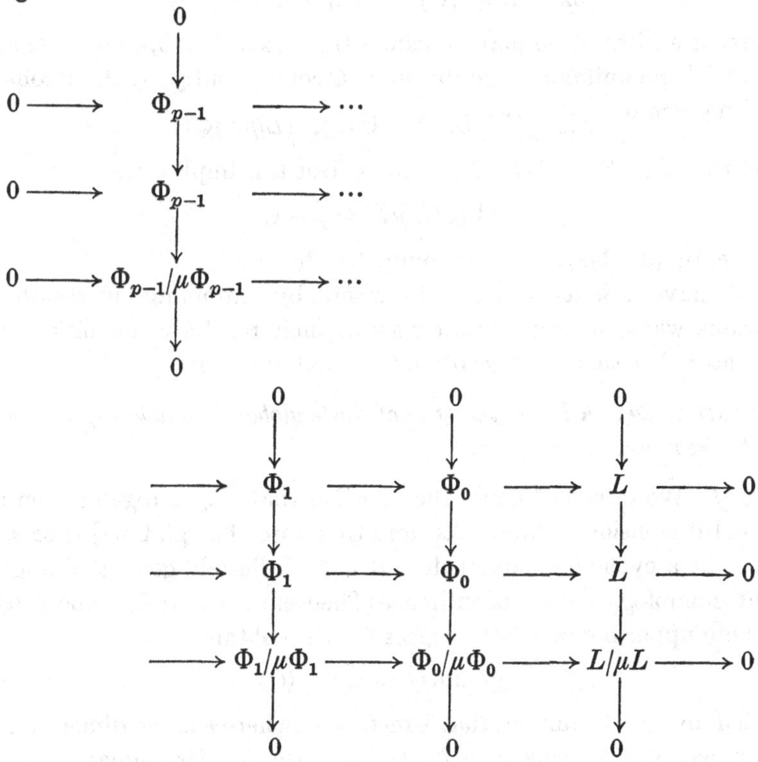

in which each vertical map from the first to the second row is μ times an identity map. Remembering that L is a submodule of a free module and that μ is not a zero-divisor, we see that the columns are all exact. The lowest row is certainly a 0-sequence,† hence, by Theorem 6 of section (4.6), the lowest row must be exact. Put $Q^* = Q/(\mu)$, then

$$0 \to \Phi_{p-1}/\mu_{p-1}\Phi_{p-1} \to \cdots \to \Phi_0/\mu\Phi_0 \to L/\mu L \to 0$$

is an exact sequence of Q^*-modules and, of course, $\Phi_i/\mu\Phi_i$ $(0 \leqslant i \leqslant p-1)$ is Q^*-free; consequently, by the definition of homological dimension, $\mathrm{dh}_{Q^*}(L/\mu L) \leqslant p-1 < \infty$. But, by Lemma 7, $\mathrm{codh}_{Q^*}(Q^*) \leqslant s-1$, hence, by the inductive hypothesis,

$$\mathrm{dh}_{Q^*}(L/\mu L) \leqslant s-1. \tag{9.6.1}$$

Now $\mathrm{dh}_Q(L) = p-1$, hence (Theorem 20) $\mathrm{dh}_Q(L/\mu L) = p$ and therefore (Theorem 15) $\mathrm{Ext}_Q^p(L/\mu L, K) \neq 0$. Further, the exact sequence

† The fact that this row has been broken, to facilitate printing, should be ignored.

$0 \to \mathfrak{m} \to Q \to K \to 0$ and the fact that $\mathrm{dh}_Q(L/\mu L) = p$, together show that

$$\mathrm{Ext}_Q^p(L/\mu L, Q) \to \mathrm{Ext}_Q^p(L/\mu L, K) \to 0$$

is exact and thus we may conclude that $\mathrm{Ext}_Q^p(L/\mu L, Q) \neq 0$. Again, μ is neither a unit nor a zero-divisor in Q consequently, by the corollary to Theorem 6,

$$\mathrm{Ext}_Q^p(L/\mu L, Q) \approx \mathrm{Ext}_{Q^*}^{p-1}(L/\mu L, Q^*),$$

and therefore $\mathrm{Ext}_{Q^*}^{p-1}(L/\mu L, Q^*) \neq 0$. But this implies that

$$\mathrm{dh}_{Q^*}(L/\mu L) \geqslant p-1,$$

hence, by (9.6.1), $p \leqslant s$. This completes the proof.

We have now reached a point where, by combining our results in various ways, we can obtain very explicit results concerning commutative Noetherian rings of finite global dimension.

Theorem 24. *A local ring Q is of finite global dimension if and only if it is regular.*

Proof. We already know (Theorem 22) that if Q is regular then its global dimension is finite. Assume therefore that $\mathrm{gl.dim}\, Q < \infty$ and let A be a cyclic Q-module; then A is both finitely generated and of finite homological dimension, hence (Theorem 23) $\mathrm{dh}_Q(A) \leqslant \mathrm{codh}_Q(Q)$. Taking upper bounds with respect to A we obtain

$$\mathrm{gl.dim}\, Q \leqslant \mathrm{codh}_Q(Q)$$

which, by (9.5.1), implies that $\mathrm{Dim}\, Q = n$, where n is the dimension of $\mathfrak{m}/\mathfrak{m}^2$ as a vector space over K. In other words, Q is regular.

Corollary. *If Q is a local ring then either*

$$\mathrm{gl.dim}\, Q = \infty \quad \textit{or} \quad \mathrm{gl.dim}\, Q = \mathrm{Dim}\, Q,$$

where $\mathrm{Dim}\, Q$ *denotes the ideal-theoretic dimension of Q.*

This follows by combining Theorems 22 and 24.

The next theorem is purely ideal-theoretic in character but is included here because it has been responsible for arousing a good deal of interest in the application of homological algebra to commutative rings.

Theorem 25. *Let Q be a regular local ring and let $\mathfrak{p}$ ($\mathfrak{p} \neq Q$) be one of its prime ideals. Then, with the usual notation for rings of fractions, $Q_{\mathfrak{p}}$ is also a regular local ring.*

Proof. The global dimension of Q is finite, consequently, by Theorem 8, so also is that of $Q_{\mathfrak{p}}$. Further, $Q_{\mathfrak{p}}$ is a local ring. That $Q_{\mathfrak{p}}$ is regular now follows from the last theorem.

From our characterization of the local rings Q for which

$$\text{gl.dim}\, Q < \infty$$

we can derive an answer to the more general question posed at the beginning of the chapter.

Theorem 26. *Let Λ be a commutative Noetherian ring and let p $(p \geqslant 0)$ be an integer. Then in order that* gl.dim Λ *should not exceed p it is necessary and sufficient that, for every maximal ideal $\mathfrak{m}$, $\Lambda_{\mathfrak{m}}$ should be a regular local ring of ideal-theoretic dimension not greater than p.*

This follows at once by combining Theorem 10 with Theorem 24 and its corollary.

We conclude with two more theorems of which the second is the one which explains the origin of the name *codimension* as used in this chapter.

Theorem 27. *Let Q be a local ring of finite global dimension and let A be a finitely generated Q-module. Then $\text{dh}_Q(A) = \text{gl.dim}\, Q$ if and only if there exists a submodule $B \neq 0$ of A such that $\mathfrak{m}B = 0$.*

Proof. Suppose first that there exists $B \subseteq A$ such that $B \neq 0$ and $\mathfrak{m}B = 0$. We can arrange (by replacing B by a suitable submodule if necessary) that B is generated by a single element x (say). The homomorphism $Q \to B$, in which an element $q \in Q$ is mapped into qx, has kernel $\mathfrak{m}$, hence B is isomorphic to $Q/\mathfrak{m} = K$. It follows that we can construct an exact sequence of the form $0 \to K \to A \to M \to 0$. Let $\text{gl.dim}\, Q = p$, then Tor^Q_{p+1} vanishes identically and therefore

$$0 \to \text{Tor}^Q_p(K, K) \to \text{Tor}^Q_p(A, K)$$

is exact. Now, by Theorem 17, $\text{Tor}^Q_p(K, K) \neq 0$, consequently $\text{Tor}^Q_p(A, K) \neq 0$, which shows that $\text{dh}_Q(A) \geqslant p$. However, by the definition of global dimension, $\text{dh}_Q(A) \leqslant p$, hence $\text{dh}_Q(A) = p$ as required.

Assume next that $\text{dh}_Q(A) = \text{gl.dim}\, Q$, then certainly $A \neq 0$. Also there must be a submodule $B \neq 0$ of A such that $\mathfrak{m}B = 0$. For assume the contrary, then, by Proposition 6*, $\mathfrak{m}$ contains an element μ which is not a zero-divisor in A. We can now apply Theorem 20 to obtain

$$\text{dh}_Q(A/\mu A) = \text{dh}_Q(A) + 1 = \text{gl.dim}\, Q + 1.$$

But this is clearly impossible.

Theorem 28. *Let Q be a local ring for which* $\text{gl.dim}\, Q < \infty$, *and let $A \neq 0$ be a finitely generated Q-module. Then*

$$\text{dh}_Q(A) + \text{codh}_Q(A) = \text{gl.dim}\, Q.$$

Proof. Let $\mu_1, \mu_2, \ldots, \mu_s$ be a normal A-sequence then, by Theorem 20,

$$s + \mathrm{dh}_Q(A) = \mathrm{dh}_Q[A/(\mu_1, \ldots, \mu_s)A] \leqslant \mathrm{gl.dim}\, Q,$$

and so $s \leqslant \mathrm{gl.dim}\, Q$. In particular, this shows that $\mathrm{codh}_Q(A)$ is finite and therefore, from now on, we may suppose that $s = \mathrm{codh}_Q(A)$. To complete the proof we need only show that

$$\mathrm{dh}_Q[A/(\mu_1, \ldots, \mu_s)A] = \mathrm{gl.dim}\, Q$$

and, by Theorem 27, this will follow if we can prove that $\mathfrak{m}$ annihilates some non-zero submodule of $A/(\mu_1, \ldots, \mu_s)A$. Assume the contrary then, by Proposition 6*, there exists $\mu_{s+1} \in \mathfrak{m}$ such that μ_{s+1} is not a zero-divisor in $A/(\mu_1, \ldots, \mu_s)A$. But this means that $\mu_1, \ldots, \mu_s, \mu_{s+1}$ is a normal A-sequence and therefore $\mathrm{codh}_Q(A) \geqslant s+1$. Thus we have obtained a contradiction and thereby proved the theorem.

10

HOMOLOGY AND COHOMOLOGY THEORIES OF GROUPS AND MONOIDS

Notation. The letter G is used to denote a monoid, that is, an associative semi-group with a neutral element. The neutral element is written as 1 or, more specifically, as 1_G. Small Greek letters σ, τ, ρ, etc., are used as symbols for the elements of the monoid. As usual, Z denotes the ring of integers. Whenever Z is regarded as a G-module, then always the elements of G operate only trivially upon it. Finally, $Z(G)$ denotes the monoid-ring of G with integer coefficients. In section (10.12) and all the later sections of the chapter, G is restricted to be a *finite* group.

10.1 General remarks concerning monoids and groups

The homology and cohomology theories of groups form a large subject which deserves a book to itself. We shall therefore attempt to describe only a few of the most important ideas and results.

At the outset, one notices that a good deal of the more general parts of the theory does not involve the particular group axiom which stipulates the existence of inverses. If we drop this axiom, then we are left with the so-called monoids, and therefore the observation means that we can go some way towards creating a theory applicable to monoids. The more general approach has several advantages and therefore we shall adopt it.

First, however, we must be clear about our definitions. A set G will be called a *monoid* if with each pair σ, τ of elements of G there is associated a further element of G called the 'product' of σ and τ and written as $\sigma\tau$. In addition, it is required that the associative law $(\sigma\tau)\rho = \sigma(\tau\rho)$ shall hold without exceptions and that there shall exist a 'neutral element'. By a neutral element is meant one, e say, such that $\sigma e = e\sigma = \sigma$ for all $\sigma \in G$.

If G is a monoid, then it is clear that the neutral element of G is *unique*. In future we shall always denote the neutral element of G by 1 or by 1_G, the latter notation being usually kept for situations involving more than one monoid. A monoid is called *abelian* or *commutative* if $\sigma\tau = \tau\sigma$ for all pairs σ, τ of elements.

Let G and G' be monoids. By a *monoid-homomorphism* of G into G' we mean a mapping $\phi : G \to G'$ such that

$$\phi(\sigma\tau) = \phi(\sigma)\phi(\tau), \quad \phi(1_G) = 1_{G'},$$

where, of course, σ and τ are arbitrary elements of G. If, in addition, ϕ is a 1-1 mapping of G on to G', we say that ϕ is an *isomorphism of G on to G'* and then ϕ^{-1} is an isomorphism of G' on to G.

Every group is clearly a monoid, and a monoid G is a group if, for each $\sigma \in G$, there exists an element σ' such that $\sigma\sigma' = \sigma'\sigma = 1$. Note too that, if $G_1, G_2, \ldots, G_n$ are monoids and we consider all ordered sets $(\sigma_1, \sigma_2, \ldots, \sigma_n)$, where $\sigma_i \in G_i$ for $1 \leqslant i \leqslant n$, then these ordered sets form a monoid if we put

$$(\sigma_1, \sigma_2, \ldots, \sigma_n)(\sigma_1', \sigma_2', \ldots, \sigma_n') = (\sigma_1\sigma_1', \sigma_2\sigma_2', \ldots, \sigma_n\sigma_n').$$

This particular monoid is called the *direct product* of $G_1, G_2, \ldots, G_n$ and it will be denoted by $G_1 \times G_2 \times \ldots \times G_n$.

We conclude these general remarks by some observations concerning *free monoids* and *free groups*. Suppose that S is a set of symbols, and let us use small letters s, s', etc., to denote typical specimens. By a *word*, or, more specifically, a *monoid-word*, will be meant a formal product $s_1 s_2 \ldots s_r$, where each of $s_1, s_2, \ldots, s_r$ belongs to S though they need not all be different. If now we have a second word, say $s_1' s_2' \ldots s_t'$, then we define their 'product' to be the word

$$s_1 s_2 \ldots s_r s_1' s_2' \ldots s_t'.$$

It is clear that these words now form a monoid which has the *empty word* as its neutral element. This particular monoid is known as the *free monoid generated by S*.

The construction of free groups is a little more complicated. As before we start with an arbitrary set S of symbols. If s belongs to S we associate with s an entirely *new* symbol s^{-1} which we call the 'inverse' of s, it being understood that different members of S have different inverses. By a *word* or, better, a *group-word*, we shall mean a formal product $x_1 x_2 \ldots x_r$, where each x_i $(1 \leqslant i \leqslant r)$ is either in S or is the inverse of a member of S. The word will be said to be *irreducible* if, in $x_1 x_2 \ldots x_r$, no symbol is adjacent to its own inverse so that neither ss^{-1} nor $s^{-1}s$ ever appear as consecutive terms. We shall now construct a group whose elements are the different irreducible words.

Let $\sigma = x_r x_{r-1} \dots x_1$ and $\sigma' = x'_1 x'_2 \dots x'_t$ be two irreducible words, then

$$x_r \dots x_2 x_1 x'_1 x'_2 \dots x'_t$$

may or may not be irreducible. However, if it is *not* irreducible then this must be because the pair x_1, x'_1 consists of an element of S and its inverse, though not necessarily in that order. In this case we remove both x_1 and x'_1 and consider

$$x_r \dots x_3 x_2 x'_2 x'_3 \dots x'_t.$$

If this, too, is not irreducible, then x_2, x'_2 comprise an element of S and its inverse and we remove them both to obtain

$$x_r \dots x_4 x_3 x'_3 x'_4 \dots x'_t.$$

Proceeding in this way we finally arrive at an irreducible word

$$x_r \dots x_{p+1} x_p x'_p x'_{p+1} \dots x'_t$$

(say), which we call the 'product' of σ and σ' and which we denote by $\sigma\sigma'$. We contend that, with this law of multiplication, *the irreducible words form a group with the empty irreducible word as the neutral element*. Indeed, this is clear provided that we establish that the associative law of multiplication is obeyed. To see that it is, let σ and σ' be irreducible words and let x be an element of S or the inverse of such an element, then it is easy to check, by enumerating the various possibilities, that

$$(\sigma\sigma')\,x = \sigma(\sigma' x). \tag{10.1.1}$$

Now let $\sigma'' = x''_1 x''_2 \dots x''_q$ be a third irreducible word, then, by (10.1.1),

$$(\sigma\sigma')\,x''_1 = \sigma(\sigma' x''_1),$$

whence $[(\sigma\sigma')\,x''_1]\,x''_2 = [\sigma(\sigma' x''_1)]\,x''_2$. Applying (10.1.1) again, we find that

$$(\sigma\sigma')\,(x''_1 x''_2) = \sigma[(\sigma' x''_1)\,x''_2] = \sigma[\sigma'(x''_1 x''_2)].$$

A further application yields

$$(\sigma\sigma')\,(x''_1 x''_2 x''_3) = \sigma[\sigma'(x''_1 x''_2 x''_3)],$$

and, in this way, we finally obtain

$$(\sigma\sigma')\,\sigma'' = \sigma(\sigma'\sigma''),$$

as required.

Definition. The group formed by the irreducible words will be called the *free group generated by S*.

Let us observe that if $s_1, s_2, \dots, s_p$ all belong to S, then $s_1 s_2 \dots s_p$ is an irreducible word because no inverses are present. Further, multiplication of these special irreducible words is the same as their multiplication in the free monoid generated by S, consequently the *free group generated by S contains the free monoid generated by the same symbols.*

10.2 Modules with respect to monoids and groups

Let G be a monoid and A a Z-module, that is to say, an additive abelian group. We say that A is a *left G-module* if, for each $\sigma \in G$ and $a \in A$, there is defined a 'product' σa which belongs to A and satisfies the following axioms:

$$\text{(i)}\quad \sigma(a_1+a_2) = \sigma a_1 + \sigma a_2 \quad (\sigma \in G;\ a_1, a_2 \in A),$$

$$\text{(ii)}\quad \sigma(\tau a) = (\sigma\tau)\, a \quad (\sigma, \tau \in G;\ a \in A),$$

$$\text{(iii)}\quad 1a = a \quad (1 = 1_G;\ a \in A).$$

Right G-modules are defined similarly save that the product of σ and a is written as $a\sigma$ and the above axioms become

$$\text{(i)}'\quad (a_1+a_2)\,\sigma = a_1\sigma + a_2\sigma,$$

$$\text{(ii)}'\quad (a\sigma)\,\tau = a(\sigma\tau),$$

$$\text{(iii)}'\quad a1 = a.$$

Usually we shall confine our attention to left G-modules it being understood that our definitions and results apply equally well to right G-modules after the obvious formal changes have been made.

If A is an arbitrary Z-module and G is an arbitrary monoid, then we can turn A into a G-module by setting $\sigma a = a$ for all $\sigma \in G$ and $a \in A$. We then say that G *operates trivially on A* and that A *has the elements of G as trivial operators.*

Suppose now that G is a *group* and that A is a left G-module. If $\sigma \in G$, then the mappings $a \to \sigma a$ and $a \to \sigma^{-1}a$ are Z-homomorphisms of A into itself, and the result of combining them, in either order, is the identity map of A. It follows that $a \to \sigma a$ is an isomorphism of A on to itself (that is an *automorphism* of the Z-module A) and $a \to \sigma^{-1}a$ is the inverse automorphism.

We shall now, very briefly, introduce some further terminology relating to G-modules where, once again, G is an arbitrary monoid. Let A and B be (left) G-modules and let f be a mapping of A into B. We say that f is a *G-homomorphism* if, first, f is a homomorphism of Z-modules and, secondly,

$$f(\sigma a) = \sigma f(a)$$

for all $\sigma \in G$ and $a \in A$. Arising out of this definition we have the concepts of *G-epimorphism*, *G-monomorphism* and *G-isomorphism*, the meanings of all of which will be evident to the reader.

Next, let G be a monoid and A a left G-module. A family $\{a_i\}_{i \in I}$ of elements of A is called a *G-base* for A if the elements σa_i ($\sigma \in G$, $a_i \in A$) form a Z-base for A. Further, any G-module which admits a G-base will be said to be *G-free*.

Finally, a G-module A is said to be *G-projective* if, whenever we have a diagram

$$\begin{array}{ccccc} & & A & & \\ & & \downarrow & & \\ B & \longrightarrow & C & \longrightarrow & 0 \end{array}$$

of G-modules and G-homomorphisms in which the row is exact, there always exists a G-homomorphism $A \to B$ such that $A \to C$ is the combined mapping $A \to B \to C$. On the other hand, A is said to be *G-injective* if, given a diagram

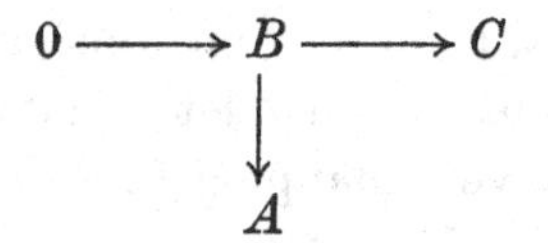

of G-modules and homomorphisms with the row exact, there always exists a G-homomorphism $C \to A$ such that the combined mapping $B \to C \to A$ coincides with the given homomorphism $B \to A$.

In the next section, the connexion between the concepts introduced here, and similar concepts employed in earlier chapters, will be made clear by using a device which associates a ring with each monoid.

10.3 Monoid-rings and group-rings

Let G be a monoid and denote by $Z(G)$ the free Z-module generated by the elements of G. If now λ and λ' belong to $Z(G)$, then

$$\lambda = \sum_{\sigma \in G} n_\sigma \sigma, \quad \lambda' = \sum_{\tau \in G} n'_\tau \tau,$$

where n_σ, n'_σ ($\sigma \in G$) are integers of which only a finite number are different from zero. We now define the 'product' of λ and λ' by means of the formula

$$\lambda\lambda' = \sum_{\sigma, \tau} n_\sigma n'_\tau (\sigma\tau).$$

A simple verification shows that this turns $Z(G)$ into a ring having the element 1_G as its identity element.

Definition. $Z(G)$, when endowed with the above ring-structure, will be called the *monoid-ring*† *of G with integer coefficients.*

We may note, in passing, that if Δ is *any* commutative ring with an identity element, and we let $\Delta(G)$ denote the free Δ-module generated by the elements of G, then the above construction (with trivial modifications) will make $\Delta(G)$ into a ring. $\Delta(G)$ is then called the monoid-ring of G with coefficients in Δ. Observe that *if G is commutative, then $\Delta(G)$, and more particularly $Z(G)$, is a commutative ring.*

Suppose now that we have a G-module A; say a left G-module for definiteness. If, with a self-explanatory notation, $\lambda = \Sigma n_\sigma \sigma$ is a typical element of $Z(G)$ and x belongs to A, we can turn A into a left $Z(G)$-module by defining the product of λ and x as

$$\lambda x = \sum_{\sigma \in G} n_\sigma(\sigma x).$$

Indeed, we can even say that the notions of a module with respect to the monoid G and of a module with respect to the ring $Z(G)$ are essentially equivalent; in fact, from now on we shall regard them as interchangeable. On this understanding, let A and B be two left G-modules and suppose that we have a mapping $f: A \to B$; then f is a G-homomorphism if, and only if, f is a $Z(G)$-homomorphism. It follows that the concept of a G-projective (injective) module is effectively the same as that of a $Z(G)$-projective (injective) module as defined in the theory of ring-modules. Yet again, it should be noted that a G-module A is G-free and has $\{a_i\}_{i \in I}$ as a G-base, if and only if it is $Z(G)$-free and has $\{a_i\}_{i \in I}$ as a $Z(G)$-base.

Although these remarks show that concepts concerning modules with respect to monoids can be replaced by analogous concepts in the theory of modules with respect to rings, it is convenient to adapt our previous notation so as to emphasize the fundamental role played by the monoid. Let us give a few examples. If $\Lambda = Z(G)$ we shall, in future, write $A \otimes_G B$, $\mathrm{Hom}_G(A, B)$, $\mathrm{Ext}_G^n(A, B)$ and $\mathrm{Tor}_n^G(A, B)$, where we would formerly have written $A \otimes_\Lambda B$, $\mathrm{Hom}_\Lambda(A, B)$, $\mathrm{Ext}_\Lambda^n(A, B)$ and $\mathrm{Tor}_n^\Lambda(A, B)$.

One final remark. The mapping

$$\sum_{\sigma \in G} n_\sigma \sigma \to \sum_{\sigma \in G} n_\sigma$$

defines a ring-epimorphism

$$\epsilon: Z(G) \to Z,$$

called the *augmentation mapping* of $Z(G)$. If $I = \mathrm{Ker}(\epsilon)$ then I is both

† The name *group-ring* is used if G is a group.

a left and a right ideal of $Z(G)$. Clearly $\sigma - 1 \in I$ for all $\sigma \in G$. On the other hand, if $\Sigma n_\sigma \sigma$ belongs to I then $\Sigma n_\sigma = 0$ and so

$$\sum_{\sigma \in G} n_\sigma \sigma = \sum_{\sigma \in G} n_\sigma (\sigma - 1).$$

It follows that *the elements* $\{\sigma - 1\}_{\sigma \neq 1}$ *form a* Z*-base for* I.

Definition. The two-sided ideal I will be called the *augmentation ideal* of $Z(G)$.

Note that if G is regarded as operating trivially on Z then

$$0 \to I \to Z(G) \xrightarrow{\epsilon} Z \to 0 \tag{10.3.1}$$

is an exact sequence of G-modules.

10.4 The functors A^G and A_G

Let G be a monoid and let A be a (left) G-module. Denote by A^G the set of all *fixed* elements of A, that is to say, the set of all elements $a \in A$ such that

$$\sigma a = a$$

for all σ belonging to G. It is clear that A^G is a Z-module; it is, indeed, a G-module on which G acts trivially but this latter aspect can be ignored. Consider now a homomorphism $\phi : A \to B$ of G-modules. Clearly if $\alpha \in A^G$ then $\phi(\alpha) \in B^G$, and so there is induced a mapping $A^G \to B^G$. In this way A^G *becomes an additive covariant functor of the* G*-module* A *taking values which are* Z*-modules.* This functor will now be expressed in a different form.

Proposition 1. *There is a* Z*-isomorphism*

$$A^G \approx \mathrm{Hom}_G(Z, A), \tag{10.4.1}$$

in which, if $\alpha \in A^G$ *corresponds to* $f \in \mathrm{Hom}_G(Z, A)$, *then*

$$f(1_Z) = \alpha. \tag{10.4.2}$$

Further, the isomorphism (10.4.1) *is a natural equivalence between the two sides, when these are regarded as functors of* A.

Remark. When A is a left (right) G-module then, in $\mathrm{Hom}_G(Z, A)$, Z is to be regarded as a left (right) G-module on which G acts trivially.

The verification of the assertions of the proposition is entirely simple and straightforward and we shall omit it. Note that, since Hom_G is a left exact functor, we have the

Corollary. A^G *is a left exact functor of* A.

We shall now construct a functor A_G which is complementary to A^G. To this end, denote by IA ,where I is the augmentation ideal, the set of all elements which can be expressed as finite sums

$$\lambda_1 a_1 + \lambda_2 a_2 + \ldots + \lambda_n a_n \quad (\lambda_r \in I,\ a_r \in A).$$

Then IA is a submodule of A and

$$A_G = A/IA, \tag{10.4.3}$$

regarded as a Z-module (its structure as G-module is trivial), may be thought of as an additive covariant functor of A. For if $\phi : A \to B$ is a homomorphism of G-modules, then $\phi(IA) \subseteq IB$ and there is induced a homomorphism of $A/IA = A_G$ into $B/IB = B_G$. The remaining functor properties are easily verified.

A_G will now be expressed in a new form.

Proposition 2. *If A is a left G-module then (using $\bar{a}$ to denote the natural image of an element $a \in A$ in $A/IA = A_G$) the mapping*

$$n \otimes a \to n\bar{a} \tag{10.4.4}$$

defines an isomorphism $\quad Z \otimes_G A \approx A_G. \qquad (10.4.5)$

Further, when A varies, this isomorphism gives a natural equivalence between the two sides of (10.4.5) *regarded as functors of A.*

Remark. As usual, Z is to be regarded as a G-module with trivial operators. In the case of right G-modules (10.4.5) has to be replaced by $A \otimes_G Z \approx A_G$.

Proof. It is clear that (10.4.4) defines an epimorphism $Z \otimes_G A \to A_G$. Suppose now that x belongs to the kernel of the mapping. Write x in the form $1_Z \otimes a$ then its image is $\bar{a}$ and this is zero, consequently we can write $a = \lambda_1 a_1 + \lambda_2 a_2 + \ldots + \lambda_n a_n$, where $\lambda_r \in I$ and $a_r \in A$. Accordingly

$$x = 1_Z \otimes a = \sum_r (1_Z \otimes \lambda_r a_r) = \sum_r ((1_Z \lambda_r) \otimes a_r) = 0,$$

since, if $\lambda \in I$, then $1_Z \lambda = 0$. This proves that $Z \otimes_G A \to A_G$ is an isomorphism and now the rest is a matter of simple verification.

Corollary. *A_G is a right exact functor of A.*

Consider next the way in which A^G and A_G depend on G. For this purpose suppose that G' is a second monoid and that we have a monoid-homomorphism

$$\phi : G' \to G. \tag{10.4.6}$$

In these circumstances, any left G-module A can be regarded as a left G'-module by writing

$$\sigma' a = \phi(\sigma') a \quad (\sigma' \in G',\ a \in A), \tag{10.4.7}$$

and then homomorphisms of G-modules are automatically also homomorphisms of G'-modules. On the understanding therefore that each G-module is regarded as being a G'-module as well, we have $A^G \subseteq A^{G'}$. The inclusion mapping now yields a monomorphism

$$A^G \to A^{G'}, \tag{10.4.8}$$

which is found to constitute a natural transformation of functors.

Again, let I' be the augmentation ideal of $Z(G')$ then, writing $1' = 1_{G'}$ and $1 = 1_G$, the elements

$$(\sigma' - 1')a = (\phi(\sigma') - 1)a$$

generate $I'A$ as a Z-module. This shows that $I'A \subseteq IA$, consequently the identity mapping of A induces an epimorphism $A/I'A \to A/IA$, that is to say, an epimorphism

$$A_{G'} \to A_G \tag{10.4.9}$$

which, like (10.4.8), is a natural transformation of functors. It should be observed, before we proceed, that when $G' = G$ and ϕ is the identity map, each of (10.4.8) and (10.4.9) is an identity transformation.

10.5 Axioms for the homology theory of monoids

As before we shall suppose that G is a monoid and that A is a variable G-module. For definiteness we shall usually assume that A varies in the category of left G-modules.

Definition. By a *homology theory* for G we shall understand an exact connected left sequence

$$\ldots, H_2(G, A), H_1(G, A), H_0(G, A)$$

of covariant functors of A (the functors having their values in the category of Z-modules) with the following two properties:

(i) $H_0(G, A) = A_G$;

(ii) *whenever A is G-projective, then $H_p(G, A) = 0$ for all $p > 0$.*

It will be recalled that the notion of a connected left sequence of covariant functors was defined in section (6.5). By saying that such a connected sequence is *exact* we mean that, whenever

$$0 \to A' \to A \to A'' \to 0$$

is an exact sequence of G-modules, the associated sequence

$$\begin{aligned} \cdots \to H_{n+1}(G, A'') \to H_n(G, A') \to H_n(G, A) \to H_n(G, A'') \to \cdots \\ \to H_1(G, A'') \to H_0(G, A') \to H_0(G, A) \to H_0(G, A'') \end{aligned} \tag{10.5.1}$$

is not merely a 0-sequence but is, in fact, *exact*. On this occasion we can indeed say a little more, because, by (i) of the definition and the corollary to Proposition 2, (10.5.1) remains exact when it is extended in the form

$$\cdots \to H_0(G, A') \to H_0(G, A) \to H_0(G, A'') \to 0 \to 0 \to \cdots. \quad (10.5.2)$$

Presently it will be shown that, for a given monoid G, the homology theory is uniquely determined to within an isomorphism of connected sequences. Anticipating this result, the Z-module $H_p(G, A)$ will be called *the p-th homology group of G with coefficients in the left G-module A*. Of course, the homology groups of G with coefficients in a right G-module are defined similarly.

Suppose now that we have a homomorphism

$$\phi : G' \to G$$

of a second monoid G' into the given monoid G. As we already know (see (10.4.7)), we can regard each G-module A as a G'-module consequently, if we have a homology theory for G', then

$$\ldots, H_2(G', A), H_1(G', A), H_0(G', A)$$

will be an exact connected left sequence of functors of A.

Theorem 1. *Given a monoid-homomorphism $G' \to G$ and homology theories $H_n(G', A')$ and $H_n(G, A)$ for G' and G respectively, then (for a variable G-module A) there exists a unique homomorphism of the connected sequence*

$$\ldots, H_2(G', A), H_1(G', A), H_0(G', A) = A_{G'}$$

into the connected sequence

$$\ldots, H_2(G, A), H_1(G, A), H_0(G, A) = A_G,$$

which extends the natural transformation $A_{G'} \to A_G$ described in (10.4.9).

This theorem is an immediate consequence of Theorem 12 of section (6.5).

Assume now that $\phi : G' \to G$ is an isomorphism and let $\psi : G \to G'$ be the inverse isomorphism. Then every G-module A is automatically a G'-module and vice versa and the theorem yields certain well-defined natural transformations

$$H_p(G', A) \to H_p(G, A) \quad \text{and} \quad H_p(G, A) \to H_p(G', A) \quad (10.5.3)$$

which commute with the connecting homomorphisms. It follows (from the uniqueness assertion in Theorem 1) that the result of combining (in either order) the transformations of (10.5.3) is an identity transformation, hence $H_p(G', A) \to H_p(G, A)$ is an isomorphism. This establishes

Corollary 1. *If $G' \to G$ is a monoid-isomorphism then the natural equivalence $A_{G'} \approx A_G$ has a unique extension to an isomorphism of the given homology theories of G' and G.*

Finally, by taking $G' = G$ and $G' \to G$ to be the identity map, we obtain

Corollary 2. *Any two homology theories, for a given monoid G, are isomorphic under an isomorphism which extends the identity transformation of the homology group A_G. Furthermore, there is only one such isomorphism.*

The existence of at least one homology theory for each monoid has still to be established, and the simplest way to do this is to observe that the left-derived functors of A_G have all the required properties. Another method, this time using the properties of torsion functors, will now be given because it provides a description of the homology theory which is important for the later sections.

Theorem 2. *If G is an arbitrary monoid then one obtains a homology theory for G by setting, for each left G-module A,*

$$H_p(G, A) = \mathrm{Tor}_p^G(Z, A) \quad (p \geqslant 0) \tag{10.5.4}$$

(Z is to be regarded as a right G-module with trivial operators, and $\mathrm{Tor}_0^G(Z, A)$ is identified with A_G by virtue of Proposition 2).

Remarks. No proof is needed because it is clear that the $H_p(G, A)$, as defined above, have all the required properties. Of course, for a homology theory in which the coefficients belong to right G-modules, one replaces (10.5.4) by

$$H_p(G, A) = \mathrm{Tor}_p^G(A, Z).$$

Corollary. *The functors $H_p(G, A)$ are all additive.*

10.6 Axioms for the cohomology theory of monoids

As usual G will denote a monoid and A a variable (left) G-module.

Definition. By a *cohomology theory* for G will be meant an *exact* connected right sequence

$$H^0(G, A),\ H^1(G, A),\ H^2(G, A),\ \ldots \tag{10.6.1}$$

of covariant functors of A (taking values in the category of Z-modules) possessing the following properties:

(i) $H^0(G, A) = A^G$,

(ii) *whenever A is G-injective, then $H^p(G, A) = 0$ for all $p > 0$.*

It is convenient to make a number of observations concerning this definition.

(*a*) By saying that (10.6.1) is an *exact* connected sequence we mean, of course, that each exact sequence $0 \to A' \to A \to A'' \to 0$ gives rise to an *exact* sequence (and not merely a 0-sequence)

$$H^0(G, A') \to H^0(G, A) \to H^0(G, A'') \to H^1(G, A') \to \cdots$$
$$\to H^n(G, A') \to H^n(G, A) \to H^n(G, A'') \to H^{n+1}(G, A') \to \cdots.$$

Further, by the corollary to Proposition 1,

$$0 \to H^0(G, A') \to H^0(G, A) \to H^0(G, A'')$$

is also exact.

(*b*) It should be noted that, in contrast to our previous practice in regard to suffixes and superfixes, $H^0(G, A)$ and $H_0(G, A)$ have quite different meanings.

(*c*) Anticipating the uniqueness theorem to be established shortly, we shall call $H^p(G, A)$ *the p-th cohomology group of G with coefficients in the left G-module A.* Of course one defines, in an entirely analogous manner, the cohomology groups with coefficients taken from a *right* G-module.

Theorem 3. (*Uniqueness theorem*). *Let $G' \to G$ be a homomorphism of a monoid G' into a monoid G and suppose that $H^p(G', A')$ and $H^p(G, A)$, where $p = 0, 1, 2, \ldots$, are connected sequences constituting cohomology theories for G' and G respectively. If now A denotes a variable G-module and we also regard it as a G'-module, then there exists a unique homomorphism of the connected sequence*

$$A^G = H^0(G, A),\ H^1(G, A),\ H^2(G, A),\ \ldots$$

into the connected sequence

$$A^{G'} = H^0(G', A),\ H^1(G', A),\ H^2(G', A),\ \ldots$$

which extends the transformation $A^G \to A^{G'}$ described in (10.4.8).

This theorem follows at once from Theorem 10 of section (6.5). Further, if $\phi : G' \to G$ is an isomorphism, then $A^G \to A^{G'}$ is a natural equivalence, hence, by the corollary to the theorem just quoted, the homomorphism described in Theorem 3 becomes an isomorphism of connected sequences. In particular, by taking $G' = G$ and $G' \to G$ to be the identity mapping, we see that *the cohomology theory of G is unique to within a (unique) isomorphism of connected sequences extending the identity transformation of A^G.*

The *existence* of a cohomology theory for an arbitrary monoid presents no problem because the right-derived functors of A^G clearly satisfy the axioms. The cohomology theory can also be described by means of extension functors as in the next theorem. This requires no proof in view of results previously obtained.

Theorem 4. *Let G be a monoid, then a cohomology theory for G is obtained by setting*

$$H^p(G, A) = \operatorname{Ext}_G^p(Z, A) \quad (p \geq 0) \tag{10.6.2}$$

and identifying $\operatorname{Ext}_G^0(Z, A)$ *with A^G by means of Proposition* 1.

Note that when A is a left (right) G-module then, in (10.6.2), Z has to be regarded as a left (right) G-module with trivial operators.

Corollary. *The functors $H^p(G, A)$ are all additive functors of A.*

10.7 Standard resolutions of Z

Throughout this section we shall use the letter A to denote a left G-module and B to denote a right G-module. G is an arbitrary monoid except where the contrary is stated.

Regarding Z as a left G-module on which G acts trivially, it is possible to construct an exact sequence

$$\cdots \to \Phi_n \to \Phi_{n-1} \to \cdots \to \Phi_1 \to \Phi_0 \to Z \to 0 \tag{10.7.1}$$

of (left) G-modules in which each Φ_n is G-free. This will give a G-free resolution of Z. Each Φ_n is, of course, Z-free as well as G-free, and so (10.7.1) *also provides a free resolution of Z regarded as a Z-module.*

Consider the complexes $B \otimes_G \mathbf{\Phi}$ and $\operatorname{Hom}_G(\mathbf{\Phi}, A)$. Their homology modules are $\operatorname{Tor}_n^G(B, Z)$ and $\operatorname{Ext}_G^n(Z, A)$ respectively, and so by Theorems 2 and 4,

$$H_n(G, B) = H_n\{B \otimes_G \mathbf{\Phi}\}, \tag{10.7.2}$$

while

$$H^n(G, A) = H^n\{\operatorname{Hom}_G(\mathbf{\Phi}, A)\}, \tag{10.7.3}$$

where, on the right-hand sides, we have used our customary notation for the homology modules of a complex.

The following terminology is frequently used. The elements of $B \otimes_G \Phi_n$ are called *n-chains*, those of $\operatorname{Ker}\{B \otimes_G \Phi_n \to B \otimes_G \Phi_{n-1}\}$ are known as *n-cycles*, while for $\operatorname{Im}\{B \otimes_G \Phi_{n+1} \to B \otimes_G \Phi_n\}$ the term used is *n-boundaries. Thus* $H_n(G, B)$ *consists of the n-cycles taken modulo the n-boundaries.* Analogously, the elements of $\mathrm{Hom}_G(\Phi_n, A)$,

$$\mathrm{Ker}\{\mathrm{Hom}_G(\Phi_n, A) \to \mathrm{Hom}_G(\Phi_{n+1}, A)\}$$

and

$$\mathrm{Im}\{\mathrm{Hom}_G(\Phi_{n-1}, A) \to \mathrm{Hom}_G(\Phi_n, A)\},$$

are known respectively as *n-cochains*, *n-cocycles* and *n-coboundaries*.

If it is required to express $H_n(G, A)$ and $H^n(G, B)$ in similar terms, we must use a G-free resolution of Z regarded as a *right* G-module with trivial operators, but clearly there is no need for a separate detailed discussion of this complementary situation.

The main purpose of the present section is to discuss free resolutions (10.7.1) of particular interest. Before embarking on this, however, it is convenient to derive two results which will be useful later.

Proposition 3. *Let* M *be a* Z*-module, put* $A = \mathrm{Hom}_Z\{Z(G), M\}$ *and use the fact that* $Z(G)$ *is a right* $Z(G)$*-module to endow* A *with the structure of a left* G*-module. Then* $H^n(G, A) = 0$ *for all* $n \geqslant 1$.

Proof. Let C be an arbitrary left G-module then, by Theorem 4 of section (8.5), we have an isomorphism

$$\mathrm{Hom}_Z\{Z(G) \otimes_G C, M\} \approx \mathrm{Hom}_G\{C, \mathrm{Hom}_Z(Z(G), M)\}.$$

But $Z(G) \otimes_G C \approx C$, consequently the isomorphism becomes

$$\mathrm{Hom}_Z(C, M) \approx \mathrm{Hom}_G(C, A), \tag{10.7.4}$$

and provides an equivalence between the two sides regarded as functors of C. It follows that the complexes

$$\cdots \to \mathrm{Hom}_G(\Phi_n, A) \to \mathrm{Hom}_G(\Phi_{n+1}, A) \to \cdots$$

and

$$\cdots \to \mathrm{Hom}_Z(\Phi_n, M) \to \mathrm{Hom}_Z(\Phi_{n+1}, M) \to \cdots,$$

derived from (10.7.1), have isomorphic homology modules. Accordingly, for $n \geqslant 1$,

$$H^n(G, A) \approx \mathrm{Ext}_Z^n(Z, M) = 0,$$

since Z is Z-free. This completes the proof.

Proposition 4. *Let* M *be a* Z*-module, put* $B = M \otimes_G Z(G)$ *and use the fact that* $Z(G)$ *is a right* $Z(G)$*-module to endow* B *with a structure as right* G*-module. Then* $H_n(G, B) = 0$ *for all* $n \geqslant 1$.

Proof. The proof is similar to that of Proposition 3 save that it uses the isomorphism

$$B \otimes_G C \approx [M \otimes_Z Z(G)] \otimes_G C \approx M \otimes_Z [Z(G) \otimes_G C] \approx M \otimes_Z C$$

in place of (10.7.4). Further details are omitted.

We turn now to the problem of constructing an explicit free resolution for the G-module Z. It will be convenient to begin with the case in which G is a *group*.

Take for Φ_n the free Z-module generated by all ordered sets $\langle\sigma_0, \sigma_1, \ldots, \sigma_n\rangle$, where $\sigma_0, \sigma_1, \ldots, \sigma_n$ are arbitrary elements of G, then Φ_n can be given the structure of a left G-module in which

$$\sigma\langle\sigma_0, \sigma_1, \ldots, \sigma_n\rangle = \langle\sigma\sigma_0, \sigma\sigma_1, \ldots, \sigma\sigma_n\rangle. \tag{10.7.5}$$

Next define homomorphisms $d_n : \Phi_n \to \Phi_{n-1}$ $(n \geqslant 1)$ and $e : \Phi_0 \to Z$ by means of the formulae

$$d_n\langle\sigma_0, \sigma_1, \ldots, \sigma_n\rangle = \sum_{r=0}^{n} (-1)^r \langle\sigma_0, \ldots, \hat{\sigma}_r, \ldots, \sigma_n\rangle \tag{10.7.6}$$

and

$$e\langle\sigma_0\rangle = 1_Z, \tag{10.7.7}$$

where the ^ over σ_r means that it is to be omitted from the bracket. It is clear from these formulae that d_n and e are not merely Z-homomorphisms but G-homomorphisms and an easy verification shows that

$$d_{n-1}d_n = 0 \quad (n \geqslant 2), \quad ed_1 = 0.$$

Further, *since G is a group*, the elements $\langle 1, \sigma_1, \sigma_2, \ldots, \sigma_n\rangle$ form a G-base for Φ_n, consequently Φ_n is *G-free*. It will be proved, very shortly, that

$$\cdots \to \Phi_n \xrightarrow{d_n} \Phi_{n-1} \xrightarrow{d_{n-1}} \Phi_{n-2} \to \cdots \to \Phi_1 \xrightarrow{d_1} \Phi_0 \xrightarrow{e} Z \to 0 \tag{10.7.8}$$

is not just a 0-sequence, but an *exact* sequence. Anticipating this result we make the following

Definition. If G is a group, the exact sequence (10.7.8), given by the above construction, is called the *standard homogeneous G-free resolution of Z.*

To establish the exactness of (10.7.8), we introduce auxiliary Z-homomorphisms (not G-homomorphisms)

$$s_{-1} : Z \to \Phi_0 \quad \text{and} \quad s_n : \Phi_n \to \Phi_{n+1} \quad (n \geqslant 0)$$

defined by

$$s_{-1}(1_Z) = \langle 1\rangle$$

and

$$s_n\langle\sigma_0, \sigma_1, \ldots, \sigma_n\rangle = \langle 1, \sigma_0, \sigma_1, \ldots, \sigma_n\rangle.$$

Then

$$es_{-1}, \quad s_{-1}e + d_1s_0, \quad d_{n+1}s_n + s_{n-1}d_n$$

are all identity maps. This is easily verified. For example, for $n \geqslant 1$,

$$(d_{n+1}s_n + s_{n-1}d_n)\langle\sigma_0, \sigma_1, \ldots, \sigma_n\rangle$$

$$= d_{n+1}\langle 1, \sigma_0, \ldots, \sigma_n\rangle + s_{n-1}\left[\sum_{r=0}^{n}(-1)^r\langle\sigma_0, \ldots, \hat{\sigma}_r, \ldots, \sigma_n\rangle\right]$$

$$= \langle\sigma_0, \sigma_1, \ldots, \sigma_n\rangle + \sum_{r=0}^{n}(-1)^{r+1}\langle 1, \sigma_0, \ldots, \hat{\sigma}_r, \ldots, \sigma_n\rangle + \sum_{r=0}^{n}(-1)^r\langle 1, \sigma_0, \ldots, \hat{\sigma}_r, \ldots, \sigma_n\rangle$$

$$= \langle\sigma_0, \sigma_1, \ldots, \sigma_n\rangle,$$

showing that $d_{n+1}s_n + s_{n-1}d_n$ is an identity map. It follows that if we regard

$$\cdots \to \Phi_n \to \cdots \to \Phi_0 \to Z \to 0 \to 0 \to \cdots$$

as a complex of Z-modules, then the identity translation is null homotopic, consequently, by Theorem 7 of section (4.7), the identity maps of the homology modules are null maps. The homology modules themselves are therefore all null, and this establishes the exactness of the sequence.

The standard homogeneous free resolution of Z can be described in different terms which, although more complicated, have the advantage that they suggest a construction valid for monoids and not merely for groups. Put

$$[\sigma_1, \sigma_2, \ldots, \sigma_n] = \langle 1, \sigma_1, \sigma_1\sigma_2, \sigma_1\sigma_2\sigma_3, \ldots, \sigma_1\sigma_2\ldots\sigma_n\rangle \quad (n \geqslant 1)$$

and

$$[\ ] = \langle 1\rangle,$$

then the $[\sigma_1, \sigma_2, \ldots, \sigma_n]$ form a G-base for Φ_n. Moreover

$$d_n[\sigma_1, \sigma_2, \ldots, \sigma_n] = \langle\sigma_1, \sigma_1\sigma_2, \ldots, \sigma_1\sigma_2\ldots\sigma_n\rangle + \sum_{r=1}^{n-1}(-1)^r\langle 1, \sigma_1, \ldots, \sigma_1\sigma_2\ldots\sigma_{r-1}, \sigma_1\sigma_2\ldots\sigma_r\sigma_{r+1}, \ldots, \sigma_1\sigma_2\ldots\sigma_n\rangle + (-1)^n\langle 1, \sigma_1, \sigma_1\sigma_2, \ldots, \sigma_1\sigma_2\ldots\sigma_{n-1}\rangle,$$

which can be rewritten as

$$d_n[\sigma_1, \sigma_2, \ldots, \sigma_n] = \sigma_1[\sigma_2, \ldots, \sigma_n] + \sum_{r=1}^{n-1}(-1)^r[\sigma_1, \ldots, \sigma_r\sigma_{r+1}, \ldots, \sigma_n] + (-1)^n[\sigma_1, \sigma_2, \ldots, \sigma_{n-1}]. \quad (10.7.9)$$

At this point we are ready to make a fresh start. Suppose, from now on, that *G is an arbitrary monoid.* For each $n \geqslant 0$, we construct a

G-free module Φ_n having the totality of all ordered sets $[\sigma_1, \sigma_2, \ldots, \sigma_n]$ as a G-base. (It is to be understood that Φ_0 is a G-free module with a G-base consisting of a single element denoted by [ ].) For each $n \geqslant 1$, a G-homomorphism $d_n : \Phi_n \to \Phi_{n-1}$ is then defined by (10.7.9) where, when $n = 1$, this is to be interpreted as

$$d_1[\sigma_1] = \sigma_1[\ ] - [\ ]. \tag{10.7.10}$$

Finally, $e : \Phi_0 \to Z$ is taken to be the G-homomorphism with the property that

$$e[\ ] = 1_Z. \tag{10.7.11}$$

The sequence $\cdots \to \Phi_n \overset{d_n}{\to} \Phi_{n-1} \to \cdots \to \Phi_0 \overset{e}{\to} Z \to 0$

is then a 0-sequence. Indeed, since by (10.7.10) $ed_1 = 0$, to substantiate this claim we need only show that when $n \geqslant 2$,

$$d_{n-1} d_n[\sigma_1, \sigma_2, \ldots, \sigma_n] = 0.$$

However, by the case already considered, this is so when $\sigma_1, \sigma_2, \ldots, \sigma_n$ are distinct generators in a *free* group and so it will also hold when they are distinct generators in a *free* monoid. From this it follows that the required relation holds for an *arbitrary* monoid.

Lastly, we contend that we have, in fact, a free resolution of Z and to prove this it will suffice to show that the sequence is exact. As before we introduce auxiliary Z-homomorphisms $s_n : \Phi_n \to \Phi_{n+1}$ $(n \geqslant 0)$ and $s_{-1} : Z \to \Phi_0$ this time defined by

$$s_n(\sigma[\sigma_1, \sigma_2, \ldots, \sigma_n]) = [\sigma, \sigma_1, \ldots, \sigma_n]$$

and

$$s_{-1}(1_Z) = [\ ].$$

In the case of a group, these are the same mappings as were used previously, and in the more general situation, at present under discussion, they retain the property that es_{-1}, $s_{-1}e + d_1 s_0$ and $d_{n+1}s_n + s_{n-1}d_n$ are all identity maps. The exactness of the sequence therefore follows on repeating the original arguments.

Definition. If G is a monoid, then the exact sequence

$$\cdots \to \Phi_n \overset{d_n}{\to} \Phi_{n-1} \to \cdots \to \Phi_1 \overset{d_1}{\to} \Phi_0 \to Z \to 0$$

(of G-modules) obtained by the above construction, is called the *standard non-homogeneous free resolution of Z.*

Consider the cochains and chains that one obtains by using the standard non-homogeneous resolution.† According to the definition,

† These will be referred to as *standard* cochains and chains

the n-cochains are the elements of $\mathrm{Hom}_G(\Phi_n, A)$. However, since the $[\sigma_1, \sigma_2, \ldots, \sigma_n]$ form a G-base for Φ_n, a typical n-cochain is obtained by associating, *in an entirely free manner*, an element of A with each ordered set $\sigma_1, \sigma_2, \ldots, \sigma_n$. (In the case $n = 0$, the cochains correspond to the different ways of associating a member of A with the base element [ ].) Thus the n-cochains may be identified with the elements of the direct product of copies of A, where there is one copy for each ordered set $\sigma_1, \sigma_2, \ldots, \sigma_n$. (For $n = 0$, the direct product reduces to A itself.) Suppose now that f is an n-cochain and let us use $f(\sigma_1, \sigma_2, \ldots, \sigma_n)$ to denote the element of A associated with $\sigma_1, \sigma_2, \ldots, \sigma_n$. Then, by (10.7.9), the image of f under the boundary homomorphism d^n is $d^n f$, where

$$(d^n f)(\sigma_1, \sigma_2, \ldots, \sigma_{n+1}) = \sigma_1 f(\sigma_2, \ldots, \sigma_{n+1})$$
$$+ \sum_{r=1}^{n} (-1)^r f(\sigma_1, \ldots, \sigma_r \sigma_{r+1}, \ldots, \sigma_{n+1}) + (-1)^{n+1} f(\sigma_1, \sigma_2, \ldots, \sigma_n), \tag{10.7.12}$$

it being understood that, when $n = 0$,

$$(d^0 f)\sigma = \sigma a - a, \tag{10.7.13}$$

where, in (10.7.13), $a \in A$ corresponds to the 0-cochain f.

Assume, for the moment, that G operates trivially on A then $\sigma_1 f(\sigma_2, \ldots, \sigma_{n+1}) = f(\sigma_2, \ldots, \sigma_{n+1})$ and (10.7.12) takes on a more symmetric form. Let us see what changes have to be made if we regard A as a *right* G-module with trivial operators. In the first place we do not change the cochains; in the second, the worst that will happen to the boundary homomorphism, acting on the cochains of a particular degree, is that its sign will be altered. Accordingly, *if G acts trivially on a module A, then $H^p(G, A)$ is the same group whether we regard G as operating on A from the left or from the right.*

Turning now to the homology theory (with coefficients in a right G-module B), the n-chains are the elements of $B \otimes_G \Phi_n$ and the boundary mapping is defined by

$$d_n(b \otimes [\sigma_1, \sigma_2, \ldots, \sigma_n])$$
$$= b\sigma_1 \otimes [\sigma_2, \ldots, \sigma_n] + \sum_{r=1}^{n-1} (-1)^r b \otimes [\sigma_1, \ldots, \sigma_r \sigma_{r+1}, \ldots, \sigma_n]$$
$$+ (-1)^n b \otimes [\sigma_1, \ldots, \sigma_{n-1}]. \tag{10.7.14}$$

Since the $[\sigma_1, \sigma_2, \ldots, \sigma_n]$ are a G-base for Φ_n, the n-chains form a Z-module which can be identified with the *direct sum* of copies of B, where there is one copy for each ordered set $\sigma_1, \sigma_2, \ldots, \sigma_n$. If G

operates trivially on B then, in (10.7.14), $b\sigma_1$ can be replaced by b so removing the lack of symmetry in the roles played by σ_1 and σ_n. From this observation we can draw the following conclusion: *when computing the homology groups of a module with trivial G-operators, one obtains the same result whether one regards G as operating on the left or on the right.*

10.8 The first homology group

By its definition, $H^0(G, A) = A^G$ is the largest submodule of A on which G operates trivially, while $H_0(G, A) = A_G$ can be described as the largest factor module of A with the same property. Of course, when G acts trivially on A itself then $H^0(G, A) = A$ and also $H_0(G, A) = A$.

In this and the next two sections, an examination will be made of some of the simpler homology and cohomology groups. We shall start with the first homology group, that is with $H_1(G, A)$. At the outset G may be any monoid and A (for definiteness) will be a left G-module.

Let I be the augmentation ideal of $Z(G)$ then, because $Z(G)$ is G-free,

$$0 \to I \to Z(G) \to Z \to 0 \tag{10.8.1}$$

induces the exact sequence

$$0 \to \operatorname{Tor}_1^G(Z, A) \to I \otimes_G A \to Z(G) \otimes_G A.$$

But, by Theorem 2, $\operatorname{Tor}_1^G(Z, A) = H_1(G, A)$, consequently

$$H_1(G, A) \approx \operatorname{Ker}\{I \otimes_G A \to Z(G) \otimes_G A\}. \tag{10.8.2}$$

When G acts trivially on A it is possible to say more for, in this situation, $I \otimes_G A \to Z(G) \otimes_G A$ maps $(\sigma - 1) \otimes a$ into

$$1(\sigma - 1) \otimes a = 1 \otimes (\sigma - 1)a = 0,$$

hence every element of $I \otimes_G A$ is mapped into zero. Accordingly

$$H_1(G, A) \approx I \otimes_G A \approx I \otimes_G (Z \otimes_Z A) \approx (I \otimes_G Z) \otimes_Z A. \tag{10.8.3}$$

But from (10.8.1) we obtain the exact sequence

$$I \otimes_G I \to I \otimes_G Z(G) \to I \otimes_G Z \to 0,$$

and, moreover, $I \otimes_G Z(G)$ can be identified with I. If this is done then $\operatorname{Im}\{I \otimes_G I \to I \otimes_G Z(G)\}$ becomes $II = I^2$, which shows that

$$I \otimes_G Z \approx I/I^2.$$

Substituting for $I \otimes_G Z$ in (10.8.3) we now obtain

Theorem 5. *If G is a monoid and A a left G-module on which G acts trivially, then*

$$H_1(G, A) \approx (I/I^2) \otimes_Z A.$$

From now on we suppose that G is a *group*. It will be recalled that the elements of the form $\sigma\tau\sigma^{-1}\tau^{-1}$ generate a normal subgroup called the *commutator subgroup* and denoted by $[G, G]$. $G/[G, G]$ is an abelian group; indeed it is the largest abelian factor group of G.

Consider the mapping $G \to I/I^2$ in which σ is mapped into the residue class, module I^2, of $\sigma - 1$. Since

$$\sigma\tau - 1 = (\sigma - 1)(\tau - 1) + (\sigma - 1) + (\tau - 1),$$

$G \to I/I^2$ is a homomorphism of G into the *abelian* group I/I^2 and therefore it induces a group-homomorphism

$$\phi : G/[G, G] \to I/I^2.$$

Again, the elements $\sigma - 1$ $(\sigma \neq 1)$ form a Z-base for I. Accordingly there exists a homomorphism $I \to G/[G, G]$, of abelian groups, in which $\sigma - 1$ is mapped into the image of σ under $G \to G/[G, G]$. It follows, from the identity $(\sigma - 1)(\tau - 1) = (\sigma\tau - 1) - (\sigma - 1) - (\tau - 1)$, that I^2 is contained in the kernel of the mapping which therefore induces a homomorphism

$$\psi : I/I^2 \to G/[G, G].$$

Now $\phi\psi$ and $\psi\phi$ are both identity maps, consequently we have proved the following theorem.

Theorem 6. *If G is a group and I is the augmentation ideal of $Z(G)$, then I/I^2 and $G/[G, G]$ are isomorphic abelian groups.*

Theorem 7. *Let G be a group and A a (left) G-module on which G acts trivially. Then*

$$H_1(G, A) \approx (G/[G, G]) \otimes_Z A \tag{10.8.4}$$

and, in particular,

$$H_1(G, Z) \approx G/[G, G]. \tag{10.8.5}$$

It should be noted that, in the right-hand side of (10.8.4), we are simply taking the tensor product of two abelian groups. For this purpose, the law of composition in $G/[G, G]$ should be thought of as *addition*.

The theorem itself needs no proof, since it merely combines Theorems 5 and 6.

10.9 The first cohomology group

Let us now turn our attention to $H^1(G, A)$, where A is a left G-module and, in the first instance, G is an arbitrary monoid. Employing the standard non-homogeneous resolution of Z, we may say that a

1-cochain is just a mapping $f : G \to A$. The boundary of such a mapping is, by (10.7.12), $df : G \times G \to A$, where

$$(df)(\sigma, \tau) = \sigma f(\tau) - f(\sigma\tau) + f(\sigma),$$

consequently, for f to be a 1-cocycle, it is necessary and sufficient that

$$f(\sigma\tau) = \sigma f(\tau) + f(\sigma) \qquad (10.9.1)$$

for all σ, τ in G.

Definition. A mapping $f : G \to A$ which has the property (10.9.1) is called a *crossed homomorphism* of G into A.

Next let ϕ be the 0-cochain associated with the element $a \in A$ then, by (10.7.13), its boundary $d\phi$ is the mapping $f : G \to A$, where

$$f(\sigma) = \sigma a - a. \qquad (10.9.2)$$

Definition. A mapping $f : G \to A$ for which there exists an element $a \in A$ such that $f(\sigma) = \sigma a - a$ for all $\sigma \in G$, is called a *principal crossed homomorphism.*

In terms of the above definitions, $H^1(G, A)$ *consists of the crossed homomorphisms modulo the principal crossed homomorphisms*, where the sum of two crossed homomorphisms f and g is given by

$$(f+g)\sigma = f(\sigma) + g(\sigma).$$

When G acts trivially on A, the principal crossed homomorphisms are all null, while (10.9.1) reduces to $f(\sigma\tau) = f(\sigma) + f(\tau)$. Hence, *for a module A with trivial G-operators,* $H^1(G, A)$ *consists of the additive group of monoid-homomorphisms of G into A.* If, in addition, G is a group then each homomorphism $G \to A$ will vanish identically on the commutator subgroup and so will induce a homomorphism

$$G/[G, G] \to A.$$

Indeed, there is an isomorphism between the group of homomorphisms of G into A and that of $G/[G, G]$ into A. This observation yields the following theorem, which is complementary to Theorem 7 and which, incidentally, could have been established by the dual argument.

Theorem 8. *Let G be a group and A a G-module on which G acts trivially. Then*

$$H^1(G, A) \approx \mathrm{Hom}_Z(G/[G, G], A). \qquad (10.9.3)$$

In interpreting (10.9.3) it will help if the law of composition, in $G/[G, G]$, is regarded as 'addition' rather than 'multiplication'.

The next result will be useful later when we come to discuss homology and cohomology for free monoids and free groups.

Theorem 9. *Let G be either the free monoid or the free group generated by a set S of symbols, let A be a left G-module and suppose that an arbitrary mapping $f : S \to A$ is given. Then there exists one and only one extension of f to a crossed homomorphism of G into A.*

Proof. We shall deal only with the case of the free group since that of the free monoid is similar but less complicated. It is clear that if such an extension exists and we denote it also by f, then

$$f(1) = 0, \quad f(s^{-1}) = -s^{-1}f(s) \quad (s \in S). \tag{10.9.4}$$

Let us therefore begin by defining f on 1 and also on s^{-1} ($s \in S$) by means of these relations. This will ensure that if x is either an element of S or is the inverse of such an element, then

$$f(x^{-1}) = -x^{-1}f(x). \tag{10.9.5}$$

Next assume that $p \geqslant 1$ and that the definition of f has been extended so as to include all irreducible words of length not exceeding p. For a typical irreducible word $x_1 x_2 \dots x_{p+1}$ of length $p+1$ now put

$$f(x_1 \dots x_p x_{p+1}) = x_1 x_2 \dots x_p f(x_{p+1}) + f(x_1 x_2 \dots x_p).$$

This construction determines f uniquely over the whole of G, and it is clear that, if there is an extension of f to a crossed homomorphism, this must be it. To complete the proof, it is therefore only necessary to show that

$$f(\sigma\tau) = \sigma f(\tau) + f(\sigma) \tag{10.9.6}$$

for all σ, τ in G.

If $\tau = 1$ this is clear. Suppose next that τ is an irreducible word of length unity and let $\sigma = y_1 y_2 \dots y_q$ be the representation of σ as an irreducible word. Should y_q and τ^{-1} be different, then (10.9.6) is just the definition of $f(\sigma\tau)$. On the other hand, if $y_q = \tau^{-1}$ then, using the definition of $f(y_1 y_2 \dots y_q)$ together with (10.9.5), we have

$$\begin{aligned} \sigma f(\tau) + f(\sigma) &= \sigma f(\tau) + f(y_1 \dots y_{q-1} y_q) \\ &= \sigma f(\tau) + (y_1 y_2 \dots y_{q-1}) f(y_q) + f(y_1 y_2 \dots y_{q-1}) \\ &= \sigma\{f(\tau) + \tau f(\tau^{-1})\} + f(\sigma\tau) \\ &= f(\sigma\tau). \end{aligned}$$

In conclusion, suppose that (10.9.6) has been established for all σ whenever τ is an irreducible word of length l, and consider the case where the length of τ is $l+1$. Then $\tau = \rho x$, where ρ has length l and x has unit length, consequently

$$f(\sigma\tau) = f(\sigma\rho x) = \sigma\rho f(x) + f(\sigma\rho)$$

by the case already considered, while

$$f(\sigma\rho) = \sigma f(\rho) + f(\sigma)$$

by the inductive hypothesis. Thus

$$f(\sigma\tau) = \sigma\{\rho f(x) + f(\rho)\} + f(\sigma) = \sigma f(\tau) + f(\sigma)$$

by the definition of $f(\tau)$. This proves the theorem.

The remainder of this section will be devoted to an application of $H^1(G, A)$ to the theory of group representations. First, however, we must set the stage for the discussion.

From now on we suppose that G is a *group* (not necessarily finite) and K will denote a *fixed* commutative field. Let A be both a K-space and also a left G-module, with the additional property that

$$\sigma(ka) = k\sigma(a)$$

whenever $\sigma \in G$, $k \in K$ and $a \in A$. In these circumstances we say that A is a *representation module for* G. Note that being a representation module is equivalent to being a bimodule with respect to $Z(G)$ and K. (In what follows, it will be more convenient to say '(G, K)-module' than 'bimodule with respect to $Z(G)$ and K', which is a little cumbersome for frequent use.) Yet again, if $K(G)$ denotes the group-ring of G with coefficients in K, then a representation module can also be described as a $K(G)$-module. In this latter characterization, multiplication by $k \in K$ and multiplication by the element $k1_G$ of $K(G)$ have the same effect.

Let A and B be representation modules for G, then a Z-homomorphism $f: A \to B$, such that

$$f(\sigma a) = \sigma f(a), \quad f(ka) = kf(a)$$

for all $\sigma \in G$, $k \in K$ and $a \in A$, will be called a *homomorphism of representation modules* or a *representation homomorphism*. When representation modules are regarded as bimodules, these mappings are simply the bihomomorphisms; while if we consider representation modules as $K(G)$-modules they are just the $K(G)$-homomorphisms.

By an *extension of the representation module A by the representation module B* will be meant (i) *a representation module E containing A as submodule* (with respect to both G and K), together with (ii) *a representation epimorphism $E \to B$ whose kernel is A*. Such an extension is described by an exact sequence

$$0 \to A \to E \to B \to 0 \qquad (10.9.7)$$

of (G, K)-modules, where $A \to E$ is an inclusion map. If now the sequence

$$0 \to A \to E^* \to B \to 0$$

describes a second extension, then the two are said to be *equivalent* if there exists a (G, K)-isomorphism $E \approx E^*$ such that the diagram

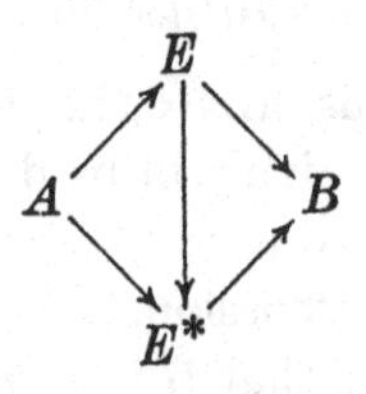

is commutative.

One particular extension of A by B can be obtained by taking for E the direct sum of A and B; and this then determines a special class of equivalent extensions all of whose members will be said to *split*. Clearly the extension (10.9.7) splits if and only if A is a direct summand of E considered as (G, K)-module.

We shall now consider the problem of finding a convenient model for the classes of equivalent extensions of A by B. As a first step towards bringing in the cohomology theory of G, observe that if C and D are representation modules, then $\operatorname{Hom}_K(C, D)$ becomes a (G, K)-module if we define σu by

$$(\sigma u)\, c = \sigma\{u(\sigma^{-1}c)\} \quad (c \in C). \tag{10.9.8}$$

Here, of course, σ and u are arbitrary elements of G and $\operatorname{Hom}_K(C, D)$ respectively. Now assume that an extension (10.9.7) has been given. Since B is K-free, the exact sequence will split if we regard it solely as a sequence of K-spaces, hence

$$0 \to \operatorname{Hom}_K(B, A) \to \operatorname{Hom}_K(B, E) \to \operatorname{Hom}_K(B, B) \to 0 \tag{10.9.9}$$

is exact. But in the latter sequence, each term is also a (G, K)-module and, as can be readily verified, each mapping is a bihomomorphism. In particular, (10.9.9) is an exact sequence of G-modules and therefore the cohomology theory yields a connecting homomorphism

$$\Delta : [\operatorname{Hom}_K(B, B)]^G \to H^1\{G, \operatorname{Hom}_K(B, A)\}.$$

Let i_B be the identity map of B then certainly $i_B \in [\operatorname{Hom}_K(B, B)]^G$ and so

$$\Delta(i_B) \in H^1\{G, \operatorname{Hom}_K(B, A)\}.$$

In this way one can associate an element of $H^1\{G, \operatorname{Hom}_K(B, A)\}$ with each extension of A by B, and it is clear that equivalent extensions have the same element associated with them.

Theorem 10. *If G is a group then the association (described above) of an element of $H^1\{G, \mathrm{Hom}_K(B, A)\}$ with each extension of A by B, sets up a one-to-one correspondence between the classes of equivalent extensions and the elements of $H^1\{G, \mathrm{Hom}_K(B, A)\}$. In this correspondence, the class of split extensions corresponds to the zero element.*

Proof. Let $0 \to A \to E \to B \to 0$ be an extension of A by B then, in particular, it yields an exact sequence of K-spaces. However B is K-free, because K is a field, and therefore we can choose a K-homomorphism $\phi : B \to E$ such that the combined mapping $B \overset{\phi}{\to} E \to B$ is the identity mapping i_B of B. In general, ϕ will not be a (G, K)-homomorphism; but in the special case where the extension belongs to the split class, we can arrange for it to be so.

Consider now the exact sequence

$$0 \to \mathrm{Hom}_K(B, A) \to \mathrm{Hom}_K(B, E) \to \mathrm{Hom}_K(B, B) \to 0$$

with the object of computing $\Delta(i_B)$. First we note that i_B is the image of ϕ. Next, the boundary of ϕ, regarded as an 0-cochain of $\mathrm{Hom}_K(B, E)$, consists of the K-homomorphisms $(\sigma\phi - \phi) : B \to E$ and this, by the general theory of the connecting homomorphism, is the image of a 1-cocycle of $\mathrm{Hom}_K(B, A)$. In other words,

$$\mathrm{Im}\,(\sigma\phi - \phi) \subseteq A$$

and, *if we regard $\sigma\phi - \phi$ as being a homomorphism $B \to A$, these homomorphisms form a 1-cocycle for the G-module $\mathrm{Hom}_K(B, A)$; furthermore, the cohomology class of this cocycle is the element of $H^1\{G, \mathrm{Hom}_K(B, A)\}$ corresponding to the equivalence class which contains the extension.* Note that, if the extension splits, then, taking ϕ to be a (G, K)-homomorphism, we have $\sigma\phi - \phi = 0$ for all σ. This proves that the zero element of $H^1\{G, \mathrm{Hom}_K(B, A)\}$ is associated with the split class.

Of course, ϕ is not uniquely determined; in fact the general K-homomorphism $B \to E$ with the same property is $\phi + \psi$, where ψ is any K-homomorphism $B \to E$ with $\mathrm{Im}\,(\psi) \subseteq A$. Now, if we replace ϕ by $\phi + \psi$, $\sigma\phi - \phi$ becomes $(\sigma\phi - \phi) + (\sigma\psi - \psi)$. But, regarded as homomorphisms $B \to A$, the $\sigma\psi - \psi$ form a typical 1-coboundary. Thus, *when ϕ is varied in all possible ways, the 1-cocycle $\{\sigma\phi - \phi\}$ of $\mathrm{Hom}_K(B, A)$ ranges through all the members of the cohomology class representing the extension.*

Now suppose that

$$0 \to A \to E \to B \to 0 \quad \text{and} \quad 0 \to A \to E' \to B \to 0$$

are two extensions which determine the *same* element of

$$H^1\{G, \operatorname{Hom}_K(B, A)\}.$$

Choose $\phi' : B \to E'$ to satisfy similar conditions to those imposed on ϕ, then the 1-cocycles $\{\sigma\phi - \phi\}$ and $\{\sigma\phi' - \phi'\}$ of $\operatorname{Hom}_K(B, A)$ belong to the same cohomology class. It is therefore possible, by adjusting the choice of ϕ', to arrange that, for each $\sigma \in G$, $\sigma\phi - \phi$ and $\sigma\phi' - \phi'$ are identical as mappings of B into A.

Each element of E has a unique representation in the form $a + \phi(b)$, where $a \in A$ and $b \in B$. By mapping $a + \phi(b)$ into $a + \phi'(b)$ we obtain a K-isomorphism $E \approx E'$. Also, if $\sigma \in G$ then

$$\sigma(a + \phi(b)) = \sigma a + [\sigma\phi](\sigma b) = \sigma a + [\sigma\phi - \phi](\sigma b) + \phi(\sigma b).$$

But $[\sigma\phi - \phi](\sigma b) \in A$, consequently $\sigma(a + \phi(b))$ corresponds to

$$\begin{aligned}\sigma a + [\sigma\phi - \phi](\sigma b) + \phi'(\sigma b)\\ = \sigma a + [\sigma\phi' - \phi'](\sigma b) + \phi'(\sigma b) = \sigma(a + \phi'(b)).\end{aligned}$$

This makes it clear that $E \approx E'$ is a (G, K)-isomorphism and now the two extensions are seen to be equivalent. Accordingly, two extensions determine the same cohomology class when and only when they are equivalent.

Finally, let $h_\sigma \in \operatorname{Hom}_K(B, A)$ and suppose that the h_σ form a 1-cocycle. Then

$$\sigma h_\tau - h_{\sigma\tau} + h_\sigma = 0 \quad (\sigma, \tau \in G), \tag{10.9.10}$$

from which it follows that $h_1 = 0$. We shall construct an extension $0 \to A \to E \to B \to 0$ and a K-homomorphism $\phi : B \to E$ such that (i) $B \to E \to B$ is an identity map and (ii) $\sigma\phi - \phi = h_\sigma$ for all σ. This will complete the proof.

Take for E the set of all pairs (a, b), where $a \in A$ and $b \in B$, and regard it as the direct sum of A and B *considered simply as K-spaces*. Next define $\sigma(a, b)$ by

$$\sigma(a, b) = (\sigma a + h_\sigma(\sigma b), \sigma b).$$

This makes E into a (G, K)-module. Of course, certain verifications are needed, but they are all trivial with one exception and this goes as follows:

$$\begin{aligned}\sigma\{\tau(a, b)\} &= \sigma\{\tau a + h_\tau(\tau b), \tau b\} = \{\sigma\tau a + [\sigma h_\tau](\sigma\tau b) + h_\sigma(\sigma\tau b), \sigma\tau b\}\\ &= \{\sigma\tau a + h_{\sigma\tau}(\sigma\tau b), \sigma\tau b\} = (\sigma\tau)(a, b).\end{aligned}$$

Note that we have used both the definition of σh_τ and the relation (10.9.10).

The mappings $a \to (a, 0)$ and $(a, b) \to b$ are the one a (G, K)-monomorphism $A \to E$ and the other a (G, K)-epimorphism $E \to B$; moreover, the sequence $0 \to A \to E \to B \to 0$ is exact. If we define the K-homomorphism $\phi : B \to E$ by $\phi(b) = (0, b)$, then the combined mapping $B \to E \to B$ is i_B and

$$(\sigma\phi - \phi)\, b = \sigma(\phi(\sigma^{-1}b)) - \phi(b) = \sigma(0, \sigma^{-1}b) - \phi(b)$$
$$= (h_\sigma(b), b) - (0, b) = (h_\sigma(b), 0).$$

Thus if we identify A with its image in E then $\sigma\phi - \phi$ coincides with h_σ. The proof is now complete.

It is interesting to deduce (from Theorem 10) a homological equivalent of Maschke's well-known theorem on completely reducible group-rings for, in this way, Maschke's result is seen against the general background of extension theory.

Theorem 11. *Let G be a finite group of order q and K a commutative field of characteristic zero or characteristic p, where p is prime to q. Then the group-ring $K(G)$ has its left and right global dimensions equal to zero.*

Proof. It will suffice to prove that $\text{l.gl.dim}\, K(G) = 0$, and this will follow if we show that an arbitrary left $K(G)$-module B is projective. To this end construct an exact sequence $0 \to A \to E \to B \to 0$ of $K(G)$-modules, where E is free. This provides an extension of A by B. If now we can establish that

$$H^1\{G, \text{Hom}_K(B, A)\} = X \text{ (say)} = 0,$$

then the extension will split (Theorem 10) showing that B is isomorphic to a direct summand of the free $K(G)$-module E. This, of course, will prove the theorem.

Now, in the first place, $\text{Hom}_K(B, A)$ is a bimodule with respect to $Z(G)$ and K consequently, by the general principles explained in section (8.2), X has a natural structure as K-space. Next observe that $qH^1(G, C) = 0$ for any G-module C. (To see this, let $f : G \to C$ be a crossed homomorphism and put $a = \sum_{\sigma \in G} f(\sigma)$. Then, summing the identity $f(\sigma\tau) = \sigma f(\tau) + f(\sigma)$ with respect to τ, one obtains

$$\sigma a - a = -qf(\sigma)$$

showing that qf is a coboundary.) It follows that $qX = 0$ and therefore $(q1_K)\, X = 0$, where 1_K is the identity element of K. But the conditions of the theorem secure that $q1_K \neq 0$ consequently $X = 0$ as required.

10.10 The second cohomology group

Throughout section (10.10), G will denote a *group* and our object will be to obtain an interpretation for the second cohomology group $H^2(G, A)$. This interpretation will be in terms of what are called 'group extensions of A by G'.

Using the standard non-homogeneous resolution of Z, a 2-cochain (for a given left G-module A) is just a mapping $f: G \times G \to A$ and (10.7.12) shows that the condition for f to be a 2-cocycle is

$$\sigma f(\tau,\rho) - f(\sigma\tau,\rho) + f(\sigma,\tau\rho) - f(\sigma,\tau) = 0. \qquad (10.10.1)$$

By custom these 2-cocycles are given a special name of their own.

Definition. A mapping $f: G \times G \to A$ for which (10.10.1) holds for all σ, τ, ρ in G is called a *factor system*.

Now observe that $f: G \times G \to A$ is the boundary $d\phi$ of a 1-cochain $\phi: G \to A$, provided that

$$f(\sigma,\tau) = \sigma\phi(\tau) - \phi(\sigma\tau) + \phi(\sigma) \qquad (10.10.2)$$

holds identically. This observation gives rise to another definition.

Definition. If $f: G \times G \to A$ and there exists a mapping $\phi: G \to A$ such that (10.10.2) holds for all σ, τ in G, then f is called a *principal factor system*.

Thus, in our new terminology, $H^2(G, A)$ is just the additive group of factor systems taken modulo the subgroup of principal factor systems.

So far we have always taken A to be an *additive* group on which G operates but, for the discussion which follows, it will be more convenient to treat the internal law of composition of A as *multiplication*. A second change of notation will consist of writing a^σ where formerly we should have written σa and so, from now on, the hypothesis that A is a left G-module means that

$$(a_1 a_2)^\sigma = a_1^\sigma a_2^\sigma, \quad (a^\sigma)^\tau = a^{\tau\sigma}, \quad a^1 = a, \qquad (10.10.3)$$

where a, a_1, a_2 are in A and σ, τ belong to G. Another change results from observing that a 2-cochain is obtained by associating an element of A with each ordered pair (σ, τ) of elements of G; and so we might equally well describe a 2-cochain as a doubly indexed family $\{a_{\sigma,\tau}\}$ of elements of A. Further, the internal law of composition of A has been used to make a group out of the cochains; but it is now more natural to describe the operation of combining $\{a'_{\sigma,\tau}\}$ and $\{a''_{\sigma,\tau}\}$ as 'multiplication' than as 'addition', which was the term previously used. Accordingly the product of these 2-cochains is $\{a^*_{\sigma,\tau}\}$, where

$$a^*_{\sigma,\tau} = a'_{\sigma,\tau}a''_{\sigma,\tau}.$$

Finally, the condition for $\{a_{\sigma,\tau}\}$ to be a factor system, see (10.10.1), takes the form

$$a^{\sigma}_{\tau,\rho} = \frac{a_{\sigma\tau,\rho}a_{\sigma,\tau}}{a_{\sigma,\tau\rho}} \quad (\sigma, \tau, \rho \in G), \tag{10.10.4}$$

and it will be a principal factor system when there exists a family $\{\alpha_\sigma\}$ of elements of A, such that

$$a_{\sigma,\tau} = \frac{\alpha^{\sigma}_{\tau}\alpha_{\sigma}}{\alpha_{\sigma\tau}} \quad (\sigma, \tau \in G). \tag{10.10.5}$$

The new multiplicative notation will be in force only for the rest of this section, after which we shall return to our previous usages.

It is now time to introduce the concepts of the theory of group extensions so far as they will concern us here. Suppose then that A is a given *commutative* group and that G is an *arbitrary* group. In rough terms, to find an extension of A by G means to find a group Π which contains A as a normal subgroup and is such that Π/A is (isomorphic to) G. This lacks the necessary precision for our purposes, so we set the matter out in formal terms as follows:

Definition. A pair (Π, χ) consisting of a group Π (containing A as a normal subgroup) and a group-epimorphism $\chi : \Pi \to G$ with kernel A, will be called an *extension of A by G*.

The notion of equivalent extensions is now easily explained. Two extensions (Π, χ) and (Π^*, χ^*) of A by G are said to be *equivalent* if there exists a group isomorphism $\Pi \approx \Pi^*$ such that the diagram

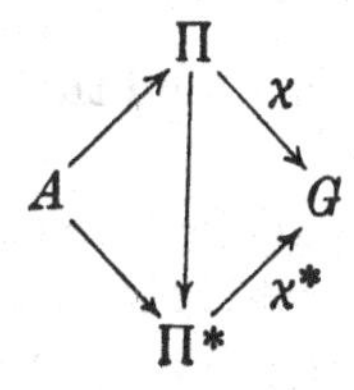

is commutative. Here it is to be understood that $A \to \Pi$ and $A \to \Pi^*$ are inclusion mappings.

Suppose now that (Π, χ) is an extension of A by G. Let $\sigma \in G$ and choose $\pi_\sigma \in \Pi$ so that $\chi(\pi_\sigma) = \sigma$ then, since A is a normal subgroup of Π, the mapping $a \to \pi_\sigma a \pi_\sigma^{-1}$ is an isomorphism of A on to itself.

Further, if π'_σ is another element of Π which χ maps into σ, then $\pi'_\sigma = \alpha\pi_\sigma$ for a suitable $\alpha \in A$. It follows that

$$\pi'_\sigma a \pi'^{-1}_\sigma = \alpha\pi_\sigma a \pi_\sigma^{-1}\alpha^{-1} = \pi_\sigma a \pi_\sigma^{-1},$$

because A is commutative and $\pi_\sigma a \pi_\sigma^{-1}$ belongs to A. Hence, if we put

$$a^\sigma = \pi_\sigma a \pi_\sigma^{-1} \quad (a \in A, \sigma \in G), \tag{10.10.6}$$

then a^σ is independent of the choice of π_σ. It is now easy to verify that the relations of (10.10.3) hold, consequently *each extension of A by G causes A to be endowed with a structure as* (*left*) *G-module* and it is clear that equivalent extensions will produce identical G-module structures. In view of this we shall reformulate our problem.

From now on, it will be supposed that A and G are given and that, in addition, there is prescribed, on A, a definite structure as (left) G-module. The problem will then be to determine the classes of equivalent extensions of A by G, *which induce, on A, the prescribed G-module structure.*

Let (Π, χ) be such an extension and, for each $\sigma \in G$, choose $\pi_\sigma \in \Pi$ so that $\chi(\pi_\sigma) = \sigma$. A family $\{\pi_\sigma\}$ of elements of Π, which possesses this property, will be called a *section* of Π. If now σ, τ are in G then, since $\chi(\pi_\sigma\pi_\tau) = \sigma\tau = \chi(\pi_{\sigma\tau})$, it follows that

$$\pi_\sigma\pi_\tau = a_{\sigma,\tau}\pi_{\sigma\tau},$$

where $a_{\sigma,\tau}$ belongs to A. Now

$$(\pi_\sigma\pi_\tau)\pi_\rho = a_{\sigma,\tau}\pi_{\sigma\tau}\pi_\rho = a_{\sigma,\tau}a_{\sigma\tau,\rho}\pi_{\sigma\tau\rho},$$

and, by (10.10.6),

$$\begin{aligned}\pi_\sigma(\pi_\tau\pi_\rho) &= \pi_\sigma a_{\tau,\rho}\pi_{\tau\rho} = \pi_\sigma a_{\tau,\rho}\pi_\sigma^{-1}\pi_\sigma\pi_{\tau\rho}\\ &= a^\sigma_{\tau,\rho}a_{\sigma,\tau\rho}\pi_{\sigma\tau\rho};\end{aligned}$$

consequently $a_{\sigma,\tau}a_{\sigma\tau,\rho} = a^\sigma_{\tau,\rho}a_{\sigma,\tau\rho}$ and therefore, by (10.10.4), the $a_{\sigma,\tau}$ form a factor system.

Next suppose that we are given a second extension (Π^*, χ^*) say; let $\{\pi^*_\sigma\}$ be a section of Π^* and let $\{a^*_{\sigma,\tau}\}$ be the factor system to which it gives rise. It is clear that if (Π, χ) and (Π^*, χ^*) are equivalent, then it is possible to choose the section $\{\pi_\sigma\}$ so that $a_{\sigma,\tau} = a^*_{\sigma,\tau}$ for all σ, τ. On the other hand, if we can choose $\{\pi_\sigma\}$ so that $a_{\sigma,\tau} = a^*_{\sigma,\tau}$ without exception, then the two extensions are equivalent. Indeed, in these circumstances, the mapping $\Pi \to \Pi^*$ defined by $\alpha\pi_\sigma \to \alpha\pi^*_\sigma$, where $\alpha \in A$, is easily seen to be an isomorphism. Moreover, to prove that it gives an equivalence we need only show that it becomes the identity map on A. Now $\pi_1\pi_1 = a_{1,1}\pi_1$ hence $\pi_1 = a_{1,1}$ and similarly $\pi_1^* = a_{1,1}$. Thus if $\alpha \in A$ then $\alpha = \alpha a_{1,1}^{-1}\pi_1 = \alpha a_{1,1}^{-1}\pi_1^*$ which makes it quite clear that α is mapped into itself.

To recapitulate, (Π, χ) *and* (Π^*, χ^*) *are equivalent if and only if it is possible to rechoose the section* $\{\pi_\sigma\}$ *so that we have* $a_{\sigma,\tau} = a^*_{\sigma,\tau}$ *for all* σ *and* τ.

Observe now that if $\{\pi_\sigma\}$ is a particular section of Π then any other section $\{\bar{\pi}_\sigma\}$ (say) is obtained by putting $\bar{\pi}_\sigma = \alpha_\sigma \pi_\sigma$, where the α_σ are arbitrary elements of A. Let $\bar{a}_{\sigma,\tau}$ be the factor system associated with $\{\bar{\pi}_\sigma\}$ then, on the one hand,

$$\bar{\pi}_\sigma \bar{\pi}_\tau = \alpha_\sigma \pi_\sigma \alpha_\tau \pi_\tau = \alpha_\sigma \pi_\sigma \alpha_\tau \pi_\sigma^{-1} \pi_\sigma \pi_\tau = \alpha_\sigma \alpha_\tau^\sigma a_{\sigma,\tau} \pi_{\sigma\tau},$$

and on the other $\quad \bar{\pi}_\sigma \bar{\pi}_\tau = \bar{a}_{\sigma,\tau} \bar{\pi}_{\sigma\tau} = \bar{a}_{\sigma,\tau} \alpha_{\sigma\tau} \pi_{\sigma\tau},$

which gives
$$\bar{a}_{\sigma,\tau} = a_{\sigma,\tau} \frac{\alpha_\tau^\sigma \alpha_\sigma}{\alpha_{\sigma\tau}}.$$

Accordingly, (Π, χ) *and* (Π^*, χ^*) *will be equivalent if and only if the original factor systems* $\{a_{\sigma,\tau}\}$ *and* $\{a^*_{\sigma,\tau}\}$ *belong to the same cohomology class*. It follows, in particular, that the cohomology class of $\{a_{\sigma,\tau}\}$ does not depend on the choice of the section $\{\pi_\sigma\}$. Thus with each extension (Π, χ) there is associated a uniquely determined element of $H^2(G, A)$; and, moreover, it has been shown that extensions are equivalent when, and only when, they determine the same element.

Theorem 12. *Let A be an abelian group, G an arbitrary group, and suppose that A has a prescribed structure as a (left) G-module. Then with each extension of A by G, which induces on A the prescribed structure, there is associated a definite element of $H^2(G, A)$. Further, this association produces a one-to-one correspondence between the classes of equivalent extensions and the elements of $H^2(G, A)$.*

Proof. In view of what has already been proved, it is enough to show that each element of $H^2(G, A)$ arises from an extension. Accordingly, let such an element be given and let $\{a_{\sigma,\tau}\}$ be a representative factor system. By multiplying the factor system by a suitable principal factor system we can arrange that $a_{1,1} = 1_A$. (This step is not essential but it makes for simplicity in what follows.) If now, in (10.10.4), we first give σ, τ, ρ the values σ, 1, 1 and then the values 1, 1, ρ we find that

$$a_{\sigma,1} = 1_A = a_{1,\sigma} \quad (\sigma \in G). \tag{10.10.7}$$

Take for Π the set of all pairs (α, σ) where $\alpha \in A$ and $\sigma \in G$; and, if (β, τ) is a second such pair, put

$$(\alpha, \sigma)(\beta, \tau) = (\alpha\beta^\sigma a_{\sigma,\tau}, \sigma\tau),$$

thereby defining multiplication. This multiplication is associative for if (γ, ρ) is a third pair, then

$$[(\alpha, \sigma)(\beta, \tau)](\gamma, \rho) = (\alpha\beta^\sigma\gamma^{\sigma\tau}a_{\sigma,\tau}a_{\sigma\tau,\rho}, \sigma\tau\rho),$$

while $$(\alpha, \sigma)[(\beta, \tau)(\gamma, \rho)] = (\alpha\beta^\sigma\gamma^{\sigma\tau}a^\sigma_{\tau,\rho}a_{\sigma,\tau\rho}, \sigma\tau\rho),$$

and these are the same because $a_{\sigma,\tau}a_{\sigma\tau,\rho} = a^\sigma_{\tau,\rho}a_{\sigma,\tau\rho}$. Put

$$e = (1_A, 1),$$

then, by (10.10.7), $$(\alpha, \sigma)e = (\alpha, \sigma) = e(\alpha, \sigma),$$

showing that e is neutral for multiplication. Further,

$$(\alpha, 1)(\alpha^{-1}, 1) = e,$$

and from this it follows that every element has both a right and a left inverse; for if $(\alpha, \sigma)(1_A, \sigma^{-1}) = (\beta, 1)$ and $(1_A, \sigma^{-1})(\alpha, \sigma) = (\gamma, 1)$ then

$$(\alpha, \sigma)(1_A, \sigma^{-1})(\beta^{-1}, 1) = e = (\gamma^{-1}, 1)(1_A, \sigma^{-1})(\alpha, \sigma).$$

The above remarks constitute a verification that Π is a group. Clearly the mapping $\chi : \Pi \to G$, in which $(\alpha, \sigma) \to \sigma$, is an epimorphism whose kernel, $\text{Ker}(\chi)$, consists of all elements of the form $(\alpha, 1)$. These elements therefore form a normal subgroup. Moreover, the mapping $\alpha \to (\alpha, 1)$ is an isomorphism $A \approx \text{Ker}(\chi)$ hence, if these are identified, A becomes a normal subgroup of Π and χ has kernel A. Next, the elements $(1_A, \sigma)$ form a section of Π and, since

$$(1_A, \sigma)(\alpha, 1)(1_A, \sigma)^{-1} = (\alpha^\sigma, \sigma)(1_A, \sigma)^{-1} = (\alpha^\sigma, 1)(1_A, \sigma)(1_A, \sigma)^{-1} = (\alpha^\sigma, 1),$$

the structure induced on A by the extension (Π, χ) is the originally prescribed structure as G-module. Finally,

$$(1_A, \sigma)(1_A, \tau) = (a_{\sigma,\tau}, \sigma\tau) = (a_{\sigma\tau}, 1)(1_A, \sigma\tau),$$

hence the factor system associated with the chosen section is just the one with which we started. The proof is thus complete.

We shall take this opportunity to establish an additional property of group extensions. As before, A will denote a commutative group and G an arbitrary group. We shall be concerned with a *fixed* extension (Π, χ) of A by G and when we speak of A as a G-module it is to be understood that we are referring to the structure induced by the extension in the manner explained above.

By an *automorphism of* Π *which is trivial for both* A *and* G will be meant a group-isomorphism $\phi : \Pi \approx \Pi$ such that

$$\phi(a) = a \quad (\text{all}\, a \in A) \quad \text{and} \quad \chi\phi = \chi. \tag{10.10.8}$$

Clearly these special automorphisms form a multiplicative group.

Now let $\alpha \in A$ and put

$$\phi_\alpha(\pi) = \alpha\pi\alpha^{-1} \quad (\pi \in \Pi);$$

then ϕ_α is an isomorphism of Π on to itself and is called the *inner automorphism of* Π *determined by* α. Since A is commutative, ϕ_α is trivial for A and, since $A = \text{Ker}(\chi)$, it is also trivial for G. Indeed, *the inner automorphisms of* Π, *determined by the elements of* A, *form a subgroup of the full group of automorphisms, which are trivial for both* A *and* G.

Let ϕ be an automorphism of Π satisfying (10.10.8), let $\sigma \in G$ and choose π_σ so that $\chi(\pi_\sigma) = \sigma$. Since $\chi(\pi_\sigma) = \chi(\phi(\pi_\sigma))$, we have $\phi(\pi_\sigma) = a\pi_\sigma$ for a suitable $a \in A$. Furthermore, if $\alpha \in A$ and we put $\pi'_\sigma = \alpha\pi_\sigma$, then

$$\phi(\pi'_\sigma) = \alpha\phi(\pi_\sigma) = a\alpha\pi_\sigma = a\pi'_\sigma;$$

consequently a depends only on σ and not on the choice of π_σ. It will therefore be denoted by a_σ. Thus ϕ determines a family $\{a_\sigma\}$ of elements of A which is such that, for any section $\{\pi_\sigma\}$ of Π,

$$\phi(\pi_\sigma) = a_\sigma\pi_\sigma \quad (\sigma \in G). \tag{10.10.9}$$

Again, $\phi(\pi_\sigma\pi_\tau) = \phi(\pi_\sigma)\,\phi(\pi_\tau) = a_\sigma\pi_\sigma a_\tau\pi_\tau = a_\sigma a_\tau^\sigma\pi_\sigma\pi_\tau,$

and therefore $a_{\sigma\tau} = a_\tau^\sigma a_\sigma$ showing that $\{a_\sigma\}$ is a standard 1-cocycle.†

The mapping $\phi \to \{a_\sigma\}$ is a monomorphism of groups and we contend that every 1-cocycle is the image of some ϕ. Note that this implies that the *group of* Π-*automorphisms, which are trivial for both* A *and* G, *is isomorphic to the group of standard* 1-*cocycles*. To fill in the proof, we have but to observe that a given 1-cocycle $\{a'_\sigma\}$ arises from ϕ', where ϕ' is defined by

$$\phi'(\alpha\pi_\sigma) = a'_\sigma\alpha\pi_\sigma \quad (\alpha \in A).$$

Let us now consider the image, under the above isomorphism, of the set of inner automorphisms of Π that are determined by elements of A. This is easily done for, if $\alpha \in A$, $\alpha\pi_\sigma\alpha^{-1} = (\alpha/\alpha^\sigma)\,\pi_\sigma$, which shows that the cocycle corresponding to ϕ_α is the typical 1-coboundary $\{\alpha/\alpha^\sigma\}$. These observations provide a proof of the following theorem.

† The reader should remember that we are using a multiplicative notation.

Theorem 13. *Let A be a commutative group, G an arbitrary group, and (Π, χ) an extension of A by G. Then the group of Π-automorphisms, which are trivial for both A and G, is commutative and has, as subgroup, the inner automorphisms of Π determined by the elements of A; furthermore, the factor-group of the former with respect to the latter, is isomorphic to $H^1(G, A)$.*

Corollary. *If $H^1(G, A)$ is trivial (i.e. consists only of a single element) then every Π-automorphism, which is trivial for both A and G, is an inner automorphism determined by an element of A.*

10.11 Homology and cohomology in special cases

In section (10.7) we introduced, for an arbitrary monoid, the standard non-homogeneous free resolution of Z. However, in the case of special monoids, there may be other free resolutions which are simpler and more useful for discussing homology and cohomology. Indeed, this section will provide a number of illustrations of this possibility.

Theorem 14. *Let S be a set of symbols, let G be either the free monoid or the free group generated by S, and let I be the augmentation ideal of $Z(G)$. Then I is a free G-module with the elements $s-1$ $(s \in S)$ as a G-base.*

Proof. We shall confine our attention to the case of the free group this being slightly the more complicated. First, it will be shown that the elements $s-1$ $(s \in S)$ generate I as a left G-module. To this end, let M be the submodule of $Z(G)$ which they do generate then, if we prove that $\sigma - 1 \in M$ for every $\sigma \in G$, it will follow that $M = I$. Let $\sigma = x_1 x_2 \ldots x_p$ be the representation of σ as an irreducible word. If $p = 1$ then $\sigma = s$ or $\sigma = s^{-1}$, where $s \in S$; and then $\sigma - 1 = s - 1$ or $\sigma - 1 = (-s^{-1})(s-1)$, so that, in any event, $\sigma - 1 \in M$. Assume now that $p > 1$ and that the relation $\sigma' - 1 \in M$ has been proved for all irreducible words σ' of length $p-1$. Then

$$\sigma - 1 = x_1 x_2 \ldots x_{p-1}(x_p - 1) + (x_1 x_2 \ldots x_{p-1} - 1),$$

and here the first of the two terms belongs to M by the case considered above and the second belongs to it by the inductive hypothesis. Accordingly $\sigma - 1 \in M$, and now it follows that $M = I$ as required. It remains to be shown that the generators are independent.

Let

$$\cdots \to \Phi_2 \to \Phi_1 \to \Phi_0 \to Z \to 0$$

be the standard non-homogeneous free resolution of Z. Then Φ_0 has as G-base the single element [ ], hence Φ_0 can be identified with $Z(G)$

so that [ ] corresponds to 1. If this is done, then $\Phi_0 \to Z$ becomes the augmentation homomorphism $Z(G) \to Z$ and therefore its kernel is I. One therefore obtains an exact sequence

$$\Phi_2 \to \Phi_1 \to I \to 0,$$

in which $\Phi_1 \to I$ maps $[\sigma]$ into $\sigma - 1$. Now let A be an arbitrary left G-module then

$$0 \to \mathrm{Hom}_G(I, A) \to \mathrm{Hom}_G(\Phi_1, A) \to \mathrm{Hom}_G(\Phi_2, A)$$

is exact, consequently

$$\mathrm{Hom}_G(I, A) \approx \mathrm{Ker}\{\mathrm{Hom}_G(\Phi_1, A) \to \mathrm{Hom}_G(\Phi_2, A)\},$$

and here the right-hand side is (by definition) the group of crossed homomorphisms of G into A. Further, if $\phi \in \mathrm{Hom}_G(I, A)$ corresponds to the crossed homomorphism ψ, then $\phi(s-1) = \psi(s)$. But, by Theorem 9, the values which a crossed homomorphism takes on the elements of S may be prescribed freely; hence there is always a G-homomorphism $I \to A$ in which the images of the elements $s - 1$ $(s \in S)$ have been assigned arbitrarily. However, since A may be *any* left G-module and the family $\{s-1\}_{s \in S}$ generates I, this clearly implies that the generators form a base.

The next result is an immediate consequence of the theorem just proved.

Theorem 15. *Let S be a set of symbols, and G either the free monoid or the free group generated by S. Then*

$$\cdots \to 0 \to 0 \to I \to Z(G) \to Z \to 0$$

is a G-free resolution of Z. Accordingly, if A is any G-module then, for all $n \geqslant 2$, both $H^n(G, A) = 0$ and $H_n(G, A) = 0$.

The next case to be considered is that of the finitely generated free abelian monoids and free abelian groups. We shall find that their homology theories are just a little more complicated.

Let $\xi_1, \xi_2, \ldots, \xi_p$ be p symbols and consider all formal power-products

$$\xi_1^{r_1} \xi_2^{r_2} \ldots \xi_p^{r_p},$$

where, to begin with, $r_1, r_2, \ldots, r_p$ are non-negative integers. These power-products can be made into an abelian monoid by multiplying any two of them, say $\xi_1^{r_1} \xi_2^{r_2} \ldots \xi_p^{r_p}$ and $\xi_1^{s_1} \xi_2^{s_2} \ldots \xi_p^{s_p}$, in the following way:

$$(\xi_1^{r_1} \xi_2^{r_2} \ldots \xi_p^{r_p})(\xi_1^{s_1} \xi_2^{s_2} \ldots \xi_p^{s_p}) = \xi_1^{r_1+s_1} \xi_2^{r_2+s_2} \ldots \xi_p^{r_p+s_p}.$$

The monoid so obtained is called *the free abelian monoid generated by* $\xi_1, \xi_2, \ldots, \xi_p$. Observe that if R is any commutative ring with an identity element and G is this monoid, then $R(G)$ is just the polynomial ring in $\xi_1, \xi_2, \ldots, \xi_p$ with coefficients in R.

The free abelian *group* generated by $\xi_1, \xi_2, \ldots, \xi_p$ is defined in the same way, save that the exponents $r_1, r_2, \ldots, r_p$ are no longer so restricted but may be any positive, negative, or zero integers.

We wish to study the homology and cohomology theories of this free abelian monoid and free abelian group. Observe, first, that for a *single* symbol ξ the *free abelian* monoid (group) and the *free* monoid (group), which it generates, coincide. Thus the case $p = 1$ has already been dealt with in Theorem 15. Next, if G_r is the free abelian monoid (group) generated by ξ_r alone and G is the free abelian monoid (group) generated by $\xi_1, \xi_2, \ldots, \xi_p$ together, then G is the direct product of $G_1, G_2, \ldots, G_p$ and we know that there exists, for each of the latter, a very simple free resolution of Z.

This focuses our attention on the following general question. Suppose that G and G' are any two given monoids and let

$$\cdots \to X_2 \to X_1 \to X_0 \to Z \to 0 \tag{10.11.1}$$

and

$$\ldots X_2' \to X_1' \to X_0' \to Z \to 0 \tag{10.11.2}$$

be G-free and G'-free resolutions of Z respectively. We ask now if there is a method of combining these two to obtain a free resolution for the direct product $G \times G'$. In fact, there is such a method and it will now be described.

We start by observing that if X is a left G-module and X' a left G'-module, then $X \otimes_Z X'$ can be made into a $(G \times G')$-module by putting

$$(\sigma, \sigma')(x \otimes x') = \sigma x \otimes \sigma' x' \quad (\sigma \in G, \sigma' \in G', x \in X, x' \in X').$$

Further, if X is G-free and X' is G'-free, then $X \otimes_Z X'$ is $(G \times G')$-free; indeed if $\{u_i\}$ is a G-base for X and $\{v_j\}$ a G'-base for X', then the elements $u_i \otimes v_j$ form a $(G \times G')$-base for the tensor product $X \otimes_Z X'$. Now denote by $\mathbf{X}$ and $\mathbf{X}'$ the complexes

$$\cdots \to X_2 \to X_1 \to X_0 \to 0 \to 0 \to \cdots$$

and

$$\cdots \to X_2' \to X_1' \to X_0' \to 0 \to 0 \to \cdots,$$

then (Chapter 6) $\mathbf{X} \otimes_Z \mathbf{X}'$ is a complex of Z-modules whose typical component module is a direct sum of the form

$$\sum_{r+s=n} X_r \otimes_Z X_s'. \tag{10.11.3}$$

Here each term, and hence the whole sum, can be given the structure of a $(G \times G')$-module and, if this is done, then all the modules of $\mathbf{X} \otimes_Z \mathbf{X}'$ will be $(G \times G')$-free. Again, the boundary homomorphisms† of $\mathbf{X} \otimes_Z \mathbf{X}'$ are easily verified as being not merely Z-homomorphisms but $(G \times G')$-homomorphisms, thus $\mathbf{X} \otimes_Z \mathbf{X}'$ has become a complex of free $(G \times G')$-modules.

Proposition 5. *The notation being as above,* $\mathbf{X} \otimes_Z \mathbf{X}'$ *is a* $(G \times G')$*-free resolution of* Z*, if the augmentation homomorphism* $X_0 \otimes_Z X'_0 \to Z$ *is defined as* $X_0 \otimes_Z X'_0 \to Z \otimes_Z Z$ *followed by the canonical identification* $Z \otimes_Z Z \approx Z$.

Proof. We need only show that the sequence

$$\cdots \to \sum_{r+s=2} X_r \otimes_Z X'_s \to \sum_{r+s=1} X_r \otimes_Z X'_s \to \sum_{r+s=0} X_r \otimes_Z X'_s \to Z \otimes_Z Z \to 0$$

is exact. Now

$$(X_1 \otimes_Z X'_0 + X_0 \otimes_Z X'_1) \to X_0 \otimes_Z X'_0 \to Z \otimes_Z Z \to 0$$

is exact by Theorem 4 of section (3.12). On the other hand, the nth homology group of $\mathbf{X} \otimes_Z \mathbf{X}'$ is just $\mathrm{Tor}_n^Z(Z, Z)$, because (10.11.1) and (10.11.2) provide Z-free resolutions of Z. But $\mathrm{Tor}_n^Z(Z, Z) = 0$ if $n \geqslant 1$ and this establishes the proposition.

Returning to the problem which led us to the last result, let G be either the free abelian group or the free abelian monoid generated by $\xi_1, \xi_2, \ldots, \xi_p$. For each s $(0 \leqslant s \leqslant p)$ we now form the free (left) G-module $X_s^{(p)}$ generated by all ordered sets $[i_1, i_2, \ldots, i_s]$, where $i_1, i_2, \ldots, i_s$ are integers satisfying $1 \leqslant i_1 < i_2 < \ldots < i_s \leqslant p$. (It is to be understood that $X_0^{(p)}$ is the G-free module generated by the single element $[\ ]$.) G-homomorphisms

$$d_s^{(p)} : X_s^{(p)} \to X_{s-1}^{(p)} \quad (1 \leqslant s \leqslant p), \quad \omega^{(p)} : X_0^{(p)} \to Z$$

are then defined by

$$d^{(p)}[i_1, i_2, \ldots, i_s] = \sum_{\nu=1}^{s} (-1)^{\nu+1} (\xi_{i_\nu} - 1) [i_1, \ldots, \hat{i}_\nu, \ldots, i_s] \tag{10.11.4}$$

and

$$\omega^{(p)}([\ ]) = 1_Z, \tag{10.11.5}$$

where, as usual, the ^ over i_ν in (10.11.4) means that i_ν is to be omitted from the bracket. The next result shows that these modules and mappings provide a particularly simple G-free resolution of Z.

† The way these are defined is explained in section (6.2).

Theorem 16. *Let G be either the free abelian group or the free abelian monoid generated by the p ($p \geqslant 1$) symbols $\xi_1, \xi_2, \ldots, \xi_p$. Then the sequence*

$$\cdots \to 0 \to 0 \to X_p^{(p)} \xrightarrow{d_p^{(p)}} X_{p-1}^{(p)} \to \cdots \to X_1^{(p)} \xrightarrow{d_1^{(p)}} X_0^{(p)} \xrightarrow{\omega^{(p)}} Z \to 0 \qquad (10.11.6)$$

(defined above) is exact and therefore yields a G-free resolution of Z.

Proof. The argument is the same for groups as it is for monoids and so, to avoid complicating the wording, we shall always refer to G as a group. First consider the case $p = 1$. In this case, G coincides with the free group generated by ξ_1 consequently, by Theorem 14, the augmentation ideal I has $\xi_1 - 1$ as a G-base. We therefore have G-isomorphisms

$$I \approx X_1^{(1)} \quad \text{and} \quad Z(G) \approx X_0^{(1)},$$

where, in the former, $\xi_1 - 1$ corresponds to [1] and, in the latter 1 corresponds to [ ]. In view of this the exact sequence

$$\cdots \to 0 \to 0 \to I \to Z(G) \to Z \to 0$$

shows that

$$\cdots \to 0 \to 0 \to X_1^{(1)} \to X_0^{(1)} \to Z \to 0$$

is exact as well and thereby establishes the theorem when $p = 1$.

Assume now that $p > 1$ and that the theorem has been proved for free abelian groups with fewer than p generators. Then

$$\cdots \to 0 \to 0 \to X_{p-1}^{(p-1)} \to \cdots \to X_0^{(p-1)} \to Z \to 0$$

is exact by the inductive hypothesis while, by the case of a single generator (which, on this occasion, we call ξ_p),

$$\cdots \to 0 \to 0 \to X_1^{(1)} \to X_0^{(1)} \to Z \to 0$$

is also exact. We can therefore apply Proposition 5 to obtain a free resolution

$$\cdots \to 0 \to 0 \to X_p \to X_{p-1} \to \cdots \to X_0 \to Z \to 0 \qquad (10.11.7)$$

of Z, for the group which is the direct product of the free abelian group generated by $\xi_1, \xi_2, \ldots, \xi_{p-1}$ with the free abelian group generated by ξ_p. In other words, (10.11.7) is a free resolution of Z for G. We shall show that it can be identified with (10.11.6).

By construction, if $0 \leqslant s \leqslant p$, then

$$X_s = X_s^{(p-1)} \otimes_Z X_0^{(1)} + X_{s-1}^{(p-1)} \otimes_Z X_1^{(1)} \quad \text{(direct sum)},$$

hence the elements

$$[j_1, j_2, \ldots, j_s] \otimes [\ ] \quad (1 \leqslant j_1 < j_2 < \ldots < j_s \leqslant p-1)$$

and

$$[k_1, k_2, \ldots, k_{s-1}] \otimes [1] \quad (1 \leqslant k_1 < k_2 < \ldots < k_{s-1} \leqslant p-1)$$

form a G-base for X_s. Further, if d_s is the boundary homomorphism $X_s \to X_{s-1}$ ($s \geqslant 1$) then, again by construction,

$$d_s\{[j_1, \ldots, j_s] \otimes [\ ]\} = d_s^{(p-1)}[j_1, \ldots, j_s] \otimes [\ ]$$
$$= \sum_{\nu=1}^{s} (-1)^{\nu+1} (\xi_{j_\nu} - 1)[j_1, \ldots, \hat{j}_\nu, \ldots, j_s] \otimes [\ ], \tag{10.11.8}$$

while

$$d_s\{[k_1, \ldots, k_{s-1}] \otimes [1]\} = d_{s-1}^{(p-1)}[k_1, \ldots, k_{s-1}] \otimes [1]$$
$$+ (-1)^{s+1}[k_1, \ldots, k_{s-1}] \otimes d_1^{(1)}[1]$$
$$= \sum_{\nu=1}^{s-1} (-1)^{\nu+1} (\xi_{k_\nu} - 1)[k_1, \ldots, \hat{k}_\nu, \ldots, k_{s-1}] \otimes [1]$$
$$+ (-1)^{s+1}[k_1, \ldots, k_{s-1}] \otimes (\xi_p - 1)[\ ]. \tag{10.11.9}$$

Now $X_s^{(p)}$ can be identified with X_s as a G-module. Indeed, suppose that $1 \leqslant i_1 < i_2 < \ldots < i_s \leqslant p$, then we make $[i_1, i_2, \ldots, i_s]$ correspond to $[i_1, i_2, \ldots, i_s] \otimes [\ ]$ if $i_s < p$ and to $[i_1, i_2, \ldots, i_{s-1}] \otimes [1]$ if $i_s = p$. When this is done, (10.11.8) and (10.11.9) show that $d_s^{(p)}$ coincides with d_s. Finally, it is obvious that $X_0 \to Z$ now coincides with $\omega^{(p)} : X_0^{(p)} \to Z$ consequently the exactness of (10.11.6) has been established.

Corollary. *Let G be either the free abelian group or the free abelian monoid generated by p symbols. Then for any G-module A, both $H_n(G, A) = 0$ and $H^n(G, A) = 0$ whenever $n > p$.*

From the point of view of homology and cohomology the next simplest groups are the cyclic groups of finite order. However, it will be more convenient to postpone their consideration until we have had some discussion of finite groups in general.

10.12 Finite groups

A special feature of the theory for finite groups is that the two sequences

$$H^0(G, A),\ H^1(G, A),\ H^2(G, A),\ \ldots$$

and

$$\ldots,\ H_2(G, A),\ H_1(G, A),\ H_0(G, A)$$

can be combined to form a single exact connected sequence of functors, which continues indefinitely in both directions. (The piecing together involves certain preliminary modifications, but these affect only $H^0(G, A)$ and $H_0(G, A)$, the other terms being left unaltered.) Once the complete sequence has been derived, it will be characterized by means of a few basic properties and this will put it on a similar footing to the original homology and cohomology theories. Finally, it will be shown how, by adapting the concept of a free resolution of Z, one can obtain the whole complete sequence at a single step. This is fundamental for the more advanced parts of the theory, but we shall leave the reader to explore, for himself, all that lies beyond.

From now on, G will denote a *finite group* and when we speak of a G-module we shall mean a *left* G-module unless there is an explicit statement to the contrary. Put

$$N = \sum_{\sigma \in G} \sigma, \tag{10.12.1}$$

so that N is an element of $Z(G)$ then, for any $\tau \in G$,

$$N\tau = N = \tau N. \tag{10.12.2}$$

If now A is a left G-module then the mapping $A \to A$ defined by $a \to Na$, that is to say by

$$a \to \sum_{\sigma \in G} \sigma a, \tag{10.12.3}$$

is a G-homomorphism of A into itself. This homomorphism, which will be denoted by

$$N : A \to A, \tag{10.12.4}$$

is called the *norm homomorphism* of A, and the image of an element of the module, under the mapping, will be called the *norm* of the element. Write

$$\operatorname{Im}(N : A \to A) = N(A) \tag{10.12.5}$$

and

$$\operatorname{Ker}(N : A \to A) = {}_N A, \tag{10.12.6}$$

then, by (10.12.2),

$$\tau(Na) = Na, \quad N[(\tau - 1)a] = 0 \quad (a \in A, \tau \in G),$$

hence $N(A) \subseteq A^G$ and $IA \subseteq {}_N A$. It follows that the norm homomorphism induces a Z-homomorphism

$$N_* : A_G \to A^G \tag{10.12.7}$$

which, as one verifies immediately, is a natural transformation

$$N_* : H_0(G, A) \to H^0(G, A). \tag{10.12.8}$$

Furthermore, by Theorem 1 of section (4.3),

$$\operatorname{Ker}(N_*) = {}_N A/IA \quad \text{and} \quad \operatorname{Coker}(N_*) = A^G/N(A) \tag{10.12.9}$$

will be additive, covariant functors of A.

Now let $$0 \to A' \to A \to A'' \to 0 \tag{10.12.10}$$

be an exact sequence of G-modules and let us use (temporarily) N'_*, N_*, N''_*, for the mappings $A'_G \to A'^G$, $A_G \to A^G$, $A''_G \to A''^G$ obtained from (10.12.7). Consider the diagram

$$\begin{array}{ccccccccc}
H_1(A) & \longrightarrow & H_1(A'') & \longrightarrow & A'_G & \longrightarrow & A_G & \longrightarrow & A''_G \\
\downarrow & & \downarrow & & \downarrow N'_* & & \downarrow N_* & & \downarrow N''_* \\
0 & \longrightarrow & 0 & \longrightarrow & A'^G & \longrightarrow & A^G & \longrightarrow & A''^G
\end{array}$$

$$\begin{array}{ccccc}
\longrightarrow & 0 & \longrightarrow & 0 \\
 & \downarrow & & \downarrow \\
\longrightarrow & H^1(A') & \longrightarrow & H^1(A)
\end{array} \tag{10.12.11}$$

where we have written, for brevity, $H_r(A)$ in place of $H_r(G, A)$ and $H^s(A)$ for $H^s(G, A)$. First the diagram is commutative (only the second and fifth squares give cause for a moment's reflection) and secondly the rows are exact.† It now follows, by applications of Propositions 3 and 4 of section (4.5), that (10.12.11) determines an exact sequence

$$\begin{aligned} H_1(A) \to H_1(A'') \to \operatorname{Ker}(N'_*) \to \operatorname{Ker}(N_*) \to \operatorname{Ker}(N''_*) \\ \to \operatorname{Coker}(N'_*) \to \operatorname{Coker}(N_*) \to \operatorname{Coker}(N''_*) \to H^1(A') \to H^1(A). \end{aligned} \tag{10.12.12}$$

Further, every translation

$$\begin{array}{ccccccccc}
0 & \longrightarrow & A' & \longrightarrow & A & \longrightarrow & A'' & \longrightarrow & 0 \\
 & & \downarrow & & \downarrow & & \downarrow & & \\
0 & \longrightarrow & \bar{A}' & \longrightarrow & \bar{A} & \longrightarrow & \bar{A}'' & \longrightarrow & 0
\end{array}$$

of (10.12.10) induces a translation of (10.12.11) and hence a translation of (10.12.12). In this way it is seen that

$$\ldots, H_2(A), H_1(A), {}_NA/IA, A^G/N(A), H^1(A), H^2(A), \ldots$$

where the sequence extends without limit in both directions, is an exact connected sequence. Accordingly, if we put

$$J^0(G, A) = A^G/N(A), \quad J^n(G, A) = H^n(G, A) = \operatorname{Ext}^n_G(Z, A) \quad (n \geq 1) \tag{10.12.13}$$

and

$$\left.\begin{aligned} J^{-1}(G, A) &= {}_NA/IA, \\ J^{-n}(G, A) = H_{n-1}(G, A) &= \operatorname{Tor}^G_{n-1}(Z, A) \quad (n \geq 2), \end{aligned}\right\} \tag{10.12.14}$$

† The fact that the rows have been broken, to facilitate printing, should be ignored.

then $\ldots, J^{-2}(G,A), J^{-1}(G,A), J^0(G,A), J^1(G,A), J^2(G,A), \ldots$

is an *exact connected sequence of additive, covariant functors.* It will be called the *complete derived sequence* of G.

10.13 The norm of a homomorphism

In this section, A, A', etc., will denote left G-modules, B, B', etc., right G-modules (G is a finite group), and we shall establish certain results of an auxiliary nature, which will help us to develop the theory of the complete derived sequence of G. It is particularly desirable that we should have knowledge about the modules A for which $J^n(G,A) = 0$ for all n, and a good deal of our effort will be devoted to this end. We shall discover later that all G-projective and G-injective modules have this property and there is a still larger class (the G-special modules), with a simple characterization, whose members exhibit the same behaviour.

Let A and A' be (left) G-modules and let $u : A \to A'$ be a Z-homomorphism (not a G-homomorphism). It is proposed to define the 'norm of u' and, for this to tie up with the notion of norm as introduced in the last section, it is necessary that we should give $\mathrm{Hom}_Z(A,A')$ a suitable structure as G-module. The relevant structure is, in fact, obtained by defining σu, where $\sigma \in G$, by means of the formula

$$(\sigma u)\, a = \sigma\{u(\sigma^{-1}a)\}. \tag{10.13.1}$$

A simple verification shows that this really does turn $\mathrm{Hom}_Z(A,A')$ into a G-module and, to distinguish this structure from other G-module structures which can be imposed on it, we shall say that the elements of G act as *diagonal operators.* By way of illustration, observe that $\sigma u = u$ if and only if $\sigma^{-1}\{u(a)\} = u(\sigma^{-1}a)$ for all $a \in A$. Thus *when* $\mathrm{Hom}_Z(A,A')$ *has the elements of* G *as diagonal operators, the fixed elements are just the* G*-homomorphisms of* A *into* A'.

It is now to be understood that, by the norm, $N(u)$ or Nu, of $u : A \to A'$ *we shall always mean the ordinary norm* $\sum_\sigma \sigma u$ *when* $\mathrm{Hom}_Z(A,A')$ *has its diagonal structure,* hence, from (10.13.1),

$$(Nu)\, a = \sum_{\sigma \in G} \sigma\{u(\sigma^{-1}a)\}. \tag{10.13.2}$$

Of course, in the case of right G-modules B and B', the norm of a Z-homomorphism $v : B \to B'$ is given by

$$(Nv)\, b = \sum_{\sigma \in G} \{v(b\sigma^{-1})\}\,\sigma. \tag{10.13.3}$$

Proposition 6. *Let A and A' be left G-modules and $u : A \to A'$ a Z-homomorphism. Then $N(u)$ is a G-homomorphism of A into A'.*

This simply interprets the fact that the norm of u is a fixed element of $\mathrm{Hom}_Z(A, A')$.

Proposition 7. *Let A_1, A_2, A_3, A_4 be left G-modules and suppose that $f : A_1 \to A_2$, $g : A_3 \to A_4$ are G-homomorphisms while $u : A_2 \to A_3$ is a Z-homomorphism. Then*

$$N(guf) = g(Nu)f. \tag{10.13.4}$$

This follows at once from (10.13.2).

In order to introduce the next definition, we draw attention to a property of $Z(G)$ regarded as a left G-module. Let $u : Z(G) \to Z(G)$ be the Z-homomorphism obtained by putting

$$u(\sum_\sigma n_\sigma \sigma) = n_1 1.$$

Then if we write $\Sigma n_\sigma \sigma = \lambda$, it is seen that $u(\tau^{-1}\lambda) = n_\tau 1$; hence $\tau\{u(\tau^{-1}\lambda)\} = n_\tau \tau$ and therefore

$$\sum_\tau \tau\{u(\tau^{-1}\lambda)\} = \sum_\tau n_\tau \tau = \lambda.$$

Thus the identity map of $Z(G)$ is a norm. There are many other modules which share this property which is a very important one for our purposes.

Definition. A G-module A whose identity map is the norm of some Z-homomorphism $u : A \to A$ will be said to be *G-special.*

Observe that, for a left G-module, the identity map is the norm of $u : A \to A$, when, and only when,

$$a = \sum_\sigma \sigma\{u(\sigma^{-1}a)\}$$

for all $a \in A$. (The obvious modifications are to be made in the case of a right G-module.) Observe, too, that $Z(G)$ is G-special both as a left and a right G-module.

Proposition 8. *Let A' be a direct summand of a G-module A and suppose that A is G-special. Then A' is also G-special.*

Proof. The identity map of A' can be expressed in the form $A' \to A \to A'$, where $A' \to A$ is an inclusion map and $A \to A'$ is a G-homomorphism. Let the identity map of A be the norm of $u : A \to A$, then, by (10.13.4), the identity map of A' is the norm of the combined map $A' \to A \xrightarrow{u} A \to A'$.

Proposition 9. *Let A be a left G-module and M a Z-module; further, let $A \otimes_Z M$ be given the structure of a left G-module for which*

$$\sigma(a \otimes x) = \sigma a \otimes x \quad (a \in A,\ \sigma \in G,\ x \in M).$$

If now A is G-special then so is $A \otimes_Z M$. In particular, $Z(G) \otimes_Z M$ is G-special.

Proof. Let $u : A \to A$ be a Z-homomorphism such that

$$a = \sum_\sigma \sigma\{u(\sigma^{-1}a)\}$$

for all $a \in A$ and put $v = u \otimes i_M$, where i_M is the identity map of M. Then, if $a \in A$, $x \in M$ and $y = a \otimes x$,

$$\sum_\sigma \sigma\{v(\sigma^{-1}y)\} = \sum_\sigma \sigma\{u(\sigma^{-1}a) \otimes x\} = \sum_\sigma \{\sigma u(\sigma^{-1}a) \otimes x\} = a \otimes x = y,$$

and the proposition follows immediately.

There is another result, complementary to the one just proved, which can be stated thus:

Proposition 10. *Let B be a right G-module and M a Z-module; further, let* $\mathrm{Hom}_Z(B, M)$ *be given the structure of a left G-module in which, for* $f \in \mathrm{Hom}_Z(B, M)$,

$$(\sigma f)b = f(b\sigma) \quad (b \in B,\ \sigma \in G).$$

If now B is G-special, then so is $\mathrm{Hom}_Z(B, M)$. *In particular,*

$$\mathrm{Hom}_Z(Z(G), M)$$

is G-special.

Proof. Let $u : B \to B$ be such that

$$b = \sum_\sigma \{u(b\sigma^{-1})\}\sigma$$

for all $b \in B$, and put $v = \mathrm{Hom}(u, i_M)$. If now $f \in \mathrm{Hom}_Z(B, M)$, then $v(\sigma^{-1}f)$ is the combined mapping $B \xrightarrow{u} B \xrightarrow{\sigma^{-1}f} M$, consequently, if we write $\sigma\{v(\sigma^{-1}f)\} = \phi_\sigma$, we shall have

$$\phi_\sigma(b) = \{v(\sigma^{-1}f)\}(b\sigma) = \{\sigma^{-1}f\}u(b\sigma) = f\{u(b\sigma)\sigma^{-1}\}.$$

Accordingly
$$(\sum_\sigma \phi_\sigma)b = f(b),$$

and therefore
$$\sum_\sigma \sigma\{v(\sigma^{-1}f)\} = \sum_\sigma \phi_\sigma = f.$$

This completes the proof.

It is now possible to give two very useful criteria for a G-module to be special. Let A be a left G-module, then $Z(G) \otimes_Z A$ has a structure as a left G-module in which

$$\sigma(\lambda \otimes a) = \sigma\lambda \otimes a \quad (\lambda \in Z(G),\ a \in A,\ \sigma \in G). \tag{10.13.5}$$

Further, there is a homomorphism

$$Z(G) \otimes_Z A \to A$$

in which

$$\lambda \otimes a \to \lambda a, \tag{10.13.6}$$

and this is clearly a G-epimorphism.

Proposition 11. *Let A be a left G-module. Then A is G-special if and only if the mapping $Z(G) \otimes_Z A \to A$, defined in* (10.13.6), *is direct*† *when regarded as an epimorphism of G-modules.*

Proof. If $Z(G) \otimes_Z A \to A$ is direct then A is isomorphic to a direct summand of the G-module $Z(G) \otimes_Z A$. But $Z(G) \otimes_Z A$ is G-special (Proposition 9) hence, by Proposition 8, A is G-special.

To prove the converse, let A be G-special then the identity map i_A of A is the norm of a Z-homomorphism $u : A \to A$. Let u^* be the Z-homomorphism $A \to Z(G) \otimes_Z A$ defined by $a \to 1 \otimes u(a)$, then $u : A \to A$ is the combined map

$$A \xrightarrow{i_A} A \xrightarrow{u^*} Z(G) \otimes_Z A \to A.$$

Taking norms and applying (10.13.4), we find that i_A can be represented as

$$A \xrightarrow{N(u^*)} Z(G) \otimes_Z A \longrightarrow A,$$

and, since Nu^* is a G-homomorphism, this completes the proof.

For the second criterion we use $\mathrm{Hom}_Z(Z(G), A)$ and endow it with the structure of a left G-module in which σf, where

$$f \in \mathrm{Hom}_Z(Z(G), A),$$

is given by

$$(\sigma f)\lambda = f(\lambda\sigma) \quad (\lambda \in Z(G),\ \sigma \in G). \tag{10.13.7}$$

Now, if $a \in A$, the mapping $\lambda \to \lambda a$ is a Z-homomorphism of $Z(G)$ into A. Denoting this homomorphism by f, we have

$$f(\lambda) = \lambda a, \tag{10.13.8}$$

and then $a \to f$ is a homomorphism

$$A \to \mathrm{Hom}_Z(Z(G), A), \tag{10.13.9}$$

which one easily verifies is a G-monomorphism.

† See section (1.9).

Proposition 12. *Let A be a left G-module. Then A is G-special if and only if the mapping $A \to \mathrm{Hom}_Z(Z(G), A)$ of* (10.13.9) *is direct, when regarded as a monomorphism of G-modules.*

Proof. If the monomorphism is direct, then A is isomorphic to a direct summand of the G-module $\mathrm{Hom}_Z(Z(G), A)$. But this is G-special (Proposition 10) consequently A is G-special.

Assume next that A is G-special and let the identity map i_A of A be the norm of $u : A \to A$. Now the Z-homomorphism

$$u^* : \mathrm{Hom}_Z(Z(G), A) \to A$$

defined by $u^*(f) = f(1)$, is such that u is the combined mapping

$$A \to \mathrm{Hom}_Z(Z(G), A) \xrightarrow{u^*} A \xrightarrow{i_A} A;$$

consequently, taking norms, i_A can be represented as

$$A \longrightarrow \mathrm{Hom}_Z(Z(G), A) \xrightarrow{Nu^*} A$$

and, since Nu^* is a G-homomorphism, this shows that

$$A \to \mathrm{Hom}_Z(Z(G), A)$$

is direct.

As an application of the last two results we shall prove

Theorem 17. *If A is either G-projective or G-injective, then A is G-special.*

Proof. If A is G-projective then the epimorphism $Z(G) \otimes_Z A \to A$ of Proposition 11 is direct by Theorem 1 of section (5.1). On the other hand, if A is G-injective then the G-monomorphism

$$A \to \mathrm{Hom}_Z(Z(G), A),$$

which occurs in Proposition 12, is direct by virtue of Theorem 6 of section (5.2). The theorem now follows.

10.14 Properties of the complete derived sequence

We are now in a position to establish some further facts about the complete derived sequence

$$\ldots, J^{-2}(G, A), J^{-1}(G, A), J^0(G, A), J^1(G, A), \ldots$$

of an arbitrary finite group G. These additional facts stem from

Theorem 18. *If A is G-special, then $J^n(G, A) = 0$ for all values of n.*

Proof. Let $u : A \to A$ be a Z-homomorphism such that

$$a = \sum_{\sigma} \{\sigma u(\sigma^{-1}a)\}$$

for all $a \in A$. If now $a \in A^G$ then $u(\sigma^{-1}a) = u(a)$ and therefore a is the norm of $u(a)$. Thus $A^G \subseteq N(A)$ consequently, by (10.12.13), it follows that $J^0(G, A) = 0$.

Assume next that $a \in {}_N A$ then $Na = 0$ and therefore

$$\sum_{\sigma} u(\sigma^{-1}a) = u(Na) = 0.$$

Accordingly $$a = \sum_{\sigma} (\sigma - 1)\{u(\sigma^{-1}a)\} \in IA,$$

hence ${}_N A \subseteq IA$ and so $J^{-1}(G, A) = 0$ by (10.12.14).

Put $A' = \mathrm{Hom}_Z\{Z(G), A\}$ and let A' have the structure of a left G-module as indicated in (10.13.7). By Proposition 12, A is isomorphic to a direct summand of the G-module A' and, by Proposition 3,

$$J^n(G, A') = H^n(G, A') = 0$$

for all $n \geqslant 1$. It follows that $J^n(G, A) = 0$ for $n \geqslant 1$.

Finally, let $A^* = Z(G) \otimes_Z A$, where A^* has the structure of a left G-module described in (10.13.5). Then, by Proposition 11, A is isomorphic to a direct summand of A^* (as G-module); furthermore, when $n \geqslant 2$, Proposition 4 shows that

$$J^{-n}(G, A^*) = H_{n-1}(G, A^*) = 0.$$

Accordingly $J^{-n}(G, A) = 0$ for $n \geqslant 2$ and with this the proof is complete.

Lemma 1. *Let $f : A' \to A$ be a homomorphism of left G-modules which is the norm of a Z-homomorphism $u : A' \to A$. Then there exists a G-homomorphism $A' \to Z(G) \otimes_Z A$ such that f is the combined mapping $A' \to Z(G) \otimes_Z A \to A$.*

In this lemma, $Z(G) \otimes_Z A$ is to have the same structure as G-module and $Z(G) \otimes_Z A \to A$ is to be the same G-homomorphism as in Proposition 11.

Proof. Let $u^* : A' \to Z(G) \otimes_Z A$ be the Z-homomorphism defined by $u^*(a') = 1 \otimes u(a')$, then u is the combined mapping

$$A' \xrightarrow{i_{A'}} A' \xrightarrow{u^*} Z(G) \otimes_Z A \to A.$$

The required result now follows from Proposition 7 on taking norms.

Proposition 13. *Let $f : A' \to A$ be a homomorphism of left G-modules and suppose that f is the norm of some Z-homomorphism of A' into A. Then $J^n(G, f) = 0$ for all values of n.*

Proof. $J^n(G,f)$ is the homomorphism $J^n(G,A') \to J^n(G,A)$ induced by f consequently, by Lemma 1, this can be represented in the form

$$J^n(G,A') \to J^n(G,Z(G) \otimes_Z A) \to J^n(G,A).$$

But, by Theorem 18, the second term is a null module because $Z(G) \otimes_Z A$ is G-special (Proposition 9). The result follows.

Now let A be any G-module and let i_A be its identity map. Since i_A is a G-homomorphism, it follows that $N(i_A) = qi_A$, where q is the order of G. However J^n is an additive functor and so

$$J^n(G,qi_A) = qJ^n(G,i_A).$$

This establishes the next theorem if one takes account of Proposition 13.

Theorem 19. *Let G be a finite group of order q and A any left G-module. Then $qJ^n(G,A) = 0$ for all values of n.*

Taking account of Theorems 17 and 18 we may observe that, *inter alia*, the complete derived sequence of G has the following properties:

(*a*) *the $J^n(G,A)$ form an exact, connected sequence of covariant functors;*

(*b*) *$J^0(G,A)$ is the functor consisting of the fixed elements of A modulo the elements which are norms;*

(*c*) *$J^n(G,A) = 0$ for all n whenever A is either G-projective or G-injective.*

These suffice to characterize the sequence to within an isomorphism of connected sequences. Indeed, one has the following more general uniqueness criterion which will be needed later.

Proposition 14. *Let A be a variable left G-module and let*

$$\ldots,\ T^{-2}(A),\ T^{-1}(A),\ T^{0}(A),\ T^{1}(A),\ \ldots$$

and

$$\ldots,\ U^{-2}(A),\ U^{-1}(A),\ U^{0}(A),\ U^{1}(A),\ \ldots$$

be exact connected sequences of covariant functors of A whose values are Z-modules. Suppose further that whenever A is either G-projective or G-injective, then $T^n(A) = 0$ and $U^n(A) = 0$ for all values of n. If now, for a particular integer r, there exists a functor equivalence $T^r(A) \approx U^r(A)$, then this equivalence has a unique extension to an isomorphism of the connected sequences.

The proposition needs no proof since it follows at once from the corollaries to Theorems 10 and 12 of section (6.5).

10.15 Complete free resolutions of Z

The method of obtaining the complete derived sequence of G, by combining together the homology and cohomology theories, has the advantage of showing how these all tie up with one another; but it is inconvenient in that the two halves of the sequence are then on different footings and therefore tend to require separate discussion. In a moment we shall describe a method by which this can be overcome, but first, in order not to interrupt the main development at an awkward moment, we shall establish a property of exact complexes of Z-free modules.

Proposition 15. *Let $T(M)$ be an additive functor of Z-modules, whose values are also Z-modules, and let* $\mathbf{X}$ *be an exact complex*

$$\cdots \to X_{n+1} \to X_n \to X_{n-1} \to \cdots$$

whose component modules are Z-free. Then $T(\mathbf{X})$ is also exact.

Proof. We shall suppose, for definiteness, that T is a covariant functor. The contravariant case can be treated similarly. Put

$$\operatorname{Im}(X_n \to X_{n-1}) = A_n$$

then, for each value of n,

$$0 \to A_{n+1} \to X_n \to A_n \to 0 \tag{10.15.1}$$

is an exact sequence. Now A_n is a submodule of the Z-free module X_{n-1} and Z is a principal ideal domain, consequently, by Theorem 3 of section (9.1), A_n is also Z-free. It follows that the exact sequence (10.15.1) splits and therefore, since T is additive,

$$0 \to T(A_{n+1}) \to T(X_n) \to T(A_n) \to 0$$

is exact for all n. But $T(X_{n+1}) \to T(X_n)$ and $T(X_n) \to T(X_{n-1})$ can be represented by

$$T(X_{n+1}) \to T(A_{n+1}) \to T(X_n) \quad \text{and} \quad T(X_n) \to T(A_n) \to T(X_{n-1})$$

respectively and now it can be seen that

$$T(X_{n+1}) \to T(X_n) \to T(X_{n-1})$$

is exact as required.

We come now to a new concept. Regarding Z as a left G-module (on which G acts trivially), we define a *complete G-free resolution of Z* as a pair of exact sequences

$$\cdots \to X_2 \to X_1 \to X_0 \to Z \to 0 \tag{10.15.2}$$

and $$0 \to Z \to X_{-1} \to X_{-2} \to \cdots, \tag{10.15.3}$$

where ..., X_1, X_0, X_{-1}, X_{-2}, ... are all G-free and the mappings are G-homomorphisms. If, in this situation, we define $X_0 \to X_{-1}$ as the combined mapping $X_0 \to Z \to X_{-1}$, then the sequence

$$\cdots \to X_2 \to X_1 \to X_0 \to X_{-1} \to X_{-2} \to \cdots \tag{10.15.4}$$

is exact. In view of this, it is convenient to represent the complete resolution by the single commutative diagram

$$\begin{array}{ccccccc} \cdots \to X_1 \to & X_0 & \to & X_{-1} & \to X_{-2} \to \cdots \\ & & \searrow \; Z \; \nearrow & & \\ & & \nearrow \quad \searrow & & \\ & 0 & & 0 & \end{array} \tag{10.15.5}$$

Suppose now that (10.15.5) is a complete G-free resolution of Z and let A be a left G-module. Denote by $\mathbf{X}$ the complex (10.15.4), then the homology module $H^n\{\mathrm{Hom}_G(\mathbf{X}, A)\}$ is a covariant functor of A on account of the fact that each G-homomorphism $A \to A'$ produces a translation $\mathrm{Hom}_G(\mathbf{X}, A) \to \mathrm{Hom}_G(\mathbf{X}, A')$. Furthermore, if

$$0 \to A^* \to A \to A' \to 0$$

is an exact sequence of G-modules then, by Theorem 3 of section (5.1),

$$\mathbf{0} \to \mathrm{Hom}_G(\mathbf{X}, A^*) \to \mathrm{Hom}_G(\mathbf{X}, A) \to \mathrm{Hom}_G(\mathbf{X}, A') \to \mathbf{0}$$

is an exact sequence of complexes. This, in turn, gives rise to the exact sequence

$$\begin{aligned} \cdots \to H^n\{\mathrm{Hom}_G(\mathbf{X}, A^*)\} \to H^n\{\mathrm{Hom}_G(\mathbf{X}, A)\} \to H^n\{\mathrm{Hom}_G(\mathbf{X}, A')\} \\ \to H^{n+1}\{\mathrm{Hom}_G(\mathbf{X}, A^*)\} \to H^{n+1}\{\mathrm{Hom}_G(\mathbf{X}, A)\} \to \cdots \end{aligned}$$

of homology modules. Indeed, we can sum up these remarks and extend them by saying briefly that

$$\ldots, H^{-1}\{\mathrm{Hom}_G(\mathbf{X}, A)\}, H^0\{\mathrm{Hom}_G(\mathbf{X}, A)\}, H^1\{\mathrm{Hom}_G(\mathbf{X}, A)\}, \ldots \tag{10.15.6}$$

is an exact connected sequence of additive covariant functors.

Theorem 20. *Let G be a finite group and* (10.15.5) *a complete G-free resolution of Z. Then the exact connected sequence* (10.15.6) *is isomorphic to the complete derived sequence of G.*

Proof. Since (10.15.2) is a G-free resolution of Z,

$$H^1\{\mathrm{Hom}_G(\mathbf{X}, A)\} = \mathrm{Ext}^1_G(Z, A) = J^1(G, A),$$

hence (Proposition 14) it is enough to show that

$$H^n\{\mathrm{Hom}_G(\mathbf{X}, A)\} = 0 \quad (-\infty < n < \infty), \tag{10.15.7}$$

whenever A is either G-projective or G-injective. By Theorem 17 this will be more than covered if (10.15.7) is established whenever A is G-special.

Assume therefore that A is G-special then (Proposition 12) it is isomorphic to a direct summand of $A' = \mathrm{Hom}_Z(Z(G), A)$, where A' has the structure of a left G-module obtained by regarding $Z(G)$ as a right G-module. Now for any left G-module C we have isomorphisms†

$$\begin{aligned}\mathrm{Hom}_G(C, A') = \mathrm{Hom}_G\{C, \mathrm{Hom}_Z(Z(G), A)\} &\approx \mathrm{Hom}_Z\{Z(G) \otimes_G C, A\} \\ &\approx \mathrm{Hom}_Z(C, A),\end{aligned}$$

and this gives a functor equivalence between $\mathrm{Hom}_G(C, A')$ and $\mathrm{Hom}_Z(C, A)$. It follows that corresponding homology modules of the two complexes

$$\cdots \to \mathrm{Hom}_G(X_{n-1}, A') \to \mathrm{Hom}_G(X_n, A') \to \mathrm{Hom}_G(X_{n+1}, A') \to \cdots$$

and

$$\cdots \to \mathrm{Hom}_Z(X_{n-1}, A) \to \mathrm{Hom}_Z(X_n, A) \to \mathrm{Hom}_Z(X_{n+1}, A) \to \cdots \tag{10.15.8}$$

are isomorphic. But (10.15.4), being an exact sequence of G-free modules, is also an exact sequence of Z-modules hence, by Proposition 15, (10.15.8) is exact. Accordingly $H^n\{\mathrm{Hom}_G(\mathbf{X}, A')\} = 0$ and therefore $H^n\{\mathrm{Hom}_G(\mathbf{X}, A)\} = 0$ for all values of n.

Before we go on to establish the existence of complete free resolutions of Z in the case of an arbitrary finite group, we shall illustrate the last theorem by considering the complete derived sequence of a finite cyclic group.

Let G be a cyclic group of order q and let σ be a generator. In this case $Z(G)$ is a commutative ring whose general element has the form

$$\sum_{\nu=0}^{q-1} n_\nu \sigma^\nu,$$

where, of course, the n_ν are integers. Put

$$N = 1 + \sigma + \ldots + \sigma^{q-1} \quad \text{and} \quad T = \sigma - 1 \tag{10.15.9}$$

and consider the mappings

$$Z(G) \xrightarrow{N} Z(G) \quad \text{and} \quad Z(G) \xrightarrow{T} Z(G),$$

† See (8.5.4).

where the former consists of multiplication by N and the latter of multiplication by T.

If $N(\Sigma n_\nu \sigma^\nu) = 0$ then $\Sigma n_\nu = 0$, because $N\sigma^\nu = N$, hence

$$\sum_{\nu=0}^{q-1} n_\nu \sigma^\nu = \sum_{\nu=0}^{q-1} n_\nu(\sigma^\nu - 1) = T\lambda$$

for a suitable $\lambda \in Z(G)$. On the other hand, $NT = 0$, consequently $Z(G) \xrightarrow{T} Z(G) \xrightarrow{N} Z(G)$ is exact.

Suppose now that $T(\Sigma n_\nu \sigma^\nu) = 0$, then

$$(n_0 + n_1\sigma + \ldots + n_{q-1}\sigma^{q-1}) - (n_0\sigma + n_1\sigma^2 + \ldots + n_{q-1}\sigma^q) = 0,$$

and therefore $n_0 = n_1 = \ldots = n_{q-1}$. Thus $\Sigma n_\nu \sigma^\nu = N\lambda'$, where $\lambda' \in Z(G)$, and so it is seen that $Z(G) \xrightarrow{N} Z(G) \xrightarrow{T} Z(G)$ is also exact.

Consider next the augmentation homomorphism $Z(G) \to Z$ as defined in section (10.3). If $\Sigma n_\nu \sigma^\nu$ belongs to its kernel, then

$$n_0 + n_1 + \ldots + n_{q-1} = 0,$$

which, as we saw above, implies that $\Sigma n_\nu \sigma^\nu$ is of the form λT. But every element of this form certainly belongs to the kernel, consequently $Z(G) \xrightarrow{T} Z(G) \to Z \to 0$ is exact. Finally, the G-homomorphism $Z \to Z(G)$, in which $1_Z \to N$, makes $0 \to Z \to Z(G) \xrightarrow{T} Z(G)$ exact and, moreover, $Z(G) \xrightarrow{N} Z(G)$ can be represented as $Z(G) \to Z \to Z(G)$. Collecting all these facts together we obtain

Theorem 21. *Let G be a cyclic group of order q then*

$$\begin{array}{ccccc} \cdots \to Z(G) \xrightarrow{N} Z(G) \xrightarrow{T} Z(G) & \xrightarrow{\quad\quad N \quad\quad} & Z(G) \xrightarrow{T} Z(G) \xrightarrow{N} Z(G) \to \cdots \\ & \searrow \quad \nearrow & \\ & Z & \\ & \nearrow \quad \searrow & \\ & 0 \qquad\quad 0 & \end{array}$$

is a complete G-free resolution of Z. Here N and T, when used to indicate mappings, signify multiplication by the elements $N = 1 + \sigma + \ldots + \sigma^{q-1}$ and $T = \sigma - 1$ respectively. $Z(G) \to Z$ is the usual augmentation mapping and, in $Z \to Z(G)$, 1_Z maps into N.

Still supposing that G is cyclic, let A be a left G-module then, by Theorems 20 and 21, the complete derived sequence of G consists of the homology groups of a complex

$$\cdots \to \mathrm{Hom}_G(Z(G), A) \to \mathrm{Hom}_G(Z(G), A) \to \mathrm{Hom}_G(Z(G), A) \to \cdots.$$

But $\text{Hom}_G(Z(G), A) \approx A$ and on identifying these two we obtain

Theorem 22. *Let G be a cyclic group of order q and A a left G-module. Then the complete derived sequence of G can be computed as the homology groups of the complex*

$$\cdots \to A \xrightarrow{T} A \xrightarrow{N} A \xrightarrow{T} A \xrightarrow{N} A \to \cdots,$$

where the mappings N consist of multiplication by $1+\sigma+\ldots+\sigma^{q-1}$ and the mappings T of multiplication by $\sigma-1$. (T operates on the component modules with even indices.) Accordingly

$$J^{2n}(G, A) = A^G/NA \quad \textit{and} \quad J^{2n+1}(G, A) = {}_N A/IA.$$

Observe that an exact sequence $0 \to A' \to A \to A'' \to 0$ of G-modules, gives rise to an exact sequence

$$\cdots \to {}_N A''/IA'' \to A'^G/NA' \to A^G/NA \to A''^G/NA''$$
$$\to {}_N A'/IA' \to {}_N A/IA \to {}_N A''/IA'' \to A'^G/NA' \to \cdots$$

and here the connecting homomorphisms are those obtained from the exact sequence

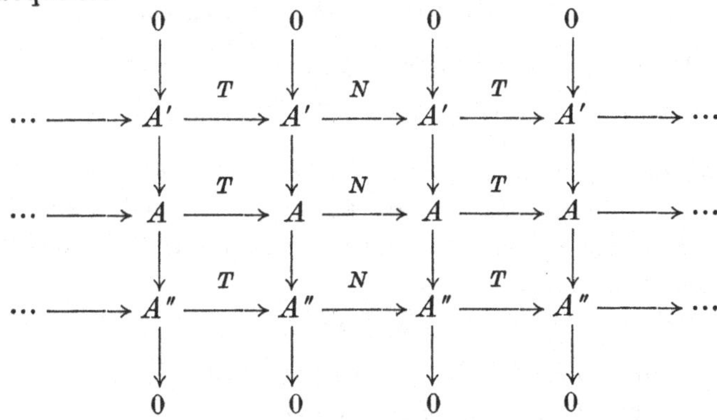

of complexes.

Let us return to the consideration of a general finite group. It will be convenient to prove two lemmas.

Lemma 2. *Let M be a Z-free module with the elements $\mu_1, \mu_2, \ldots, \mu_s$ as a base and let $\phi_i : M \to Z$ $(1 \leqslant i \leqslant s)$ be the Z-homomorphism defined by*

$$\phi_i(\mu_j) = \begin{cases} 1_Z & (i=j), \\ 0 & (i \neq j). \end{cases}$$

Then $\phi_1, \phi_2, \ldots, \phi_s$ are a Z-base for $\text{Hom}_Z(M, Z)$.

The verification is immediate.

Lemma 3. *Let Y be a right G-free module with a finite base. Then $\mathrm{Hom}_Z(Y, Z)$ has a natural structure as a left G-module and as such it is G-free and has a finite base.*

Proof. Let $y_1, y_2, \ldots, y_t$ be a G-base for Y, then the elements $y_i \sigma^{-1}$ $(1 \leqslant i \leqslant t, \sigma \in G)$ form a Z-base for Y. For each i $(1 \leqslant i \leqslant t)$, let ψ_i be the Z-homomorphism $Y \to Z$ in which all the base elements, with the exception of $y_i 1$, are mapped into zero, while $y_i 1$ itself is mapped into 1_Z. Put $\phi_{i,\sigma} = \sigma\psi_i$, then $\phi_{i,\sigma}(y_j \tau^{-1}) = \psi_i(y_j \tau^{-1}\sigma)$ and this is zero unless both $\sigma = \tau$ and $i = j$ when it has the value 1_Z. By Lemma 2, the $\phi_{i,\sigma} = \sigma\psi_i$ are a Z-base for $\mathrm{Hom}_Z(Y, Z)$ hence the ψ_i are a G-base for the module.

We come now to the existence theorem for complete resolutions.

Theorem 23. *If G is an arbitrary finite group then there exist complete G-free resolutions*

$$\begin{array}{c} \cdots \to X_1 \to X_0 \to X_{-1} \to X_{-2} \to \cdots \\ \searrow \; \nearrow \\ Z \\ \nearrow \; \searrow \\ 0 \qquad 0 \end{array}$$

of Z and, moreover, it can be arranged that each X_r has a finite G-base.

Proof. We want first to construct an exact sequence

$$\cdots \to X_2 \to X_1 \to X_0 \to Z \to 0$$

of left G-modules where each X_r $(r \geqslant 0)$ has a finite G-base. This step presents no difficulty for we are simply concerned here with the ordinary free resolutions of Z. Secondly, an exact sequence of left G-modules of the form

$$0 \to Z \to X_{-1} \to X_{-2} \to \cdots$$

is required in which $X_{-1}, X_{-2}, \ldots$ all have finite G-bases. To obtain this we start by building up an exact sequence

$$\cdots \to Y_1 \to Y_0 \to Z \to 0$$

of *right* G-modules where Y_0, Y_1, etc., all have finite G-bases. In this sequence *all* the modules are Z-free hence, by Proposition 15,

$$0 \to \mathrm{Hom}_Z(Z, Z) \to \mathrm{Hom}_Z(Y_0, Z) \to \mathrm{Hom}_Z(Y_1, Z) \to \cdots$$

is exact. Now the right G-module structures of $Z, Y_0, Y_1, \ldots$ induce left G-module structures on the terms of the new sequence and, for these, the mappings are all G-homomorphisms.† Further, by Lemma 3,

† Here we employ the general principles explained in section (8.2).

$\operatorname{Hom}_Z(Y_r, Z)$ is G-free and has a finite base. To complete the proof we have only to observe that $\operatorname{Hom}_Z(Z, Z)$ is isomorphic to Z and that G acts on it trivially.

Theorem 24. *Let G be a finite group and A a finitely generated left $Z(G)$-module. Then the groups $J^n(G, A)$, of the complete derived sequence, are all finite.*

Proof. Let

$$\begin{array}{ccccccc} \cdots \to X_1 \to & X_0 & \to & X_{-1} & \to X_{-2} \to \cdots \\ & & \searrow & \nearrow & \\ & & & Z & \\ & & \nearrow & \searrow & \\ & 0 & & & 0 \end{array}$$

be a complete G-free resolution of Z in which all the X_r have finite G-bases. By Theorem 20, $J^n(G, A)$ is effectively a factor module of a submodule of $\operatorname{Hom}_G(X_n, A)$ and hence, *a fortiori*, a factor module of a submodule of $\operatorname{Hom}_Z(X_n, A)$. Now X_n and A will both be finitely generated as Z-modules; consequently $\operatorname{Hom}_Z(X_n, A)$, and therefore also $J^n(G, A)$, will be finitely generated abelian groups. But, if the order of G is q, then, by Theorem 19, $qJ^n(G, A) = 0$. It follows that $J^n(G, A)$ must be finite.

NOTES

The following notes on the chapters of this book are intended to serve two main purposes. First, they are meant to give the reader some help in getting his orientation. This help includes suggestions for further reading, particularly at those points where the attempt to keep the account reasonably self-contained seems a little in jeopardy. Secondly, the notes give particulars of the books and papers on which I have drawn freely. To their authors I now gratefully acknowledge my indebtedness.

The figures in bold type, which occur from time to time, refer to the list of references which comes at the end. The inclusion of a publication in this list means simply that an explicit reference is made to it in these Notes.

Chapter 1

In choosing a starting point, it seemed reasonable to suppose that the reader would have had some previous contact with modern algebra, and this is made precise in the assumption that he is familiar with the notions of *ring* and *group*. However, it is worth while observing that ideas about what constitute the basic concepts of algebra are continually being modified. This constant revision, on what is a rather substantial scale, may well continue for some time to come, and it produces a situation in which what is familiar is frequently more complex than some of the topics which require to be considered *ab initio*. An account of most of the fundamental notions, showing how they stand in relation to one another so far as their primitiveness as logical concepts is concerned, will be found in Chevalley (**14**).

In section (1.8) we encounter the idea of a *free module* generated by an arbitrary set of symbols, together with the result (Theorem 2) that every module is a homomorphic image of a free module. This is an example of a point of view, in which one regards free algebraic structures as the prototypes of other and more complicated structures. This point of view is of great importance in homological algebra and, when carried to the limit in the case of modules, leads on to the concept of a resolution (see Chapter 5).

Finally, a word about notations. The *abbreviated notation* introduced in section (1.10) is a suggestion of Yoneda (**33**). In certain circumstances it is very economical and easy to follow, and this is particularly so when one wants to investigate the *connecting homomorphism* associated with a certain kind of diagram (see section (4.5)). However, a notation is only a means to an end, and the author did not feel under any compulsion to limit himself to just one among the possible alternatives. Ease in following an argument is always desirable, and therefore the choice of notation has been made to depend, to some extent, on the features of the situation under discussion.

Chapter 2

The method of defining tensor products given in the text has, in the opinion of the author, a number of features to recommend it. However, some mathematicians prefer to enumerate certain properties which tensor products are required to possess; to show that these selected properties provide an effectively complete characterization, and finally, to establish the existence of objects which behave in the required way. Such a treatment will be found in Chevalley (**14**). To illustrate the point involved, let A and B be Λ-modules and assume that Λ is a *commutative* ring. Denote by $\Lambda(A, B)$ the free Λ-module generated by all ordered pairs (a, b), where $a \in A$ and $b \in B$, and by N the submodule generated by all elements having one or other of the following four forms:

$$(a_1 + a_2, b) - (a_1, b) - (a_2, b), \quad (a, b_1 + b_2) - (a, b_1) - (a, b_2),$$
$$\lambda(a, b) - (\lambda a, b), \quad \lambda(a, b) - (a, \lambda b),$$

it being understood that λ denotes an arbitrary element of Λ. If now $A \otimes_\Lambda B$ (as defined in section (2.1)) is given its normal structure as Λ-module (see section (2.2)), then it is easy to see that $A \otimes_\Lambda B$ and $\Lambda(A, B)/N$ are Λ-isomorphic. Now it can be argued that, in the situation under discussion, it is more 'natural' to take $\Lambda(A, B)/N$ as the *definition* of the tensor product than to say it is isomorphic to it. This objection would not have arisen if we had avoided making a definition out of a particular construction.

The special features that occur in the case of a commutative ring are noted as we go along because they become important in Chapter 9, though it is the case that some of these features stem from certain general principles which are described in section (8.2). For commutative rings, tensor products obey commutative and associative laws which are developed in Chapter 8, but the reader should be aware that what is said in the text merely covers what is needed for our immediate purposes. If he wishes to extend his general knowledge he will do well to consult Bourbaki (**8**).

Among the results concerning tensor products which we do not treat at all, there is one which should be mentioned, since it is particularly likely to be needed in applications of homological algebra. The result in question asserts that the formation of tensor products commutes with the operation of taking 'direct limits', and a proof will be found in Cartan and Eilenberg (**12**), where it appears as Proposition 9.2* of Chapter 5. Considerations of space led to the omission of all discussion of direct limits, but an excellent account will be found in Eilenberg and Steenrod (**18**).

Chapter 3

The purpose of this chapter is to introduce the language of functor theory which becomes increasingly important as our subject develops. If one surveys the modern trends in mathematical ideas, it now seems remarkable that what may be termed the 'functor point of view' has taken so long to crystallize; because once it is grasped there is something extremely natural and almost inevitable about it. In the course of this book, we have occasion to consider functors on three types of category: categories of modules; translation categories of diagrams; and, in Chapter 8, categories of bimodules. These resemble one another rather closely, but their minor differences make it worth while to start with some very general considerations.

The theory of categories and functors, as touched on here, is due originally to Eilenberg and MacLane. Their account will be found in (**16**) and, of course, it contains much more than the present exposition. Indeed, as the reader will observe, once the notions of *functor*, *natural transformation* and *equivalence* have been defined, his attention is directed towards categories and functors of modules. This is because a category of modules possesses a number of special properties which are needed in the construction of *derived functors*, this being the central topic of homological algebra. Once the nature of these special properties is recognized, a natural step is to try to isolate them in order to obtain an abstract type of category capable of supporting a generalized theory of derived functors. This line of reasoning has led Buchsbaum (**9**) to the concept of an *exact category* which, besides broadening the scope of homological methods, has a second and less obvious advantage. It turns out that with each exact category one can associate a *dual category* and thereby set up a precise *principle of duality*. In this way, the tendency of our results to group themselves in pairs is explained and in such a way that, once one result of a pair has been established, the second is an immediate consequence.

Finally, in section (**3.11**), when discussing certain kinds of partial exactness for functors of modules, it was remarked that there is a further kind called *half exactness*. This is important in the theory of *satellites* an account of which will be found in chapter III of (**12**).

Chapter 4

In introducing categories and functors of diagrams, and in treating homology as a functor on the category of three-term 0-sequences, the author has been considerably influenced by the exposition of Yoneda (**33**). An alternative is to consider *modules with a differentiation* and to treat a complex as a *graded module with a differentiation of degree unity*. We shall not stop to explain this other procedure, a detailed account of which will be found in chapter IV of (**12**). The reader should be aware, however, that the concept of a graded module is an important one and, if he has not encountered it elsewhere, he is advised to take an early opportunity of becoming familiar with it.

The motivation for a good deal of the contents of Chapter 4 is provided by well-known techniques in the theory of the homology groups of topological spaces. Indeed, this is how homological algebra gets its name. For the reader who has some acquaintance with algebraic topology, we add a few remarks concerning the points of contact of the two theories.

Suppose that E is a topological space, and for each integer $p \geqslant 0$ let $C_p(E)$ be the free Z-module generated by the singular p-simplexes, so that the elements of $C_p(E)$ are what are called p-chains. If $p \geqslant 1$, the boundary operator produces a homomorphism $d_p : C_p(E) \to C_{p-1}(E)$ whose special properties make

$$\cdots \to C_{p+1}(E) \xrightarrow{d_{p+1}} C_p(E) \xrightarrow{d_p} C_{p-1}(E) \to \cdots$$

a complex of Z-modules. The homology modules $H_p(E)$ of the complex are just the *homology groups* of E and, by the method of construction, they are topological invariants.

Let us assume that we have a continuous mapping f of E into a second space E'. This mapping converts each singular p-simplex of E into a singular p-simplex of E' and therefore determines a homomorphism $C_p(E) \to C_p(E')$ for each value of p. It is easy to see that the diagram

$$\begin{array}{ccccccccc} \cdots & \longrightarrow & C_{p+1}(E) & \longrightarrow & C_p(E) & \longrightarrow & C_{p-1}(E) & \longrightarrow & \cdots \\ & & \downarrow & & \downarrow & & \downarrow & & \\ \cdots & \longrightarrow & C_{p+1}(E') & \longrightarrow & C_p(E') & \longrightarrow & C_{p-1}(E') & \longrightarrow & \cdots \end{array}$$

is commutative, consequently the continuous mapping produces a translation of the chain complexes and hence homomorphisms $H_p(E) \to H_p(E')$.

Consider now a second continuous mapping $g : E \to E'$ and let us suppose that f and g are homotopic. This means, in crude intuitive terms, that f can be changed into g by continuous variation. The usual topological analysis of this situation then shows that the translations induced by f and g are homotopic in the purely algebraic sense defined in section (4.7).

The connecting homomorphism considered in section (4.5) is encountered naturally in the development of what is called the *relative homology theory*.† Let F be a subspace of E then, since a p-simplex on F can be regarded as a p-simplex on E, $C_p(F)$ is a submodule of $C_p(E)$; further, the boundary operator d_p on $C_p(E)$ becomes, on restriction, the corresponding operator on $C_p(F)$. Put $C_p(E, F) = C_p(E)/C_p(F)$ then d_p induces a homomorphism

$$C_p(E, F) \to C_{p-1}(E, F)$$

and in this way we obtain a complex

$$\cdots \to C_{p+1}(E, F) \to C_p(E, F) \to C_{p-1}(E, F) \to \cdots.$$

Denote its pth homology group by $H_p(E, F)$. This is what is called the *relative pth homology group of E modulo F* though, in topology, it is more usual to define it in terms of subgroups of $C_p(E)$. The obvious mappings

$$0 \to C_p(F) \to C_p(E) \to C_p(E, F) \to 0$$

make an exact sequence out of our complexes consequently there arises, by Theorem 5 of section (4.6), an exact sequence

$$\cdots \to H_p(F) \to H_p(E) \to H_p(E, F) \to H_{p-1}(F) \to \cdots,$$

and this is known as the homology sequence of the pair (E, F). For a discussion from the topological point of view, together with illustrations of some of the ways in which it can be applied, the reader is referred to Wallace (**32**).

CHAPTER 5

Injective modules were introduced by Baer (**5**) (under another name, and with a different definition from the one now customarily given) who also established the fundamental result that every module is a submodule of an injective module.

We make some observations on Z-injective modules. Let Θ be a Z-module then Θ is said to be *divisible* if, given any $\theta \in \Theta$ and any integer $n \neq 0$, there always exists $\theta' \in \Theta$ such that $n\theta' = \theta$. One has then the following result: *a Z-module is injective if and only if it is divisible.* Indeed a divisible module can be proved injective in essentially the same way as we established the injective property of the group Ω of the rational numbers modulo the integers. To see the converse, suppose that Θ is Z-injective and let $\theta \in \Theta$. One obtains a homomorphism $f : Z \to \Theta$ by putting $f(m) = m\theta$. Since Θ is injective, f can be extended to a homomorphism $\bar{f} : R \to \Theta$, where R is the field of rational numbers here considered as an additive group. Clearly, if $m \neq 0$ is an integer, then

$$m\bar{f}\left(\frac{1}{m}\right) = f(1) = \theta,$$

and this shows that Θ is divisible.

The above characterization of Z-injective modules makes possible a simple proof that an arbitrary Z-module D is a submodule of a Z-injective module.

† See, for example, Wallace (**32**), ch. v, § 9.

For we can suppose that $D = F/H$, where F is a free Z-module and H is one of its submodules. Let $\{u_i\}_{i\in I}$ be a base for F and let F' be the free R-module with the same base. F', regarded as a Z-module, has F as one of its submodules and now $D = F/H$ is a submodule of F'/H. But the latter is clearly divisible and therefore, by the previous remarks, injective.

Eckmann and Schopf (**15**) have used these ideas to give a particularly neat proof of the corresponding result in the general case. We sketch their argument very briefly. Let A be a Λ-module (where Λ is an arbitrary ring with an identity element) then, as Z-module, A can be regarded as a submodule of a Z-injective module Θ. Consider $\mathrm{Hom}_Z(\Lambda, A)$ and $\mathrm{Hom}_Z(\Lambda, \Theta)$. By regarding Λ as a right Λ-module these acquire the structure of left Λ-modules and then $\mathrm{Hom}_Z(\Lambda, A)$ is a submodule (with respect to Λ) of $\mathrm{Hom}_Z(\Lambda, \Theta)$. Now the latter is Λ-injective by the same reasoning as was used to prove Lemma 4 of section (5.3). Finally, if $a \in \Lambda$ and we denote by f_a the element of $\mathrm{Hom}_Z(\Lambda, A)$ which maps λ of Λ into λa, then $a \to f_a$ is a Λ-monomorphism $A \to \mathrm{Hom}_Z(\Lambda, A)$. Combining these observations, we see that A can be considered as a submodule of the Λ-injective module $\mathrm{Hom}_Z(\Lambda, \Theta)$.

The treatment in the text combines arguments to be found in (**15**) and (**33**).

There is one further result of Eckmann and Schopf which will be described. Suppose that A is a Λ-module and X is a Λ-injective module containing A. X is called a *minimal injective extension* of A if, whenever X' is any other injective extension, there exists a Λ-monomorphism $X \to X'$ leaving each element of A fixed. By a most elegant argument, it is shown that *every Λ-module has a minimal injective extension and this is unique to within a Λ-isomorphism over A.*

Our discussion of injective modules makes use of Zorn's lemma. Since this and certain other related results are (in the opinion of the author) not as widely known among young mathematicians as is desirable, a little will be said here about the forms of transfinite induction which will be needed. These are easily explained and, if not already familiar, accounts of them can be read at leisure afterwards, for example, in chapter 3 of Birkhoff (**7**).

The basic notion is that of a *partially ordered* set. Let E be a set of objects and let us use x, y, z, etc., to denote elements of E. Suppose now that we have a relation which holds between some pairs of elements of E and let us write $x \leqslant y$ if it holds between x and y. This relation is said to define a *partial ordering* if the following conditions are all satisfied:

(i) *$x \leqslant x$ for each x in E;*

(ii) *$x \leqslant y$ and $y \leqslant x$ together imply that $x = y$;*

(iii) *whenever $x \leqslant y$ and $y \leqslant z$ then $x \leqslant z$.*

For example, if E consists of all the subsets of a set X and we write $x \leqslant y$ if the subset x is contained in the subset y, then this gives a partial ordering on E.

We make one general observation, namely, that each subset of a partially ordered set is itself partially ordered by the original ordering relation. This is called the *induced ordering.*

Let $\leqslant$ be a partial ordering on a set E, then E is said to be *totally ordered* or *simply ordered* if, whenever $x, y \in E$, then either $x \leqslant y$ or $y \leqslant x$. For instance, the real numbers are totally ordered with respect to the relation 'less than or equal to'.

Definition. A totally ordered set E is said to be *well ordered* if every non-empty subset of E has a *first element.*

Thus to say that E is well ordered is to assert that, given any non-empty subset Y, there exists $\eta \in Y$ such that $\eta \leqslant y$ for all $y \in Y$. The *well-ordering principle,* which will be used on a small number of occasions, can now be stated.

It asserts that *any given set can be well ordered by means of a suitable ordering relation.*

Before stating Zorn's lemma, it will be convenient to define one further concept.

Definition. A partially ordered set E is called an *inductive system* if, for every subset Y which is totally ordered with respect to the induced ordering relation, there exists an element ω (belonging to E and depending on Y) such that $y \leqslant \omega$ for all $y \in Y$.

It is customary to express this by saying that E is an inductive system when every totally ordered subset is bounded above in E.

The result known as *Zorn's lemma* asserts that a non-empty inductive system possesses at least one maximal element. In other words (and it is important to make the significance of *maximal* quite clear) *if E is a non-empty inductive system, then there exists $\xi \in E$ such that $\xi \leqslant x$ with $x \in E$, only when $x = \xi$.*

CHAPTER 6

An essential part of the theory of derived functors is the construction whereby one passes from an additive functor $T(A_1, A_2, \ldots, A_k)$ of modules to an additive functor $T(\mathbf{X}_1, \mathbf{X}_2, \ldots, \mathbf{X}_k)$ of complexes. In the text, only the cases $k = 1, 2$ are considered and, indeed, these are sufficient for the sequel. However, the reader will wish to know how the construction is to be carried out in the general case.

Let the covariant variables in $T(A_1, A_2, \ldots, A_k)$ be A_i $(i \in I)$ and the contravariant ones A_j $(j \in J)$. Further, when discussing complexes $\mathbf{X}_1, \mathbf{X}_2, \ldots, \mathbf{X}_k$, let us agree (for the remainder of the notes on Chapter 6) to use superscripts only to specify the component modules, and to keep subscripts for distinguishing between the different complexes.

Define $e_1, e_2, \ldots, e_k$ by

$$e_i = +1 \quad (i \in I), \qquad e_j = -1 \quad (j \in J).$$

Put
$$T^{n_1, n_2, \ldots, n_k}(\mathbf{X}_1, \mathbf{X}_2, \ldots, \mathbf{X}_k) = T(X_1^{e_1 n_1}, X_2^{e_2 n_2}, \ldots, X_k^{e_k n_k}),$$
and then set

$$T^n(\mathbf{X}_1, \mathbf{X}_2, \ldots, \mathbf{X}_k) = \sum_{n_1 + \ldots + n_k = n} T^{n_1, \ldots, n_k}(\mathbf{X}_1, \ldots, \mathbf{X}_k),$$

where the sum is direct. The $T^n(\mathbf{X}_1, \ldots, \mathbf{X}_k)$ will be the modules of the complex we are seeking to construct.

Suppose now that $\mathbf{X}_1', \mathbf{X}_2', \ldots, \mathbf{X}_k'$ is a second set of complexes, then the next thing to be done is to define $T(\boldsymbol{\varphi}_1, \boldsymbol{\varphi}_2, \ldots, \boldsymbol{\varphi}_k)$, where the $\boldsymbol{\varphi}_m = [\phi_m^r]$ are families of homomorphisms which operate so that, with an obvious notation,

$$\phi_i^r : X_i^r \to X_i'^{r+p_i} \quad (i \in I, -\infty < r < \infty),$$

$$\phi_j^r : X_j'^r \to X_j^{r+p_j} \quad (j \in J, -\infty < r < \infty).$$

The degrees of $\boldsymbol{\varphi}_1, \boldsymbol{\varphi}_2, \ldots, \boldsymbol{\varphi}_k$ are therefore $p_1, p_2, \ldots, p_k$ respectively. If now we write
$$l_i = e_i n_i \quad (i \in I), \qquad l_j = e_j(n_j + p_j) \quad (j \in J)$$

then $T(\phi_1^{l_1}, \phi_2^{l_2}, \ldots, \phi_k^{l_k})$ is a homomorphism of $T^{n_1, \ldots, n_k}(\mathbf{X}_1, \ldots, \mathbf{X}_k)$ into $T^{n_1+p_1, \ldots, n_k+p_k}(\mathbf{X}_1', \ldots, \mathbf{X}_k')$. Put

$$T^{n_1, \ldots, n_k}(\boldsymbol{\varphi}_1, \ldots, \boldsymbol{\varphi}_k) = (-1)^{\sum_{r<s} n_r p_s} T(\phi_1^{l_1}, \ldots, \phi_k^{l_k}),$$

then this too is a homomorphism between these same modules. Next, combining all the $T^{n_1,\ldots,n_k}(\boldsymbol{\varphi}_1, \ldots, \boldsymbol{\varphi}_k)$ for which $n_1 + \ldots + n_k = n$, we obtain a homomorphism

$$T^n(\boldsymbol{\varphi}_1, \boldsymbol{\varphi}_2, \ldots, \boldsymbol{\varphi}_k) : T^n(\mathbf{X}_1, \ldots, \mathbf{X}_k) \to T^{n+p}(\mathbf{X}'_1, \ldots, \mathbf{X}'_k),$$

where $p = p_1 + p_2 + \ldots + p_k$. In other words, the $T^n(\boldsymbol{\varphi}_1, \boldsymbol{\varphi}_2, \ldots, \boldsymbol{\varphi}_k)$ form a family $T(\boldsymbol{\varphi}_1, \boldsymbol{\varphi}_2, \ldots, \boldsymbol{\varphi}_k)$ of homomorphisms and the family is of degree p.

It is now necessary to describe the multiplicative properties of

$$T(\boldsymbol{\varphi}_1, \boldsymbol{\varphi}_2, \ldots, \boldsymbol{\varphi}_k).$$

To this end suppose that $\boldsymbol{\psi}_1, \boldsymbol{\psi}_2, \ldots, \boldsymbol{\psi}_k$ are further families of homomorphisms, where

$$\psi_i^r : X_i'^r \to X_i''^{r+q_i}, \quad \psi_j^r : X_j''^r \to X_j'^{r+q_j}.$$

A straightforward computation shows that

$$T^{n_1,\ldots,n_k}(\ldots, \boldsymbol{\psi}_i \boldsymbol{\varphi}_i, \ldots, \boldsymbol{\varphi}_j \boldsymbol{\psi}_j, \ldots)$$
$$= (-1)^{\sum_{r<s} p_r q_s} T^{n_1+p_1,\ldots,n_k+p_k}(\boldsymbol{\psi}_1, \ldots, \boldsymbol{\psi}_k) T^{n_1,\ldots,n_k}(\boldsymbol{\varphi}_1, \ldots, \boldsymbol{\varphi}_k)$$

and this, by combination, yields the important relation

$$T(\ldots, \boldsymbol{\psi}_i \boldsymbol{\varphi}_i, \ldots, \boldsymbol{\varphi}_j \boldsymbol{\psi}_j, \ldots) = (-1)^{\sum_{r<s} p_r q_s} T(\boldsymbol{\psi}_1, \ldots, \boldsymbol{\psi}_k) T(\boldsymbol{\varphi}_1, \ldots, \boldsymbol{\varphi}_k).$$

The construction of $T(\mathbf{X}_1, \ldots, \mathbf{X}_k)$ is now easily completed. With a notation analogous to that used in (6.2.15), put

$$\boldsymbol{\delta}_r = T(\mathbf{i}_{X_1}, \ldots, \mathbf{d}_{X_r}, \ldots, \mathbf{i}_{X_k}) \quad (1 \leqslant r \leqslant k),$$

then $\boldsymbol{\delta}_r$ is a family of mappings of degree unity and it is clear that

$$\boldsymbol{\delta}_r \boldsymbol{\delta}_r = \mathbf{0}, \quad \boldsymbol{\delta}_r \boldsymbol{\delta}_s + \boldsymbol{\delta}_s \boldsymbol{\delta}_r = \mathbf{0} \quad (r \neq s).$$

Accordingly, $\mathbf{d} = \boldsymbol{\delta}_1 + \boldsymbol{\delta}_2 + \ldots + \boldsymbol{\delta}_k$ is also a family $[d^n]$ of homomorphisms and it has the property that $\mathbf{dd} = \mathbf{0}$. Thus the modules $T^n(\mathbf{X}_1, \ldots, \mathbf{X}_k)$ and the homomorphisms

$$d^n : T^n(\mathbf{X}_1, \ldots, \mathbf{X}_k) \to T^{n+1}(\mathbf{X}_1, \ldots, \mathbf{X}_k)$$

together constitute a complex, and it is this complex that we denote by $T(\mathbf{X}_1, \ldots, \mathbf{X}_k)$. Finally, if $\mathbf{f}_1, \mathbf{f}_2, \ldots, \mathbf{f}_k$ are translations of complexes, then it is not difficult to check that $T(\mathbf{f}_1, \mathbf{f}_2, \ldots, \mathbf{f}_k)$ is itself such a translation, after which it is clear how $T(\mathbf{X}_1, \mathbf{X}_2, \ldots, \mathbf{X}_k)$ is to be regarded as a functor.

One last comment. Theorem 1 of section (6.1) concerns the homology modules of $T(\mathbf{X})$ when $T(A)$ is an exact functor of a single variable. The reader should be aware that this result has extensions to functors of several variables. He will find full particulars in (**12**) (chapter IV, §§ 6, 7).

CHAPTER 7

An account of the theory from which torsion functors derive their name is to be found in Cartan and Eilenberg ((**12**), chapter VII, § 4). We shall say no more on this point. The origin of the nomenclature for extension functors is of wider interest and will be the subject of a few remarks.

The general extension problem for modules can be stated as follows. Let Λ be an arbitrary ring with an identity element and let A and B be left Λ-modules. By an *extension of B by A* one understands a pair (E, χ) consisting of a Λ-module E containing B as submodule, together with a Λ-epimorphism $\chi : E \to A$ whose kernel is B. Two such extensions (E, χ) and (E', χ') are said to be *equivalent* if there exists a Λ-isomorphism $E \approx E'$ for which the diagram

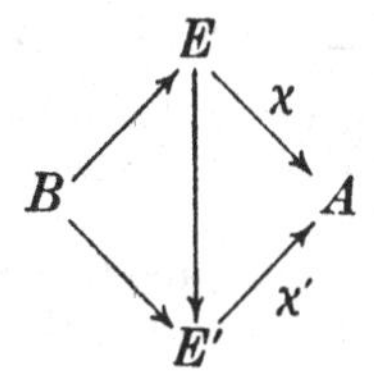

is commutative. In this way, the extensions of B by A become organized into equivalence classes.

Let (E, χ) be an extension of B by A then

$$0 \to B \to E \xrightarrow{\chi} A \to 0$$

is an exact sequence and so gives rise to a connecting homomorphism

$$\Delta : \mathrm{Hom}_\Lambda(B, B) \to \mathrm{Ext}^1_\Lambda(A, B).$$

Now the image $\Delta(i_B)$, of the identity map of B, depends only on the class of the extension and so, in this way, a class of equivalent extensions determines an element of $\mathrm{Ext}^1_\Lambda(A, B)$. The fundamental fact is that this correspondence is one-to-one ((**12**), chapter XIV, § 1). In addition, it may be noted that $\mathrm{Ext}^1_\Lambda(A, B)$ is an abelian group, consequently the correspondence enables its group structure to be transferred to the set of classes of equivalent extensions. However it is of interest to describe the latter structure directly and the reference given above provides the full details. When the necessary analysis is carried out, it is found that the law of composition for classes of extensions is the product rule previously given by Baer (**6**).

The result that the left (right) global dimension of a ring is the upper bound of the homological dimensions of its cyclic left (right) modules, is due to Auslander (**3**). The same paper contains the proof that, for Noetherian rings, global dimension and weak global dimension coincide. This, the reader will appreciate, was an important step in the investigation of Noetherian rings by homological methods. It should be kept in mind that a ring may be left Noetherian without being right Noetherian and an example, due to J. Dieudonné, to illustrate this will be found in Cartan and Eilenberg ((**12**), chapter I, § 7).

As observed in the text, the structure theory of semi-simple rings enables us to conclude (by an extremely indirect argument) that a ring has its left global dimension equal to zero when, and only when, its right global dimension is zero. The relevant structure theory will be found in van der Waerden ((**31**), vol. 2, chapter XVI, § 120).

Chapter 8

We make no comments on the contents of this chapter.

Chapter 9

The theorem which asserts that, if K is a commutative field then the polynomial ring $\Lambda = K[X_1, X_2, \ldots, X_n]$ has global dimension equal to n, has an interesting background. In his famous paper on algebraic forms, Hilbert (**22**) considered an ideal I generated by a set $\phi_1, \phi_2, \ldots, \phi_p$ of homogeneous polynomials (forms) and showed the existence of an exact sequence

$$0 \to F_n \to F_{n-1} \to \cdots \to F_1 \to \Lambda \to \Lambda/I \to 0,$$

where $F_1, F_2, \ldots, F_n$ are finitely generated free Λ-modules. He did not, of course, state his result in quite these terms. To explain Hilbert's point of view, let us suppose that we wish to construct $F_1 \to \Lambda$. This may be done by setting up a free module F_1 with a base $u_1, u_2, \ldots, u_p$ and then mapping u_j into $\phi_j(X_1, \ldots, X_n)$. An element $\Sigma \psi_j(X_1, \ldots, X_n) u_j$ of F_1 is then in the kernel of $F_1 \to \Lambda$ provided that

$$\sum_{j=1}^{p} \psi_j(X_1, \ldots, X_n) \phi_j(X_1, \ldots, X_n)$$

vanishes identically. Now a set $\psi_1, \psi_2, \ldots, \psi_p$ of polynomials having this property is called a *syzygy* of $(\phi_1, \phi_2, \ldots, \phi_p)$ and Hilbert described his theory in terms of syzygies. It would take too much space to fill in the details, but the reader who would like to know more about the classical theory will find a most enjoyable account in Gröbner (**20**). That one gets a *free* and not merely a *projective* resolution in the situation discussed by Hilbert, is due to the fact that, since $\phi_1, \phi_2, \ldots, \phi_p$ are forms, we are really concerned with graded modules over $K[X_1, X_2, \ldots, X_n]$ where the latter is to be regarded as a graded ring. In these circumstances free and projective turn out to be the same thing (see (**12**), chapter VIII, theorem 6.1).

Attention was first drawn to connexions between syzygies and homology by Koszul (**25**) and later the connexion was explored further by Cartan (**10**) who extended Hilbert's results to some other rings. More recently Eilenberg, Rosenberg and Zelinsky (**19**) have shown that, for any ring R (commutative or not),

$$\text{l.gl.dim}\, R[X_1, X_2, \ldots, X_n] = n + \text{l.gl.dim}\, R,$$

which, of course, is enormously more general than Theorem 7 of section (9.1). The text represents a compromise between generality and the desire to prove a result of interest with the techniques readily available. The 'reduction principle', which is used for this purpose, is due to Rees (**27**).

A proof of the 'Basis theorem' will be found in van der Waerden ((**31**), vol. 2, chapter XII, § 84) while the 'Zeros theorem' is established in (**31**) (vol. 2, chapter XI, § 79). However, it is necessary to explain how the latter is connected with our assertions about maximal ideals.

Suppose that K is an algebraically closed commutative field and that $f_j(X_1, X_2, \ldots, X_n)$, where $1 \leqslant j \leqslant r$, are polynomials with coefficients in K. If now $h(X_1, X_2, \ldots, X_n)$ is another such polynomial and $h(\alpha_1, \alpha_2, \ldots, \alpha_n) = 0$ whenever $\alpha_1, \alpha_2, \ldots, \alpha_n$ are in K and $f_j(\alpha_1, \alpha_2, \ldots, \alpha_n) = 0$ for all j, then there exists an integer p and polynomials $g_j(X_1, X_2, \ldots, X_n)$ such that

$$h^p(X_1, X_2, \ldots, X_n) \equiv \sum_{j=i}^{r} f_j(X_1, \ldots, X_n)\, g_j(X_1, \ldots, X_n).$$

This is the most familiar form of the Zero's theorem.

Now let M be a maximal ideal then, by the Basis theorem, it is generated by a finite set of polynomials, say by $f_j(X_1, \ldots, X_n)$ where $1 \leqslant j \leqslant r$. We contend that there exist $\alpha_1, \alpha_2, \ldots, \alpha_n$ in K which make all the $f_j(X_1, \ldots, X_n)$ vanish. For otherwise it would be possible to apply the Zero's theorem with $h(X_1, \ldots, X_n) \equiv 1$ and this would show that $1 \in M$ thereby contradicting the fact that M is a proper ideal. Next let M' be the ideal generated by $X_1-\alpha_1, X_2-\alpha_2, \ldots, X_n-\alpha_n$. M' is clearly proper and, since $f_j(\alpha_1, \alpha_2, \ldots, \alpha_n) = 0$, we see that

$$f_j(X_1, \ldots, X_n) \in M'.$$

Accordingly $M \subseteq M'$ and therefore, by the maximal property of M, $M = M'$. Thus M is generated by $X_1-\alpha_1, X_2-\alpha_2, \ldots, X_n-\alpha_n$. (It is not difficult to reason in the opposite direction and so deduce the Zero's theorem from this property of the maximal ideals of $K[X_1, X_2, \ldots, X_n]$.)

The theory, which culminates in the ideal-theoretic characterization of commutative Noetherian rings of finite global dimension, is due jointly to Auslander and Buchsbaum (**4**), who initiated it, and to Serre (**29**) who provided the proof of Theorem 21 of section (9.3). We add a few remarks about this last result.

Let Q be a local ring with maximal ideal $\mathfrak{m}$ and residue field K. For each value of p, $\mathrm{Tor}_p^Q(K, K)$ is a vector space over K of dimension B_p (say). B_p is called the pth *Betti number* of the ring Q. Let $\mathfrak{m}/\mathfrak{m}^2$ be of dimension n as a K space then Serre's result asserts that $B_p \geqslant \binom{n}{p}$ for $0 \leqslant p \leqslant n$. Now *when Q is regular* the sequence

$$\cdots \to 0 \to 0 \to F_n \to F_{n-1} \to \cdots \to F_1 \to Q \to K \to 0$$

is exact,† the notation being the same as in (9.3.7). From this it follows that, in the regular case,

$$B_p = \binom{n}{p} \quad (0 \leqslant p \leqslant n), \qquad B_p = 0 \quad (p > n).$$

For non-regular local rings the situation is quite different and, as Tate (**30**) has shown, one then has

$$B_p \geqslant \binom{n}{p} + \binom{n}{p-2} + \binom{n}{p-4} + \cdots$$

for all p, this being, in a certain sense, a best possible result.

Proofs of two of the auxiliary results, namely Propositions 4* and 5* of section (9.4), will be found in chapter IV of Northcott (**26**), where the former proposition appears as Theorem 2, Cor. 2 and the latter as Theorem 6. The remaining auxiliary result belongs to the theory of Noetherian modules, a subject which was first studied extensively by Grundy (**21**) though his work seems to have been rather overlooked until recently. We shall outline the relevant facts. For complete details, the reader is referred to Grundy's paper or to the rather condensed discussion given in Zariski and Samuel ((**35**), chpater IV, appendix).

Suppose that Λ is a commutative Noetherian ring (with an identity element) and that A is a finitely generated Λ-module. A is to be regarded as *fixed* and we consider its submodules. One can associate with such a submodule B (say) a finite set $\mathfrak{p}_1, \mathfrak{p}_2, \ldots, \mathfrak{p}_s$ of prime ideals. Just how this is done need not concern us, but these prime ideals have one property which is important for our purposes and which may be stated thus: *if $\mathfrak{a}$ is an ideal, then $B : \mathfrak{a} = B$ if and only if $\mathfrak{a}$ is not contained in any of the prime ideals associated with B.* Here by $B : \mathfrak{a}$ is meant the set of all elements x in A such that $\mathfrak{a}x \subseteq B$.

To deduce Proposition 6* of section (9.4), we observe that its hypotheses imply that $0 : \mathfrak{m} = 0$, where 0 denotes the zero submodule. Consequently $\mathfrak{m}$ is

† The reader who wishes to have proof of this should consult Cartan and Eilenberg ((**12**), ch. VIII, § 4).

not contained in any of the prime ideals $\mathfrak{p}_1, \mathfrak{p}_2, \ldots, \mathfrak{p}_s$ associated with 0, and therefore† there exists $\mu \in \mathfrak{m}$ which is also not contained in any $\mathfrak{p}_i$. Accordingly $0 : (\mu) = 0$ and this means that μ is not a zero-divisor in A.

Consider, once again, a polynomial ring $\Lambda = K[X_1, X_2, \ldots, X_n]$, where K is a commutative field. It is a comparatively classical result that the ring of fractions of Λ, with respect to an arbitrary maximal ideal, is a regular local ring whose ideal-theoretic dimension is n. Indeed, as Zariski (**34**) has shown, the geometric content of this result is simply that n-dimensional affine space is free from multiple points. In view, therefore, of our general criterion (Theorem 26 of section (9.6)), the relation gl.dim $\Lambda = n$ and Hilbert's theorem on syzygies both stem ultimately from the fact that affine space is a non-singular variety.

Perhaps the most outstanding achievement of homological methods in the theory of Noetherian rings is the proof that the ring of fractions of a regular local ring, with respect to an arbitrary prime ideal, is again regular; but many other interesting applications will be found in Auslander and Buchsbaum (**4**), Rees (**28**) and Serre (**29**).

Chapter 10

The original treatment, due to Eilenberg and MacLane (**17**), of the homological algebra of abstract groups, defined cohomology groups directly in terms of standard cocycles and coboundaries. This is a natural procedure in the light of certain familiar constructions in algebraic topology. That these ideas could prove of value in purely algebraic situations was also recognized at an early stage by Hochschild (**23, 24**), whose application of them to the theory of algebras did a great deal to stimulate interest in the early days of our subject.

Still earlier, 1-cocycles and factor systems had been encountered in connexion with Galois theory. To be more explicit let K be a field and G a finite group of automorphisms of K. The non-zero elements of K form a multiplicative G-module which will be denoted by K^*; while, with respect to addition, K itself is a G-module. In order to make our intention clear, the latter G-module will be specified by the symbol K^+.

In the case of K^*, a 1-cocycle is a family $\{\alpha_\sigma\}$ of non-zero elements of K which satisfies the relations

$$\alpha_\tau^\sigma \alpha_\sigma = \alpha_{\sigma\tau} \quad (\sigma, \tau \in G),$$

where, of course, we understand by α^σ the image of α under the automorphism σ. These relations are known as *Noether's equations* and they can be solved. In fact, if $\beta \in K^*$ and we set $\alpha_\sigma = \beta/\beta^\sigma$ for all σ in G, then we obtain a solution and, moreover, all solutions are obtainable in this way. For it is known (see (**1**), th. 12, p. 34) that the elements of G are linearly independent with respect to K; this means that if $\{x_\sigma\}_{\sigma \in G}$ is a family of elements of K and

$$\sum_{\sigma \in G} x_\sigma \sigma(\gamma) = 0$$

for all $\gamma \in K$, then $x_\sigma = 0$ for all σ. Suppose now that we are given a 1-cocycle $\{\alpha_\sigma\}$ of K^* then, for a suitable $\gamma \in K$, $\Sigma_\tau \alpha_\tau \tau(\gamma) = \beta$ (say) is not zero. We now have

$$\beta^\sigma = \sum_\tau \alpha_\tau^\sigma \sigma\tau(\gamma) = \sum_\tau \alpha_{\sigma\tau} \sigma\tau(\gamma)/\alpha_\sigma = \beta/\alpha_\sigma,$$

so that $\alpha_\sigma = \beta/\beta^\sigma$ as required. Translating these observations into homological terms, we see that they amount to a proof that $H^1(G, K^*) = 0$.

† See Northcott ((**26**), ch. I, prop. 6).

The factor systems for K^* arise in the theory of algebras in connexion with what are called *crossed products*. For further particulars, the reader is referred to Artin, Nesbit and Thrall ((**2**), chapter 8).

In the case of K^+, homology and cohomology are trivial. For, by the linear independence of the automorphisms, we can find $\alpha \in K$ such that $\Sigma_\sigma \sigma(\alpha) = a$ (say) is not zero. Now denote by $u : K^+ \to K^+$ the Z-homomorphism which consists in multiplication by α/a then, since a is left fixed by the elements of G, we have

$$\Sigma_\sigma \sigma\{u(\sigma^{-1}\beta)\} = \Sigma_\sigma \sigma\left\{\frac{\alpha}{a}(\sigma^{-1}\beta)\right\} = \frac{1}{a}\Sigma_\sigma \sigma(\alpha)\,\beta = \beta$$

for all $\beta \in K$ and so we see that K^+ is G-special. Accordingly, by Theorem 18 of section (10.14), all the terms of the complete derived sequence vanish for K^+.

The main method used for discussing homology and cohomology in the case of special groups (free groups, free abelian groups, cyclic groups) is to obtain a particularly simple free resolution of Z. This method is also useful for arbitrary finitely generated abelian groups and a convenient resolution, suitable for this purpose, has been constructed by Tate ((**30**), application 1, p. 20).

For further general reading on the homological theory of groups, the reader may consult the Cartan Seminar Notes (**11**) for 1950–1. The rather more specialized theory for finite groups has been developed with the object of applying it to Class Field Theory.† An account of much of what is needed for this purpose (but without the applications themselves) will be found in Cartan and Eilenberg ((**12**), chapter XII).

† See Chevalley (**13**). The new Class Field Theory, which arises from these applications, is largely the creation of Artin, Hochschild, Nakayama and Tate.

REFERENCES

(1) Artin, E. *Galois Theory.* Notre Dame Mathematical Lectures, No. 2. Notre Dame, Indiana, 1946.

(2) Artin, E., Nesbitt, C. J. and Thrall, R. M. *Rings with Minimum Condition.* University of Michigan Press, 1944.

(3) Auslander, M. On the dimension of modules and algebras (III). *Nagoya Math. J.* **9** (1955), 67–77.

(4) Auslander, M. and Buchsbaum, D. A. Homological dimension in local rings. *Trans. Amer. Math. Soc.* **85** (1957), 390–405.

(5) Baer, R. Abelian groups that are direct summands of every containing abelian group. *Bull. Amer. Math. Soc.* **46** (1940), 800–6.

(6) Baer, R. Erweiterung von Gruppen und ihren Isomorphismen. *Math. Z.* **38** (1934), 375–416.

(7) Birkhoff, G. *Lattice Theory.* Amer. Math. Soc. Colloquium Publications, no. 25 (revised edition, 1948).

(8) Bourbaki, N. *Algèbre Multilinéaire.* Actualités Scientifiques et Industrielles, no. 1044. Hermann, Paris, 1948.

(9) Buchsbaum, D. A. Exact categories and duality. *Trans. Amer. Math. Soc.* **80** (1955), 1–34.

(10) Cartan, H. Extensions du théorème des 'chaines de syzygies'. *Univ. Roma 1st Naz. Alta. Mat. Rend. Mat e Appl.* (**5**), **11** (1952), 156–66.

(11) Cartan, H. *Séminaire* 1950/51. (*Cohomologie des groupes, suite spectrale, faisceaux.*) Secrétariat mathematique, 11 rue Pierre Curie, Paris.

(12) Cartan, H. and Eilenberg, S. *Homological Algebra.* Princeton University Press, 1956.

(13) Chevalley, C. *Class Field Theory.* Nagoya University, 1954.

(14) Chevalley, C. *Fundamental Concepts of Algebra.* Academic Press, 1956.

(15) Eckmann, B. and Schopf, A. Über injektive Moduln. *Arch. Math.* **4** (1953), 75–8.

(16) Eilenberg, S. and MacLane S. General theory of natural equivalences. *Trans. Amer. Math. Soc.* **58** (1945), 231–94.

(17) Eilenberg, S. and MacLane, S. Cohomology theory in abstract groups. I. *Ann. Math.* **46** (1945), 58–67.

(18) Eilenberg, S. and Steenrod, N. *Foundations of Algebraic Topology.* Princeton University Press, 1952.

(19) Eilenberg, S., Rosenberg, A. and Zelinsky, D. On the Dimensions of Modules and Algebras (VIII). Dimension of Tensor Products. *Nagoya Math. J.* **12** (1957), 71–93.

(20) Gröbner, W. *Moderne algebraische Geometrie.* Springer, 1949.

(21) Grundy, P. M. A generalization of additive ideal theory. *Proc. Camb. Phil. Soc.* **38** (1942), 241–79.

(22) Hilbert, D. Über die Theorie der algebraischen Formen. *Math. Ann.* **36** (1898), 473–534.

(23) Hochschild, G. P. On the cohomology groups of an associative algebra. *Ann. Math.* **46** (1945), 58–67.

(24) Hochschild, G. P. On the cohomology theory for associative algebras. *Ann. Math.* **47** (1946), 568–79.

(25) KOSZUL, L. Sur un type d'algèbres differentielles en rapport avec la transgression. Colloque de topologie (espaces fibrés), Bruxelles, 1950. Georges Thone, Liége; Masson et Cie, Paris, 1951.

(26) NORTHCOTT, D. G. *Ideal Theory.* Cambridge Tracts no. 42, 1953.

(27) REES, D. A theorem of homological algebra. *Proc. Camb. Phil. Soc.* **52** (1956), 605–10.

(28) REES, D. The grade of an ideal or module. *Proc. Camb. Phil. Soc.* **53** (1957), 28–42.

(29) SERRE, J. P. Sur la dimension des anneaux et des modules noethériens. Proceedings of the International Symposium on Algebraic Number Theory. Tokyo and Nikko, 1955; Science Council of Japan, Tokyo, 1956.

(30) TATE, J. Homology of Noetherian rings and local rings. *Illinois J. Math.* **1** (1957), 14–27.

(31) VAN DER WAERDEN, B. L. *Modern Algebra.* Frederick Ungar Publ. Co. 1950.

(32) WALLACE, A. H. *An Introduction to Algebraic Topology.* Pergamon Press, 1957.

(33) YONEDA, N. On the homology theory of modules. *J. Fac. Sci. Univ. Tokyo*, **7** (1954), 193–227.

(34) ZARISKI, O. Foundations of a general theory of birational correspondences. *Trans. Amer. Math. Soc.* **53** (1943), 490–542.

(35) ZARISKI, O. and SAMUEL, P. *Commutative Algebra*, vol. 1. Van Nostrand, 1958.

INDEX

The numbers refer to pages